ADVANCED DESIGN TECHNIQUES AND REALIZATIONS OF MICROWAVE AND RF FILTERS

ADVANCED DESIGN TECHNIQUES AND REALIZATIONS OF MICROWAVE AND RF FILTERS

PIERRE JARRY
JACQUES BENEAT

IEEE PRESS

WILEY
A JOHN WILEY & SONS, INC., PUBLICATION

Published by John Wiley & Sons, Inc., Hoboken, New Jersey.
Published simultaneously in Canada.

Library of Congress Cataloging-in-Publication Data:

Jarry, Pierre, 1946–
Advanced design techniques and realizations of microwave and RF filters / Pierre Jarry, Jacques Beneat.
p. cm.
Includes index.
ISBN 978-0-470-18310-6 (cloth)
1. Microwave filters. I. Beneat, Jacques, 1964- II. Title.
TK7872.F5J37 2008
621.381′3224—dc22

2007050393

Printed in the United States of America

10 9 8 7 6 5 4 3 2 1

CONTENTS

FOREWORD

Being asked to review the manuscript of *Advanced Design Techniques and Realizations of Microwave and RF filters* was an honor. The title truly represents the book's focus and its contents.

Filters are the most important passive components used in RF and microwave subsystems and instruments to obtain a precise frequency response. In the early years of filter development, significant progress was made in waveguide and planar TEM filters. During the past two decades, filter technology has advanced in the area of emerging applications for both military and commercial markets. Several major developmental categories in filter technology are included: performance improvement, development of CAD tools, full-wave analysis, new structures and configurations, and advanced materials and associated technologies. Advanced materials/technologies such as high-temperature superconductor substrates, micromachining, multilayer monolithic, low-temperature co-fired ceramic, and liquid-crystal polymer are commonly used in the development of advanced filters. Some recent applications of filters include dual-band communication, such as wireless local area networks and ultrawideband communication and imaging.

This book treats the subject to meet the needs for advanced filter design based on planar and waveguide structures that can satisfy the ever-increasing demand for design accuracy, reliability, fast development times, and cost-effective solutions. The topics discussed include analyses, design, modeling, fabrication, and practical considerations for both ladder and bridged filters. Modern design techniques are discussed for a wide variety of microwave filters, including comprehensive analyses and modeling of structures. These topics are self-contained, with practical aspects addressed in detail. Extensive design information in the form of equations, tables, graphs, and solved examples are included. To aid in solving filter-related design problems from specifications to realization of the end-product, the book provides a unique integration of theory and practical aspects of filters. Simple design equations and numerous practical examples are included which simplify the concepts of advanced filter design. With emphasis on theory, design, and practical aspects geared toward day-to-day applications, the book is suitable for students, teachers, scientists, and practicing engineers.

Overall, the book is well balanced and includes exhaustive treatment of relevant topics important to a filter designer. I congratulate the authors on an outstanding book that I am confident will be very well received in the RF and microwave community.

Dr. Inder Bahl

Roanoke, Virginia
November 2007

PREFACE

Microwave and RF filters play an important role in communication systems, and due to the proliferation of radar, satellite, and mobile wireless systems there is a need for design methods that can satisfy the ever-increasing demand for accuracy, reliability, and fast development times. This book, which provides modern design techniques for a wide variety of microwave filters operating over the frequency range 1 to 35 GHz, has grown out of the authors' own research and teaching and thus can present a unity of methodology and style, essential for a smooth reading.

The book is intended for researchers and for radio-frequency (RF) and microwave engineers, but is also suitable for an advanced graduate course in the subject area. Furthermore, it is possible to choose material from the book to supplement traditional courses in microwave filter design. The fundamental principles that can be applied to the synthesis and design of microwave filters are first recalled in a concise manner. Each of the 10 design chapters provides a complete analysis and modeling of the microwave structure used for filtering, as well as the design methodology. We hope that this will provide researchers with a set of approaches that can be used for current and future microwave filter designs. We also emphasize the practical nature of the subject by summarizing the design steps and giving numerous examples of filter realizations and measured responses so that RF and microwave engineers can have an appreciation of each filter in view of their needs. This approach, we believe, has produced a coherent, practical, and real-life treatment of the subject. The book is therefore theoretical but also experimental, with over 20 microwave filter realizations.

The book is divided into four parts. In Part I comprising the first four chapters, fundamental concepts and equations for microwave filter analysis and design are provided. Chapter 1 covers definitions and examples of the scattering and *ABCD* parameters of two-port systems. Classical elements used in microwave filter design, such as impedance and admittance inverters, are reviewed. The bisection theorem, which is often very useful to simplify the synthesis of microwave filters, is presented. Chapter 2 summarizes filter approximations and synthesis. Several functions, such as the Butterworth, Chebyshev, elliptic, and pseudoelliptic, are given for amplitude-oriented filters. The Bessel and Rhodes equidistant linear-phase functions are provided for phase-oriented filters. The general synthesis method for some of these functions is discussed in terms of lumped ladder realizations. The chapter ends with useful properties and equations related to realizations based on impedance and admittance inverters. In Chapter 3 we recall the fundamental equations for waveguides, striplines, suspended substrate striplines, and distributed circuits. The general design approaches used for the filters presented in the book are given in Chapter 4. The filters are regrouped in terms of minimum or non-minimum-phase microwave filters. The latter being regrouped according to the low-pass symmetry of their magnitude response around 0 Hz, referred to as non-minimum-phase symmetrical or asymmetrical microwave filters.

Part II consists of Chapters 5 through 8, corresponding to the analysis and design of four minimum-phase filters. In Chapter 5 we describe the analysis and modeling of capacitive gap filters implemented using suspended substrate stripline. A straightforward design technique is presented and actual realizations and measured responses are given for narrow- and wideband filters in the range 8 to 16 GHz. The chapter concludes with preliminary results showing that this technique can be extended to the design of millimeter-wave filters. An example of a 35-GHz suspended substrate stripline filter is given. It becomes clear, however, that more rigorous electromagnetic analysis and treatment of problems will be needed to improve the performance of filters at these frequencies. In Chapter 6 we present a design technique for evanescent-mode waveguide filters, which can have either rectangular or circular cross sections and include a number of dielectric inserts. These filters can be smaller than traditional waveguide filters, since the cutoff frequency of the guide can be increased even if this means that the fundamental mode will be evanescent in part of the structure.

Several realizations showing the reliability of the technique are given at 14 GHz. In an attempt to reduce the form factor of these filters, the design of folded evanescent mode waveguide filters is introduced. Two realizations with different angles of curvature around 8.4 and 14.6 GHz are described. In Chapter 7 we present the design of interdigital filters, structures modeled using the method of graphs rather than the admittance matrix approach. The design is based on the use of a low-pass prototype circuit, due to the periodic nature in frequency of interdigital line segments. Two realizations around 1 and 2 GHz made using suspended substrate stripline are shown. In Chapter 8 an exact design procedure for combline filters is provided. The technique is applied to a filter at 1.2 GHz using suspended stripline.

Part III which includes Chapters 9 and 10, provides two techniques for designing non-minimum-phase symmetrical response filters. In the case of symmetrical responses, the bisection theorem can be used to reduce the complexity of the design method. The filters presented in this part are concerned primarily with phase characteristics. The non-minimum-phase condition leads to additional degrees of freedom that can be used to shape the frequency response of the filter. However, this also requires microwave structures that can accommodate the non-minimum-phase condition, and the design techniques are usually more complex. Chapter 9 makes use of the additional degrees of freedom to design filters that have good group delay characteristics. The filters are implemented using a generalized interdigital structure. A realization at 2.7 GHz based on interdigital bars is shown to give a group delay variation of 2 ns in the passband. In Chapter 10 we present very narrowband (e.g., 1%) circular monomode TE_{011} cavity filters. A method for optimizing the group delay is introduced. A monomode filter design and realization is given at 14.5 GHz. These filters are also suited for temperature variations. The stability of filter responses from -10 to $+20^\circ$C is demonstrated.

Part IV groups Chapters 11 through 14 and deals with non-minimum-phase asymmetrical response filters. The filters in this chapter present generalized Chebyshev or pseudoelliptic bandpass responses with a given number of transmission zeros. In the case of asymmetrical responses, the lowpass prototype is imaginary, and neither the bisection theorem nor traditional frequency transform techniques can be used. The microwave structures must also be able to generate zeros of transmission. In Chapter 11 we describe the design technique for capacitive-gap coupled line filters that have generalized Chebyshev asymmetrical responses. A third-order realization in suspended stripline at 10 GHz shows an asymmetrical response with a zero of transmission at a frequency above the passband. Chapter 12 deals with a state-of-the-art family of in-line dual-mode rectangular structures using resonant TE_{102} and TE_{301} modes. Two realizations are given for frequencies around 12 GHz. In Chapter 13 we describe a technique used to design cylindrical dual-mode cavity filters with asymmetrical responses. Good amplitude characteristics are obtained using the TE_{113} resonant mode. Several realizations using a dual-mode cavity are given around 8.5 GHz. The responses show that various combinations of the zeros of transmission can be obtained.

In Chapter 14 we introduce a new concept for the design of non-minimum-phase microwave filters. The filters are made of basic rectangular waveguide building blocks, and the filters can be designed by using a powerful optimization algorithm called the genetic algorithm. Use of this method makes it possible to design rectangular multimode cavity filters with generalized Chebyshev responses. The realization of a fourth-order filter at 14 GHz with two zeros of transmission using one building block (TE_{100}, TM_{120}, and TM_{210} modes), and a seventh-order filter at 20 GHz with four zeros of transmission using two building blocks are given. Due to the simplicity of the building blocks, these filters are also easy to manufacture.

The plan of the book is summarized in Figure P1. It shows that we begin in Part I with fundamental concepts and equations useful for designing microwave filters. The reader can stop after Part II which provides a synthesis of minimum-phase microwave filters. The reader could also go directly to more advanced filters known as non-minimum-phase filters. Part III provides the design techniques

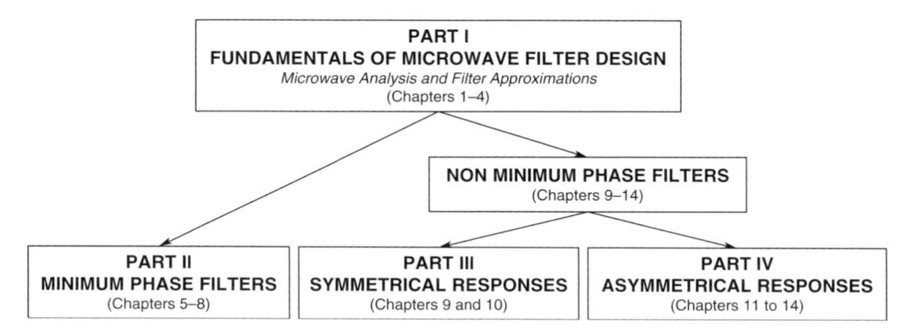

Figure P1 Summary of the organization of the book.

used for non-minimum-phase filters with symmetrical responses, and Part IV covers approaches for designing non-minimum-phase filters with asymmetrical responses. The difference between the latter two types of filters depends on the low-pass network being considered (real or imaginary).

We would like to acknowledge the contributions of our past and present research students whose collaboration has resulted in much of the material in the book. In particular, we would like to mention Professor Humberto Abdalla Junior and Professor Horacio Tertuliano from Brazil, Professor G. Tanné and Associate Professors J.F. Favennec and F. Le Pennec from France, Professor C. Djoub from Ivory Coast, Engineer M. Lyakoubi from Canada, and Engineers N. Boutheiller, CL. Guichaoua, E. Hanna, M. Lecouve, D. Lo Line Tong and O. Roquebrun from France.

The book is based on the authors' research under the sponsorship of the European Space Agency (ESA), France Telecommunications Research & Development (CNET), National Center of Spatial Studies (CNES), ALCATEL SPACE, PHILIPS and THOMSON, French Ministry of Research and Technology (MRT), French Delegation of Research in Science and Techniques (DGRST).

The work resulted in 2 patents with ESA, 3 patents with ALCATEL SPACE and approximately 16 contracts with the different agencies and companies.

ACKNOWLEDGMENTS

The authors are deeply indebted to Dr. Inder J. Bahl of Tyco Electronics M/A-COM (USA), editor of the *International Journal of RF and Microwave Computer-Aided Engineering*. The book couldn't have been written without his help, and he is acknowledged with gratitude.

Our sincere thanks extend to George Telecki of Wiley-Interscience, and the reviewers, for their support in writing the book. The help provided by Rachel Witmer and Melissa Valentine of Wiley is very much appreciated.

Pierre Jarry wishes to thank his colleagues at the University of Bordeaux, including Professors Pascal Fouillat, Eric Kerherve, and André Touboul, as well as Professor Yves Garault at the University of Limoges. Finally, he would like to express his deep appreciation to his wife, Roselyne, and his son, Jean-Pierre, for their tolerance and support.

Jacques Beneat is very grateful to Norwich University, a place conducive to trying new endeavors and succeeding. He particularly wants to thank Professors Ronald Lessard and Steve Fitzhugh for their support during the writing of the book; Bjong Wolf Yeigh, Vice President for Academic Affairs and Dean of the Faculty; and Bruce Bowman, Dean of the David Crawford School of Engineering, for their encouragement in such a difficult enterprise.

PIERRE JARRY
JACQUES BENEAT

PART I

MICROWAVE FILTER FUNDAMENTALS

1

SCATTERING PARAMETERS AND *ABCD* MATRICES

1.1 INTRODUCTION

In this chapter we recall the most important characterization techniques used in the design of microwave filters [1.1]. These consist of the scattering parameters, which are often based on electromagnetic analysis of the microwave structures, and the *ABCD* parameters, which are useful to make the link with two-port systems and have been studied exhaustively over the years. Several examples are presented to better understand the relations between the two formalisms. The bisection (or Bartlett) theorem is also reviewed and proves to be very useful in the case of symmetrical networks.

Advanced Design Techniques and Realizations of Microwave and RF Filters,
By Pierre Jarry and Jacques Beneat

1.2 SCATTERING MATRIX OF A TWO-PORT SYSTEM

1.2.1 Definitions

The scattering matrix [1.2] of a two-port system provides relations between the input and output reflected waves b_1 and b_2 and the input and output incident waves a_1 and a_2 when the structure is to be connected to a source resistance R_G and a load resistance R_L, as depicted in Figure 1.1. The notion of waves rather than voltages and currents is better suited for microwave structures.

For a two-port system, the equations relating the incident and reflected waves and the S parameters are given by

$$b_1 = S_{11}a_1 + S_{12}a_2$$
$$b_2 = S_{21}a_1 + S_{22}a_2$$

These equations can be summarized in the matrix form $(b) = (S)(a)$, where

$$\begin{pmatrix} b_1 \\ b_2 \end{pmatrix} = \begin{pmatrix} S_{11} & S_{12} \\ S_{21} & S_{22} \end{pmatrix} \begin{pmatrix} a_1 \\ a_2 \end{pmatrix}$$

The parameter S_{11}, called the *input reflection coefficient*, can be computed by setting the output incident wave a_2 to zero and taking the ratio of the input reflected wave over the input incident wave:

$$S_{11} = \left.\frac{b_1}{a_1}\right|_{a_2=0}$$

The output incident wave a_2 is set to zero by connecting the output of the system to the reference resistor R_L. The parameter S_{11} provides a measure of how much of the input incident wave does not reach the output of the system and is reflected back at the input. For microwave filters, ideally, S_{11} should be equal to zero in the passband of the filter.

The parameter S_{21}, called the *forward transmission coefficient*, can be computed by setting the output incident wave a_2 to zero and taking the ratio of the output reflected wave over the input incident wave:

$$S_{21} = \left.\frac{b_2}{a_1}\right|_{a_2=0}$$

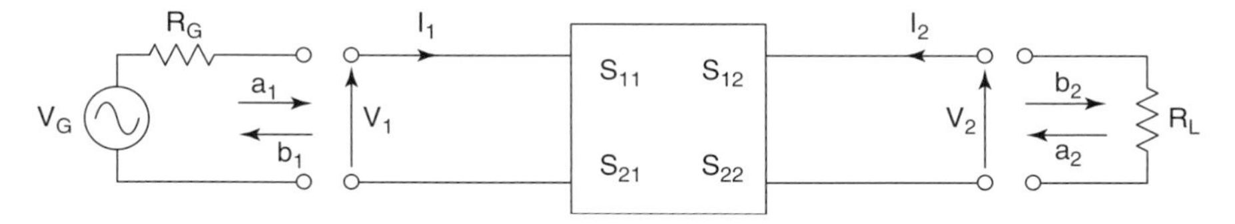

Figure 1.1 Notation used in defining the scattering matrix of a two-port system.

The output incident wave a_2 is set to zero by connecting the output to the reference resistor R_L. The parameter S_{21} provides a measure of how much of the input incident wave reaches the output of the system. For microwave filters, ideally, S_{21} should be equal to 1 in the passband of the filter.

The parameter S_{22}, called the *reflection coefficient at the output of the system*, can be computed by setting the input incident wave a_1 to zero and taking the ratio of the output reflected wave over the output incident wave:

$$S_{22} = \left.\frac{b_2}{a_2}\right|_{a_1=0}$$

The input incident wave a_1 is set to zero by connecting the input of the system to the reference resistor R_G. As in the case of S_{11}, it is desirable that S_{22} be kept close to zero in the passband of the filter. S_{11} and S_{22} provide a measure of how well the system impedances are matched to the reference terminations.

The parameter S_{12}, called the *reverse transmission coefficient*, can be computed by setting the input incident wave a_1 to zero and taking the ratio of the input reflected wave over the output incident wave:

$$S_{12} = \left.\frac{b_1}{a_2}\right|_{a_1=0}$$

The input incident wave a_1 is set to zero by connecting the input of the system to the reference resistor R_G. The parameter S_{12} provides a measure of how much of an incident wave set at the output of the system would reach the input. Due to symmetries in the system, S_{21} and S_{12} can have similar values. Since there are no generators at the output of the system, an output incident wave could appear due to a poor S_{22}.

The scattering parameters can be illustrated using a graph, as shown in Figure 1.2. The graph shows that part of the incident wave a_1 results in a reflected wave b_1 through the parameter S_{11}, and in a transmitted wave b_1 through the parameter S_{21}. Similar descriptions can be given for a_2 and the parameters S_{22} and S_{12}. It is always important to remember that the S-parameter values are associated with a given set of termination values. Changing the termination values will change the S-parameter values.

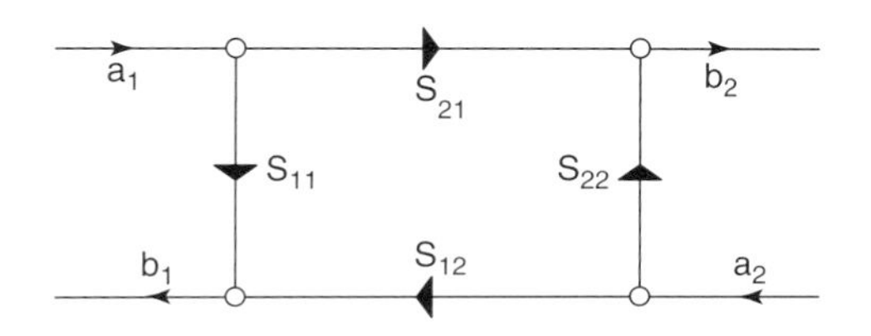

Figure 1.2 Graph of a two-port scattering matrix.

1.2.2 Computing the *S* Parameters

A common example of a scattering matrix in microwave is that of a waveguide of length l_0 and characteristic impedance Z_0, as shown in Figure 1.3. When the structure is to be connected to a source and load resistance equal to the characteristic impedance of the waveguide, the scattering matrix is given by

$$(S) = \begin{pmatrix} 0 & e^{-j\beta l_0} \\ e^{-j\beta l_0} & 0 \end{pmatrix}$$

where $j\beta$ is the propagation function of a given mode above the cutoff frequency of the waveguide. This matrix tells us that the structure will be perfectly matched to the terminations since S_{11} and S_{22} are equal to zero. It also tells us that b_1 the wave transmitted, will simply be a delayed version of a_1, the incident wave, since the forward transmission coefficient, S_{21}, has a magnitude of 1 and a linear phase of $-\beta l_0$, and the longer the length, the longer the delay. Since we cannot differentiate one end of a waveguide from the other, we would have similar results if connecting the source to the output and the load to the input (e.g., $S_{12} = S_{21}$).

As will be seen, microwave structures will at times have discontinuities that result in the apparition of "scattered" and unwanted electromagnetic fields. For these cases, matching the electromagnetic fields on each side of the discontinuity will provide relations that can be used for defining the scattering parameters of the discontinuity. In that case, the scattering parameters will be defined directly from electromagnetic wave equations. It should be noted, however, that scattering parameters are not limited to microwave structures and electromagnetic field equations.

The incident and reflected waves can be expressed in terms of voltages and currents, as shown in Figure 1.1.

$$a_1 = \frac{V_1 + R_G I_1}{2\sqrt{R_G}} \qquad a_2 = \frac{V_2 + R_L I_2}{2\sqrt{R_L}}$$

$$b_1 = \frac{V_1 - R_G I_1}{2\sqrt{R_G}} \qquad b_2 = \frac{V_2 - R_L I_2}{2\sqrt{R_L}}$$

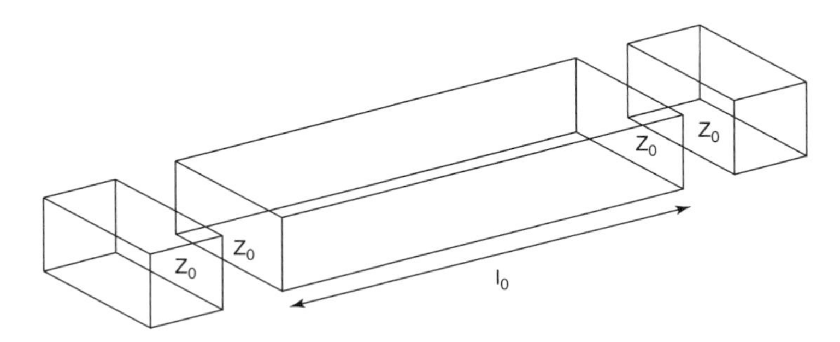

Figure 1.3 Waveguide of length l_0 and characteristic impedance Z_0.

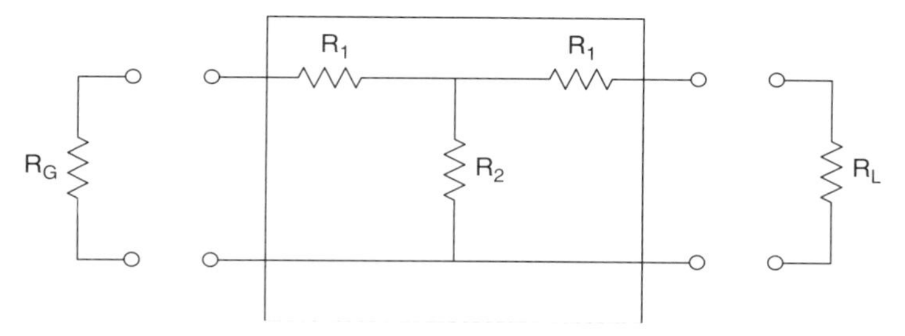

Figure 1.4 Defining the scattering parameters of a resistive two-port system.

For example, the scattering parameters of the resistive two-port system in Figure 1.4 can be defined from these voltages and currents.

The input reflection coefficient S_{11} is defined from the input incident and reflected waves when the system is connected to the reference resistor R_L, as shown in Figure 1.5. Also shown in the figure, the system connected to resistor R_L can be modeled as Z_{in}, an input impedance of the system. In this case, $V_1 = Z_{\text{in}} I_1$ and S_{11} is given by

$$S_{11} = \frac{V_1 - R_G I_1}{V_1 + R_G I_1} = \frac{Z_{\text{in}} I_1 - R_G I_1}{Z_{\text{in}} I_1 + R_G I_1} = \frac{Z_{\text{in}} - R_G}{Z_{\text{in}} + R_G}$$

This gives $Z_{\text{in}} = R_1 + R_2 || (R_1 + R_L)$ for the input impedance of the system. For S_{11} to be equal to zero, the input impedance Z_{in} should be equal to the source resistor R_G.

The forward transmission coefficient S_{21} is defined from the input incident wave and output reflected wave when the system is connected to the reference resistor R_L, as shown in Figure 1.6. Replacing the incident and reflected waves by their voltage and current expressions, S_{21} is given by

$$S_{21} = \sqrt{\frac{R_G}{R_L}} \frac{V_2 - R_L I_2}{V_1 + R_G I_1}$$

Also from the computations of S_{11}, $V_1 = Z_{\text{in}} I_1$ when the system is connected to R_L. From Figure 1.6 it is also seen that $V_2 = -R_L I_2$. Therefore, S_{21} will be

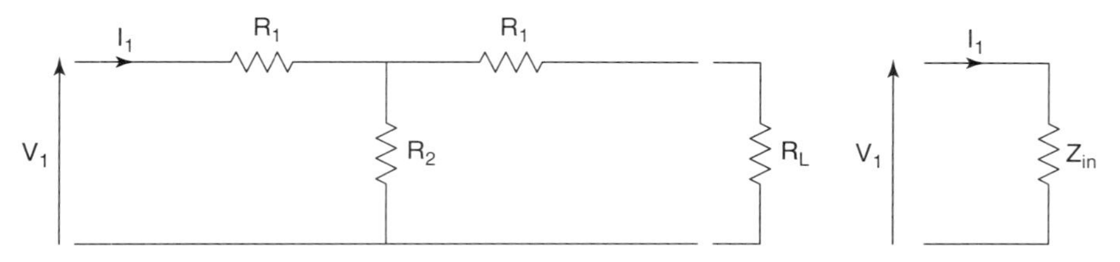

Figure 1.5 Defining the input reflection coefficient S_{11}.

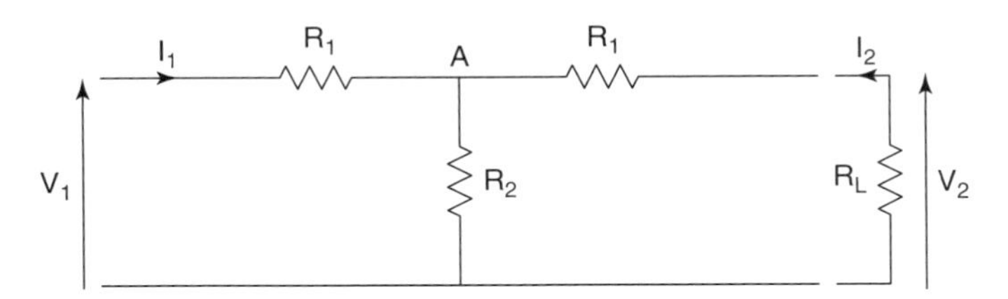

Figure 1.6 Defining the forward transmission coefficient S_{21}.

given by

$$S_{21} = \sqrt{\frac{R_G}{R_L}} \frac{V_2 - (-V_2)}{V_1 + (R_G/Z_{\text{in}})V_1} = 2\sqrt{\frac{R_G}{R_L}} \frac{1}{1 + R_G/Z_{\text{in}}} \frac{V_2}{V_1}$$

Note that in the case where $Z_{\text{in}} = R_G$ (input impedance matching) and $R_G = R_L$ (similar source and load terminations), the forward coefficient reduces to

$$S_{21} = \frac{V_2}{V_1}$$

In Figure 1.6,

$$V_2 = \frac{R_L}{R_L + R_1} V_A \quad \text{and} \quad V_A = \frac{R}{R + R_1} V_1$$

where $R = R_2 || (R_1 + R_L)$, so that

$$V_2 = \frac{R_L}{R_L + R_1} \frac{R}{R + R_1} V_1$$

and a general expression for S_{21} is given by

$$S_{21} = 2\sqrt{\frac{R_G}{R_L}} \frac{1}{1 + R_G/Z_{\text{in}}} \frac{R_L}{R_L + R_1} \frac{R}{R + R_1}$$

The S_{22} and S_{12} parameters can be defined using a similar process, where the input is now connected to the reference resistor R_G. In the resistive example above, the S parameters are independent of frequency since the impedances of the resistors are independent of frequency. However, the results can be used to define the S parameters of a more general case, as shown in Figure 1.7. The input reflection coefficient S_{11} will now be a function of the impedances of the system and therefore depend on the frequency of application through the Laplace variable s:

$$S_{11}(s) = \frac{Z_{\text{in}}(s) - R_G}{Z_{\text{in}}(s) + R_G}$$

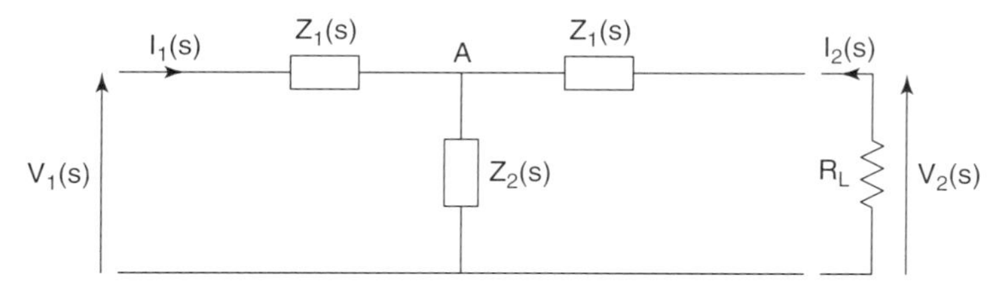

Figure 1.7 Defining the scattering parameters of a general two-port system.

where the input impedance $Z_{\text{in}}(s)$ is given by $Z_{\text{in}}(s) = Z_1(s) + Z_2(s)||[Z_1(s)+R_L]$.

The forward transmission coefficient S_{21} will also depend on the frequency of operation and is given by

$$S_{21}(s) = 2\sqrt{\frac{R_G}{R_L}}\frac{1}{1 + R_G/Z_{\text{in}}(s)}\frac{R_L}{R_L + Z_1(s)}\frac{Z(s)}{Z(s) + Z_1(s)}$$

where the impedance $Z(s) = Z_2(s)||[Z_1(s) + R_L]$. This means that the S parameters will generally have different values depending on the frequency at which they are being evaluated. In the case of a microwave filter, ideally, $S_{21}(f)$ should be equal to 1 at the frequencies of the passband of the filter and be equal to zero at the frequencies of the stopband of the filter. Similarly, $S_{11}(f)$ should be equal to zero at the frequencies of the passband of the filter and be equal to 1 at the frequencies of the stopband of the filter. Figure 1.8 shows the forward transmission coefficient S_{21} and the input reflection coefficient S_{11} versus frequency for a fifth-order Chebyshev filter with passband edges at 10 and 11 GHz and specified a −20-dB return loss.

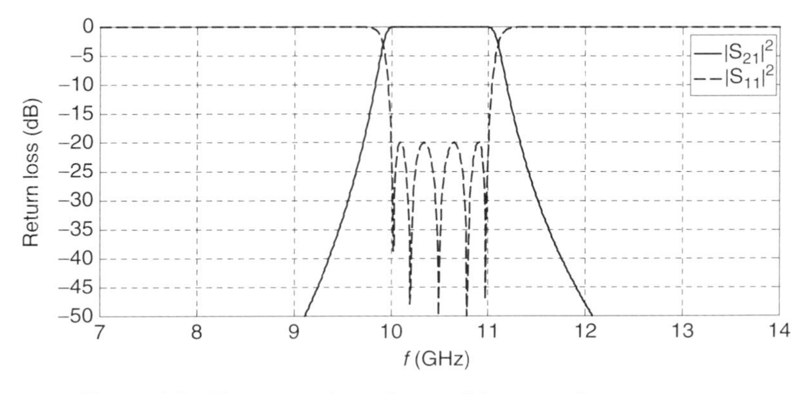

Figure 1.8 Frequency dependence of the scattering parameters.

1.2.3 *S*-Parameter Properties

Depending on the properties of the *S* parameters, the structures can be classified into the following categories:

- *Reciprocity*. A two-port system is said to be *reciprocal* if the *S* matrix is equal to its transpose:

$$(S) = \begin{pmatrix} S_{11} & S_{12} \\ S_{21} & S_{22} \end{pmatrix} = (S)^{\mathrm{T}} = \begin{pmatrix} S_{11} & S_{21} \\ S_{12} & S_{22} \end{pmatrix}$$

  In other words, the forward transmission coefficient and the reverse transmission coefficient are equal (e.g., $S_{21} = S_{12}$).
- *Symmetry*. A two-port system is said to be *symmetrical* if in addition to the reciprocity, the input and output reflection coefficients are identical (e.g., $S_{11} = S_{22}$), and *antisymmetrical* if they are opposite in sign (e.g., $S_{11} = -S_{22}$).
- *Lossless*. A two-port system is said to be *lossless* if power is conserved. In this case the complex conjugate of the *S* matrix is equal to its transpose:

$$(S)^* = \begin{pmatrix} S_{11}^* & S_{12}^* \\ S_{21}^* & S_{22}^* \end{pmatrix} = (S)^{\mathrm{T}} = \begin{pmatrix} S_{11} & S_{21} \\ S_{12} & S_{22} \end{pmatrix}$$

In the case of lossless structures, additional relations exist between the transmission and reflection coefficients:

$$\begin{aligned} |S_{11}|^2 + |S_{12}|^2 &= 1 \\ |S_{21}|^2 + |S_{22}|^2 &= 1 \\ S_{11}S_{21}^* + S_{22}^*S_{12} &= 0 \end{aligned}$$

In the case of a lossless and reciprocal two-port system, the input reflection coefficient and the forward transmission coefficient are such that $|S_{11}|^2 = 1 - |S_{21}|^2$. An example of this is shown in Figure 1.8. Additional results concerning lossless systems are given in Appendix 1.

1.3 *ABCD* MATRIX OF A TWO-PORT SYSTEM

There are several benefits of using the *ABCD* matrix representation when designing microwave filters [1.3,1.4]. They allow simulating entire structures made of a cascade of lumped elements, such as capacitors, inductors, and transformers. Lumped ladder structures are available in the literature for providing specified filtering responses. In addition, *S* parameters can be converted to *ABCD* parameters, and vice versa. When analyzing a microwave structure element, it is easier to

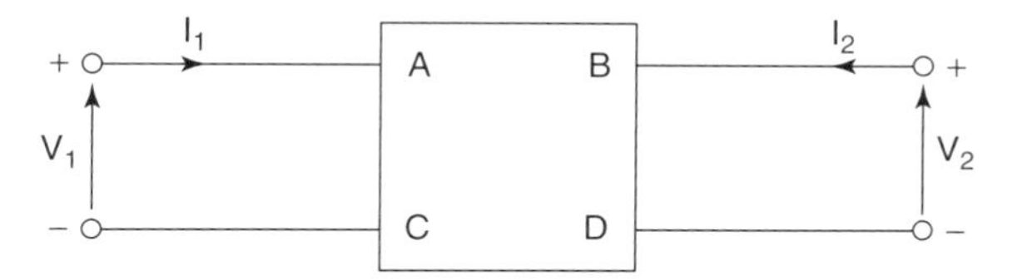

Figure 1.9 Notation used in defining the *ABCD* matrix of a two-port system.

model the element as a combination of lumped elements rather than interpreting the *S* parameters directly.

The *ABCD* matrix of a two-port is defined using voltages and currents as shown in Figure 1.9. The *ACBD* matrix is defined by

$$\begin{aligned} V_1 &= AV_2 + B(-I_2) \\ I_1 &= CV_2 + D(-I_2) \end{aligned}$$

In matrix form this provides the relation

$$\begin{pmatrix} V_1 \\ I_1 \end{pmatrix} = \begin{pmatrix} A & B \\ C & D \end{pmatrix} \begin{pmatrix} V_2 \\ (-I_2) \end{pmatrix}$$

The *ABCD* matrices of some of the most basic elements found in microwave structures are described next.

1.3.1 *ABCD* Matrix of Basic Elements

A common element found in many microwave filter design problems consists of a single impedance *Z* placed in series, as shown in Figure 1.10. The equations and the *ABCD* matrix of a series impedance are

$$\begin{aligned} I_1 &= (-I_2) \\ V_1 &= V_2 + Z(-I_2) \end{aligned} \quad \text{or} \quad \begin{pmatrix} V_1 \\ I_1 \end{pmatrix} = \begin{pmatrix} 1 & Z \\ 0 & 1 \end{pmatrix} \begin{pmatrix} V_2 \\ (-I_2) \end{pmatrix}$$

Another common element consists of an admittance *Y* placed in parallel, as shown in Figure 1.11. The equations and *ABCD* matrix of a parallel admittance are

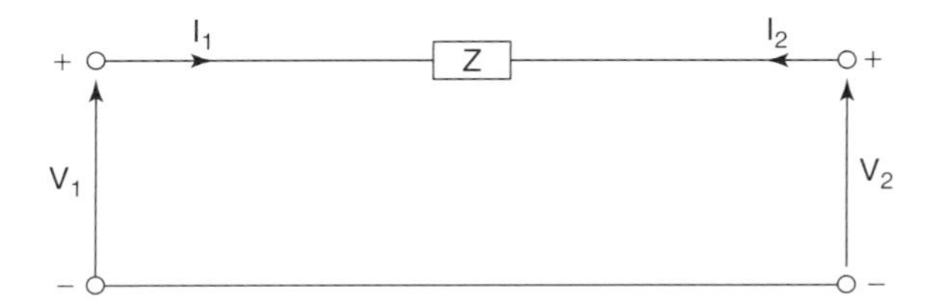

Figure 1.10 *ABCD* matrix of a series impedance.

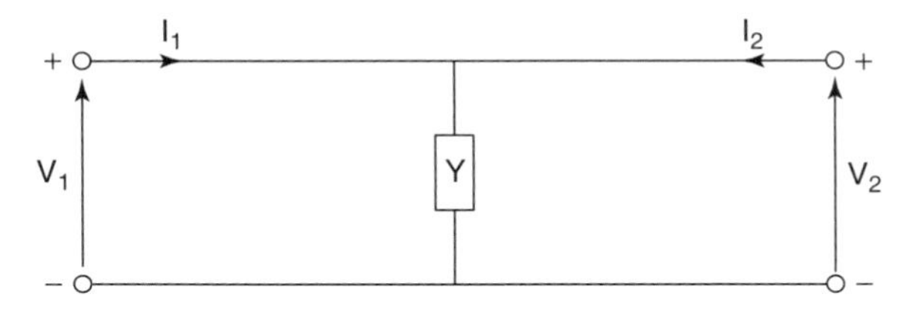

Figure 1.11 *ABCD* matrix of a parallel admittance.

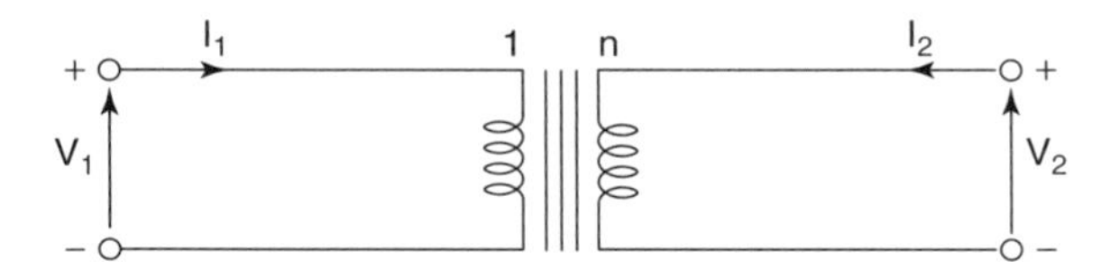

Figure 1.12 *ABCD* matrix of an ideal transformer.

$$\begin{aligned} V_1 &= V_2 \\ Y V_2 &= I_1 + I_2 \end{aligned} \quad \text{or} \quad \begin{pmatrix} V_1 \\ I_1 \end{pmatrix} = \begin{pmatrix} 1 & 0 \\ Y & 1 \end{pmatrix} \begin{pmatrix} V_2 \\ (-I_2) \end{pmatrix}$$

Another common element encountered in microwave systems is the ideal transformer shown in Figure 1.12. The equations and *ABCD* matrix of the ideal transformer are

$$\begin{aligned} V_2 &= n V_1 \\ (-I_2) &= \frac{I_1}{n} \end{aligned} \quad \text{or} \quad \begin{pmatrix} V_1 \\ I_1 \end{pmatrix} = \begin{pmatrix} \frac{1}{n} & 0 \\ 0 & n \end{pmatrix} \begin{pmatrix} V_2 \\ (-I_2) \end{pmatrix}$$

1.3.2 Cascade and Multiplication Property

One of the main advantages of the *ABCD* representation is that the *ABCD* matrix of a system made of the cascade of two systems, as shown in Figure 1.13, is equal to the multiplication of the individual *ABCD* matrices. The equations of this system are given by

$$\begin{aligned} V_1 &= A_1 V_2 + B_1(-I_2) \\ I_1 &= C_1 V_2 + D_1(-I_2) \end{aligned} \quad \text{and} \quad \begin{aligned} V_1' &= A_2 V_2' + B_2(-I_2') \\ I_1' &= C_2 V_2' + D_2(-I_2') \end{aligned}$$

but since $V_2 = V_1'$ and $(-I_2) = I_1'$ we have

$$\begin{cases} V_1 = A_1(A_2 V_2' + B_2(-I_2')) + B_1(C_2 V_2' + D_2(-I_2')) \\ \quad = (A_1 A_2 + B_1 C_2) V_2' + (A_1 B_2 + B_1 D_2)(-I_2') \\ I_1 = C_1(A_2 V_2' + B_2(-I_2')) + D_1(C_2 V_2' + D_2(-I_2')) \\ \quad = (C_1 A_2 + D_1 C_2) V_2' + (C_1 B_2 + D_1 D_2)(-I_2') \end{cases}$$

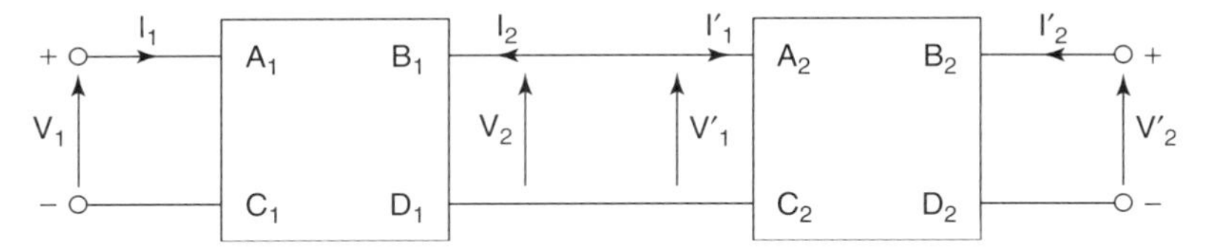

Figure 1.13 Cascade of two *ABCD* systems.

On the other side, if we multiply the *ABCD* matrices, we find that

$$\begin{pmatrix} A_1 & B_1 \\ C_1 & D_1 \end{pmatrix} \begin{pmatrix} A_2 & B_2 \\ C_2 & D_2 \end{pmatrix} = \begin{pmatrix} A_1A_2 + B_1C_2 & A_1B_2 + B_1D_2 \\ C_1A_2 + D_1C_2 & C_1B_2 + D_1D_2 \end{pmatrix}$$

Since both techniques provide the same answers, in the future the *ABCD* matrix of the cascade of two systems will be given by multiplication of their *ABCD* matrices. This is very powerful since it will be easier to define individual *ABCD* matrices of a microwave structure and then multiply these matrices for simulating the entire structure.

Application to a T Network A *T network* is often described as three impedances forming a structure that looks like a T, as shown in Figure 1.14. In this case the *ABCD* matrix is given by the multiplication of three *ABCD* matrices (impedance Z_1, admittance $1/Z_2$, and impedance Z_3):

$$\begin{pmatrix} A & B \\ C & D \end{pmatrix} = \begin{pmatrix} 1 & Z_1 \\ 0 & 1 \end{pmatrix} \begin{pmatrix} 1 & 0 \\ \dfrac{1}{Z_2} & 1 \end{pmatrix} \begin{pmatrix} 1 & Z_3 \\ 0 & 1 \end{pmatrix} = \begin{pmatrix} 1 + \dfrac{Z_1}{Z_2} & Z_1 + Z_3 + \dfrac{Z_1Z_3}{Z_2} \\ \dfrac{1}{Z_2} & 1 + \dfrac{Z_3}{Z_2} \end{pmatrix}$$

Application to a Π Network A Π *network* is often described as three admittances forming a structure that looks like a Π, as shown in Figure 1.15. In this case, the *ABCD* matrix is given by the multiplication of three *ABCD* matrices

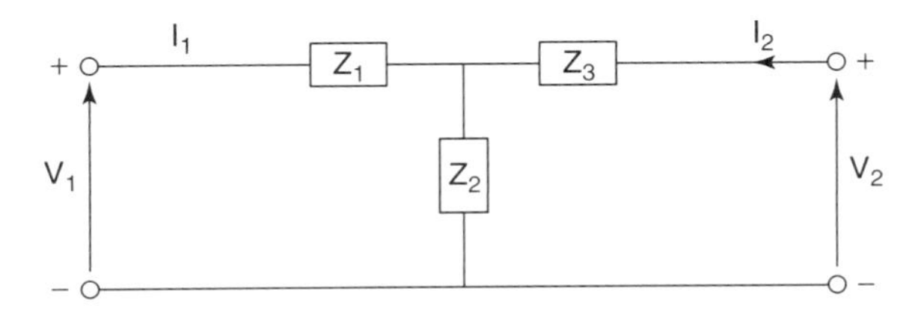

Figure 1.14 T network.

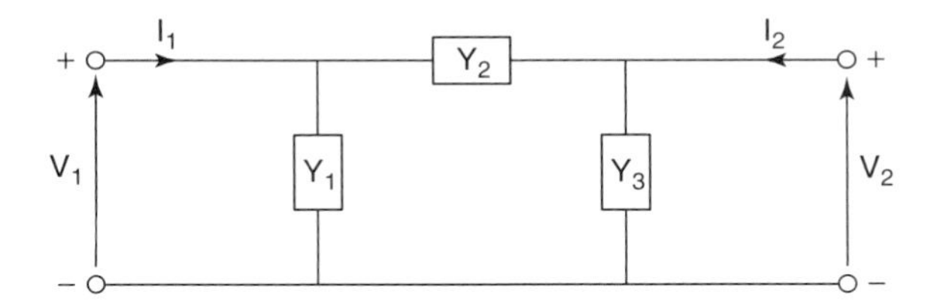

Figure 1.15 Π network.

(admittance Y_1, impedance $1/Y_2$, and admittance Y_3):

$$\begin{pmatrix} A & B \\ C & D \end{pmatrix} = \begin{pmatrix} 1 & 0 \\ Y_1 & 1 \end{pmatrix} \begin{pmatrix} 1 & \dfrac{1}{Y_2} \\ 0 & 1 \end{pmatrix} \begin{pmatrix} 1 & 0 \\ Y_3 & 1 \end{pmatrix} = \begin{pmatrix} 1 + \dfrac{Y_3}{Y_2} & \dfrac{1}{Y_2} \\ Y_1 + Y_3 + \dfrac{Y_1 Y_3}{Y_2} & 1 + \dfrac{Y_1}{Y_2} \end{pmatrix}$$

1.3.3 Input Impedence of a Loaded Two-Port

When a two-port is connected to a load $Z_L(\mathrm{s})$ as shown in Figure 1.16, the output voltage V_2 and current I_2 will be such that $V_2 = Z_L(-I_2)$. The *ABCD* equations then become

$$V_1 = AV_2 + B(-I_2) = AZ_L(-I_2) + B(-I_2)$$
$$I_1 = CV_2 + D(-I_2) = CZ_L(-I_2) + D(-I_2)$$

from which it is straightforward to extract the input impedance of the two-port network:

$$Z_{\mathrm{in}} = \left.\frac{V_1}{I_1}\right|_{Z_L} = \frac{AZ_L + B}{CZ_L + D}$$

1.3.4 Impedance and Admittance Inverters

An *impedance inverter* is a two-port network that can provide an input impedance that is the inverse of the load impedance. This property is illustrated in

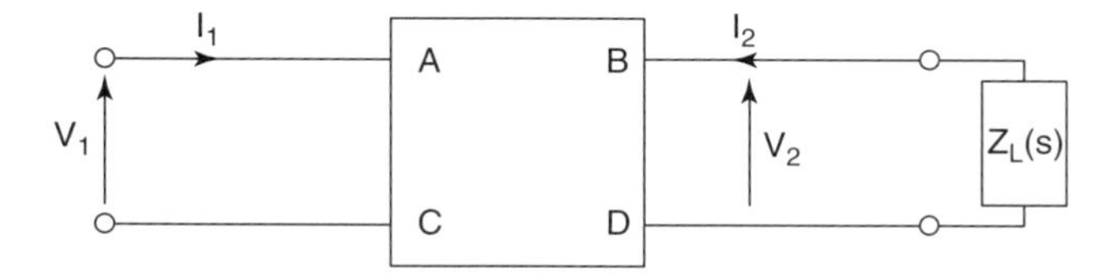

Figure 1.16 Two-port terminated on a load impedance $Z_L(s)$.

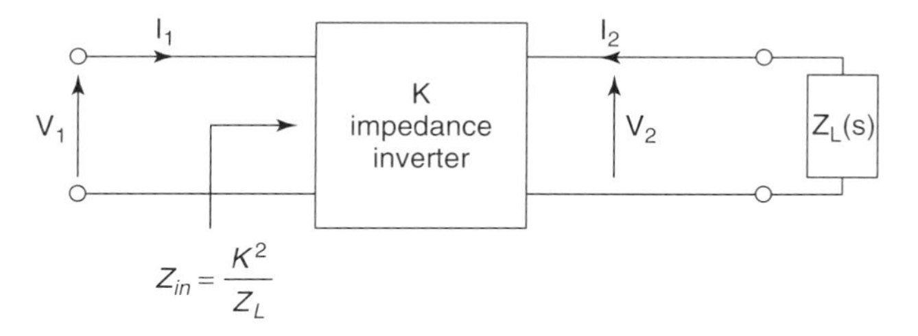

Figure 1.17 Impedance inverter principles.

Figure 1.17. An *ideal impedance inverter* will have an *ABCD* matrix of the form

$$\begin{pmatrix} V_1 \\ I_1 \end{pmatrix} = \begin{pmatrix} 0 & jK \\ \dfrac{j}{K} & 0 \end{pmatrix} \begin{pmatrix} V_2 \\ (-I_2) \end{pmatrix}$$

The input impedance of the two-port network when connected to a load impedance $Z_L(s)$ will be given by

$$Z_{\text{in}} = \left.\frac{V_1}{I_1}\right|_{Z_L} = \frac{AZ_L + B}{CZ_L + D} = \frac{0Z_L + jK}{(j/K)Z_L + 0} = \frac{K^2}{Z_L}$$

An *admittance inverter* is a two-port network that can provide an input admittance that is the inverse of the load admittance. This property is illustrated in Figure 1.18. An *ideal admittance inverter* will have an *ABCD* matrix of the form

$$\begin{pmatrix} V_1 \\ I_1 \end{pmatrix} = \begin{pmatrix} 0 & \dfrac{j}{J} \\ jJ & 0 \end{pmatrix} \begin{pmatrix} V_2 \\ (-I_2) \end{pmatrix}$$

The input admittance of the two-port network when connected to a load admittance $Y_L(s)$ will be given by

$$Y_{\text{in}} = \left.\frac{I_1}{V_1}\right|_{Y_L} = \frac{CV_2 + DY_LV_2}{AV_2 + BY_LV_2} = \frac{C + DY_L}{A + BY_L} = \frac{jJ + 0Y_L}{0 + (j/J)Y_L} = \frac{J^2}{Y_L}$$

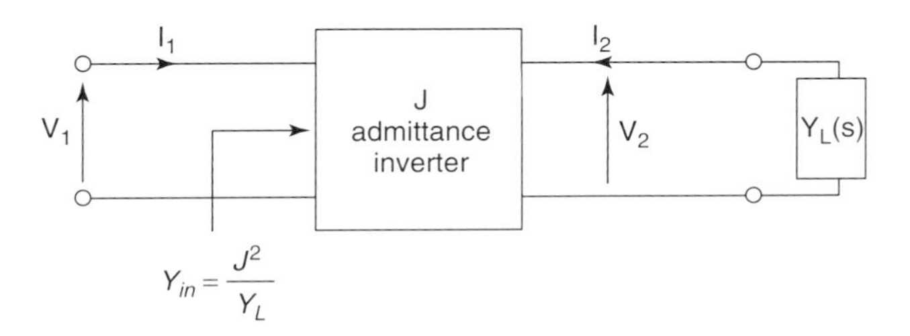

Figure 1.18 Admittance inverter principles.

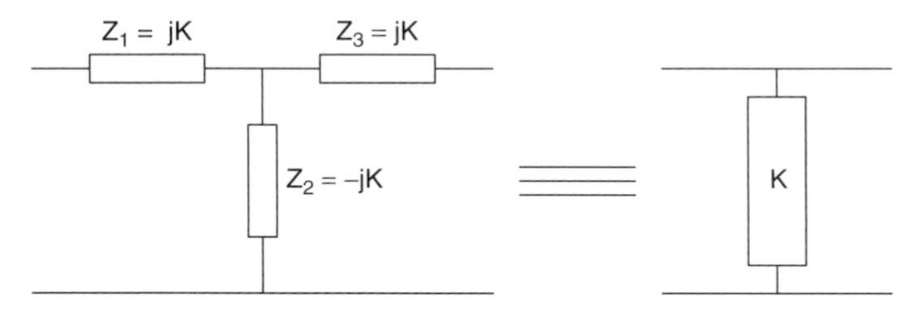

Figure 1.19 Realization of an ideal impedance inverter.

An ideal impedance inverter can be realized as a T network, as shown in Figure 1.19. This is shown by checking that the *ABCD* matrix of the T network, when $Z_1 = Z_3 = jK$ and $Z_2 = -jK$ is that of an impedance inverter:

$$\begin{pmatrix} A & B \\ C & D \end{pmatrix} = \begin{pmatrix} 1 + \dfrac{Z_1}{Z_2} & Z_1 + Z_3 + \dfrac{Z_1 Z_3}{Z_2} \\ \dfrac{1}{Z_2} & 1 + \dfrac{Z_3}{Z_2} \end{pmatrix}$$

$$= \begin{pmatrix} 1 + \dfrac{jK}{-jK} & jK + jK + \dfrac{(jK)(jK)}{-jK} \\ \dfrac{1}{-jK} & 1 + \dfrac{jK}{-JK} \end{pmatrix} = \begin{pmatrix} 0 & jK \\ \dfrac{j}{K} & 0 \end{pmatrix}$$

An ideal admittance inverter can be constructed using the Π network of Figure 1.20. This is shown by checking that the *ABCD* matrix of a Π network when $Y_1 = Y_3 = jJ$ and $Y_2 = -jJ$ is that of an admittance inverter:

$$\begin{pmatrix} A & B \\ C & D \end{pmatrix} = \begin{pmatrix} 1 + \dfrac{Y_3}{Y_2} & \dfrac{1}{Y_2} \\ Y_1 + Y_3 + \dfrac{Y_1 Y_3}{Y_2} & 1 + \dfrac{Y_1}{Y_2} \end{pmatrix}$$

$$= \begin{pmatrix} 1 + \dfrac{jJ}{-jJ} & \dfrac{1}{-jJ} \\ jJ + jJ + \dfrac{(jJ)(jJ)}{-jJ} & 1 + \dfrac{jJ}{-jJ} \end{pmatrix} = \begin{pmatrix} 0 & \dfrac{j}{J} \\ jJ & 0 \end{pmatrix}$$

An impedance inverter can be approximated using two identical inductors and a capacitor as shown in Figure 1.21. The *ABCD* matrix of this T network is given by

$$\begin{pmatrix} A & B \\ C & D \end{pmatrix} = \begin{pmatrix} 1 - LC\omega^2 & jL\omega(2 - LC\omega^2) \\ jC\omega & 1 - LC\omega^2 \end{pmatrix}$$

At the frequency ω_0 where $LC\omega_0^2 = 1$, this T network behaves as an ideal impedance inverter $K = L\omega_0$.

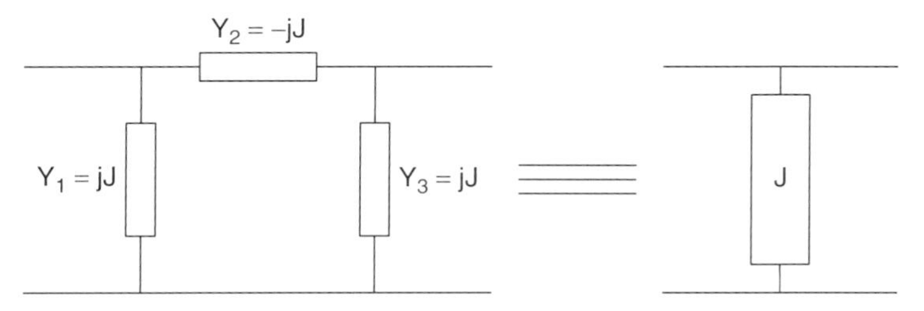

Figure 1.20 Realization of an ideal admittance inverter.

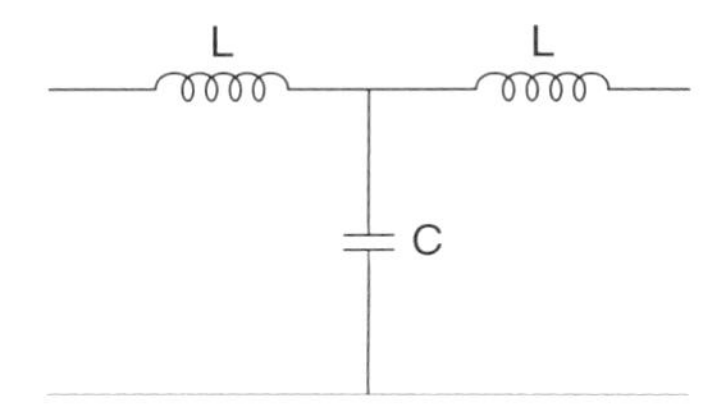

Figure 1.21 Approximate realization of an impedance inverter.

$$\begin{pmatrix} A & B \\ C & D \end{pmatrix}\bigg|_{\omega=\omega_0} = \begin{pmatrix} 0 & jL\omega_0 \\ \dfrac{j}{L\omega_0} & 0 \end{pmatrix} = \begin{pmatrix} 0 & jK \\ \dfrac{j}{K} & 0 \end{pmatrix}$$

As will be seen in the filter design chapters, one will often represent a microwave structure as an equivalent circuit based on impedance or admittance inverters. In some cases it will be possible to greatly reduce or even remove the frequency dependence of the inverter. For other cases, the ideal inverter behavior can only be assumed around the center frequency of the filter (narrowband designs).

1.3.5 *ABCD*-Parameter Properties

Depending on the properties of the *ABCD* matrix, the structures can be classified into the following categories:

- *Reciprocal*. In this case, the determinant of the ABCD matrix is equal to unity:

$$AD - BC = 1$$

- *Symmetrical*. In this case, the parameters A and D are equal:

$$A = D$$

1.4 CONVERSION FROM FORMULATION *S* TO *ABCD* AND *ABCD* TO *S*

The *ABCD* matrix can be defined from the *S* parameters and termination conditions using the following conversion equations:

$$\begin{pmatrix} A & B \\ C & D \end{pmatrix} = \begin{pmatrix} \sqrt{\dfrac{R_G}{R_L}}\dfrac{(1+S_{11})(1-S_{22})+S_{21}S_{12}}{2S_{21}} & \sqrt{R_G R_L}\dfrac{(1+S_{11})(1+S_{22})-S_{21}S_{12}}{2S_{21}} \\ \dfrac{1}{\sqrt{R_G R_L}}\dfrac{(1-S_{11})(1-S_{22})-S_{21}S_{12}}{2S_{21}} & \sqrt{\dfrac{R_L}{R_G}}\dfrac{(1-S_{11})(1+S_{22})+S_{21}S_{12}}{2S_{21}} \end{pmatrix}$$

The *S* matrix can be defined from the *ABCD* parameters and termination conditions using the following conversion equations:

$$\begin{pmatrix} S_{11} & S_{12} \\ S_{21} & S_{22} \end{pmatrix} = \begin{pmatrix} \dfrac{AR_L + B - CR_G R_L - DR_G}{AR_L + B + CR_G R_L + DR_G} & \dfrac{2\sqrt{R_G R_L}(AD - BC)}{AR_L + B + CR_G R_L + DR_G} \\ \dfrac{2\sqrt{R_G R_L}}{AR_L + B + CR_G R_L + DR_G} & \dfrac{-AR_L + B - CR_G R_L + DR_G}{AR_L + B + CR_G R_L + DR_G} \end{pmatrix}$$

From these definitions it is possible to check the effects of adding redundant elements such as impedance inverters at the input and output of a system on the scattering parameters. This is done in Appendix 2.

1.5 BISECTION THEOREM FOR SYMMETRICAL NETWORKS

When a network is symmetrical, it can be modeled as the cascade of a half network and a reverse half network, as shown in Figure 1.22. In this case, the *ABCD* matrix of the symmetrical network is given by the product of the half network matrices:

$$\begin{pmatrix} A & B \\ C & D \end{pmatrix} = \begin{pmatrix} A_1 & B_1 \\ C_1 & D_1 \end{pmatrix}\begin{pmatrix} D_1 & B_1 \\ C_1 & A_1 \end{pmatrix} = \begin{pmatrix} A_1D_1 + B_1C_1 & 2A_1B_1 \\ 2C_1D_1 & A_1D_1 + B_1C_1 \end{pmatrix}$$

This matrix corresponds to a symmetrical network since $A = D$. Furthermore, the symmetrical network is reciprocal [e.g., $AD - BC = (A_1D_1 - B_1C_1)^2 = 1$] when the half network is reciprocal [e.g., $A_1D_1 - B_1C_1 = 1$].

If we define the even-mode input impedance of the half network Z_e as

$$Z_e = \left.\frac{V_1}{I_1}\right|_{\text{half-circuit open}} = \frac{A_1 \times \infty + B_1}{C_1 \times \infty + D_1} = \frac{A_1}{C_1}$$

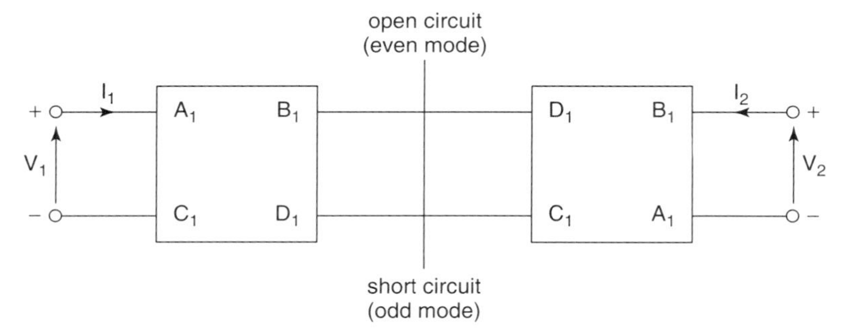

Figure 1.22 Decomposition of a symmetrical network.

and the odd-mode input impedance of the half network Z_o as

$$Z_o = \left.\frac{V_1}{I_1}\right|_{\text{half-circuit shorted}} = \frac{A_1 \times 0 + B_1}{C_1 \times 0 + D_1} = \frac{B_1}{D_1}$$

then for a reciprocal network (e.g., $A_1 D_1 - B_1 C_1 = 1$), the matrix of a symmetrical network can be expressed as

$$\begin{pmatrix} A & B \\ C & D \end{pmatrix} = \frac{1}{Z_e - Z_o} \begin{pmatrix} Z_e + Z_o & 2Z_e Z_o \\ 2 & Z_e + Z_o \end{pmatrix}$$

Using the conversion formulas from *ABCD* to *S* parameters, the *S* matrix of a symmetrical and a reciprocal network can be represented as

$$\begin{pmatrix} S_{11} & S_{12} \\ S_{21} & S_{22} \end{pmatrix} = \begin{pmatrix} \dfrac{2(Z_e Z_o - R_G R_L) - (Z_e + Z_o)(R_G - R_L)}{2(Z_e + R_G)(Z_o + R_L) + (Z_e - Z_o)(R_G - R_L)} & \dfrac{2\sqrt{R_G R_L}(Z_e - Z_o)}{2(Z_e + R_G)(Z_o + R_L) + (Z_e - Z_o)(R_G - R_L)} \\ \dfrac{2\sqrt{R_G R_L}(Z_e - Z_o)}{2(Z_e + R_G)(Z_o + R_L) + (Z_e - Z_o)(R_G - R_L)} & \dfrac{2(Z_e Z_o - R_G R_L) + (Z_e + Z_o)(R_G - R_L)}{2(Z_e + R_G)(Z_o + R_L) + (Z_e - Z_o)(R_G - R_L)} \end{pmatrix}$$

When the terminations are equal, $R_G = R_L = R_0$, and using the normalized even and odd input impedances $z_e = Z_e/R_0$ and $z_o = Z_o/R_0$, the *S* matrix reduces to

$$\begin{pmatrix} S_{11} & S_{12} \\ S_{21} & S_{22} \end{pmatrix} = \begin{pmatrix} \dfrac{z_e z_o - 1}{(z_e + 1)(z_o + 1)} & \dfrac{z_e - z_o}{(z_e + 1)(z_o + 1)} \\ \dfrac{z_e - z_o}{(z_e + 1)(z_o + 1)} & \dfrac{z_e z_o - 1}{(z_e + 1)(z_o + 1)} \end{pmatrix}$$

In addition, if we note that

$$S_{21} = S_{12} = \frac{1}{2}\frac{z_e - 1}{z_e + 1} - \frac{1}{2}\frac{z_o - 1}{z_o + 1} = \frac{1}{2}(S_{11e} - S_{11o})$$

$$S_{22} = S_{11} = \frac{1}{2}\frac{z_e - 1}{z_e + 1} + \frac{1}{2}\frac{z_o - 1}{z_o + 1} = \frac{1}{2}(S_{11e} + S_{11o})$$

the S parameters of a symmetrical network can be expressed in terms of the input reflection coefficient of the half network under open or shorted conditions:

$$\begin{pmatrix} S_{11} & S_{12} \\ S_{21} & S_{22} \end{pmatrix} = \frac{1}{2}\begin{pmatrix} S_{11e} + S_{11o} & S_{11e} - S_{11o} \\ S_{11e} - S_{11o} & S_{11e} + S_{11o} \end{pmatrix}$$

Similar formulations can be found using normalized even and odd input admittances $y_e = 1/z_e$ and $y_o = 1/z_o$.

The bisection theorem will be used in the case of symmetrical networks. Often, these networks will be composed of impedance or admittance inverters. The half network and even- and odd-mode impedances of an impedance inverter are given below.

In the case of an ideal impedance inverter K, the *ABCD* matrix can be written as

$$\begin{pmatrix} A & B \\ C & D \end{pmatrix} = \begin{pmatrix} 0 & jK \\ \dfrac{j}{K} & 0 \end{pmatrix} = \frac{1}{2}\begin{pmatrix} 1 & jK \\ \dfrac{j}{K} & 1 \end{pmatrix}^2$$

The half network of an impedance inverter K and its reverse half network can then be given by

$$\begin{pmatrix} A_1 & B_1 \\ C_1 & D_1 \end{pmatrix} = \frac{1}{\sqrt{2}}\begin{pmatrix} 1 & jK \\ \dfrac{j}{K} & 1 \end{pmatrix} \quad \text{and} \quad \begin{pmatrix} D_1 & B_1 \\ C_1 & A_1 \end{pmatrix} = \frac{1}{\sqrt{2}}\begin{pmatrix} 1 & jK \\ \dfrac{j}{K} & 1 \end{pmatrix}$$

We can check that the product of *ABCD* matrices of the half networks provides the *ABCD* matrix of the ideal inverter:

$$\begin{aligned}
\begin{pmatrix} A_1 & B_1 \\ C_1 & D_1 \end{pmatrix}\begin{pmatrix} D_1 & B_1 \\ C_1 & A_1 \end{pmatrix} &= \frac{1}{2}\begin{pmatrix} 1 & jK \\ \dfrac{j}{K} & 1 \end{pmatrix}\begin{pmatrix} 1 & jK \\ \dfrac{j}{K} & 1 \end{pmatrix} \\
&= \frac{1}{2}\begin{pmatrix} (1)(1) + (jK)\left(\dfrac{j}{K}\right) & (1)(jK) + (jK)(1) \\ \left(\dfrac{j}{K}\right)(1) + (1)\left(\dfrac{j}{K}\right) & \left(\dfrac{j}{K}\right)(jK) + (1)(1) \end{pmatrix} \\
&= \begin{pmatrix} 0 & jK \\ \dfrac{j}{K} & 0 \end{pmatrix}
\end{aligned}$$

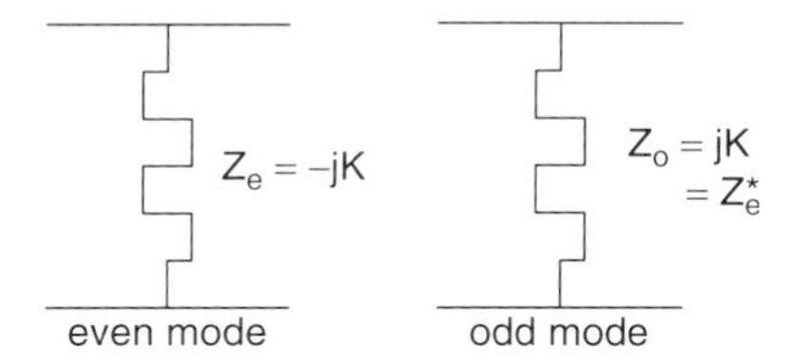

Figure 1.23 Even- and odd-mode models of an impedance inverter K.

The even-mode input impedance of the half network is given by $Z_e = A_1/C_1 = -jK$, and the odd-mode input impedance of the half network is given by $Z_o = B_1/D_1 = +jK$. The even- and odd-mode models of an impedance inverter are shown in Figure 1.23. For an admittance inverter J it can be shown that the even-mode input admittance of the half network is given by $Y_e = 1/Z_e = jJ$ and the odd-mode input admittance is given by $Y_o = Y_e^* = -jJ$.

1.6 CONCLUSIONS

In this chapter we recalled some of the fundamental relations for scattering and *ABCD* characterizations of two-port systems. We have also introduced the notion of impedance and admittance inverters, which are not easily realized using lumped elements but will be key in designing microwave filters. The bisection theorem has introduced the concept of even and odd modes that can be used to reduce the complexity of microwave filter design [1.5].

REFERENCES

[1.1] J. D. Rhodes, *Theory of Electrical Filters* Wiley, New York, 1976.

[1.2] H. Baher, *Synthesis of Electrical Networks*, Wiley, New York, 1984.

[1.3] J. S. Hong and M. J. Lancaster, *Microstrip Filters for RF/Microwave Applications*, Wiley, New York 2001.

[1.4] G. L. Matthaei, L. Young, and E. M. T. Jones, *Microwave Filters, Impedance-Matching Networks, and Coupling Structures*, McGraw-Hill, New York, 1964.

[1.5] P. Jarry, ed., Special Volume on Microwave Filters and Amplifiers, *Research Signpost*, 2005.

2

APPROXIMATIONS AND SYNTHESIS

2.1 INTRODUCTION

In this chapter we present functions that approximate ideal filter characteristics and the synthesis of these functions under the form of lumped *LC* ladders and

---

Advanced Design Techniques and Realizations of Microwave and RF Filters,
By Pierre Jarry and Jacques Beneat

ladders based on impedance and admittance inverters. First, the ideal characteristics of a low-pass filter are identified. Then several functions that approximate the ideal low-pass magnitude response as well as the ideal low-pass phase response are reviewed. A general synthesis method for a category of these functions is provided. The synthesis results are given under the form of *LC* prototype ladders. Since the prototype ladders are normalized to given termination and frequency conditions, it is important to provide impedance and frequency scaling techniques to meet the actual termination and frequency requirements. The chapter concludes with prototype ladders composed of impedance or admittance inverters. These ladders are more flexible and generally more adequate to define a microwave filter design method for the microwave structures encountered throughout this book.

2.2 IDEAL LOW-PASS FILTERING CHARACTERISTICS

The responses of an ideal low-pass filter are shown in Figure 2.1. The magnitude response is equal to 1 for frequencies starting at zero to a frequency ω_c. This band of frequencies is referred to as the *passband of the filter* and ω_c is called the *passband edge* or *cutoff frequency*. The magnitude response is then equal to zero for frequencies from ω_c to infinity. This band of frequencies is referred to as the *stopband of the filter*. In addition, the phase response of the filter should be linear with frequency in the passband and can have any value in the stopband since the magnitude there is zero. The conditions can be expressed mathematically as

$$
\begin{aligned}
|S_{21}(j\omega)|^2 &= 1 \qquad && \omega \le \omega_c \\
|S_{21}(j\omega)|^2 &= 0 && \omega > \omega_c \\
\phi(\omega) &= -k\omega && \omega \le \omega_c
\end{aligned}
$$

The ideal low-pass filter is unfortunately not physically realizable. There exist several functions that can approximate the ideal low-pass magnitude response that are realizable. The most common functions are the Butterworth, Chebyshev,

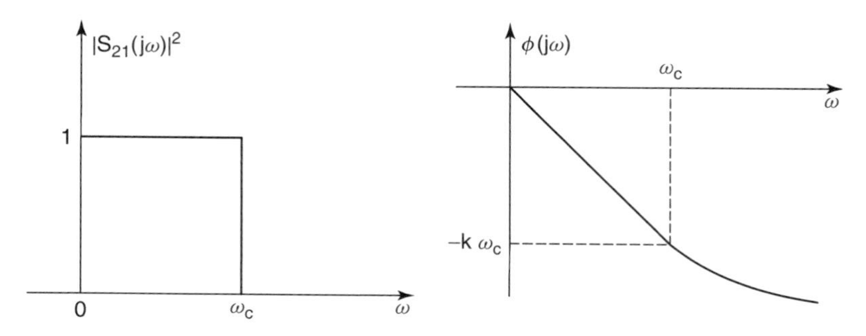

Figure 2.1 Ideal low-pass filtering characteristics.

and the elliptic functions. The generalized Chebyshev function, also called the pseudoelliptic function, requires special attention since the low-pass function can be asymmetric around zero leading to complex-valued transfer functions. There also exist a few functions that can approximate the ideal low-pass phase response that are realizable. These are the Bessel and Rhodes equidistant linear-phase functions.

Generally, these functions can be synthesized as a lumped *LC* circuit or as an inverter-based circuit. These "prototype" circuits are most often defined for a cutoff frequency of 1-rad/s and 1-Ω terminations. It will therefore be necessary to use transformation techniques that modify the elements of the low-pass filter prototype to provide low-pass or bandpass microwave filters with specified cutoffs and terminations. The frequency transformation techniques preserve the magnitude but not the phase characteristics. This makes the design of bandpass linear-phase microwave filters a particularly difficult problem.

2.3 FUNCTIONS APPROXIMATING THE IDEAL LOW-PASS MAGNITUDE RESPONSE

2.3.1 Butterworth Function

The Butterworth function [2.1] approximates the ideal low-pass magnitude response by providing a maximally flat behavior at $\omega = 0$ and $\omega = \infty$. The Butterworth function is given by

$$|S_{21}(j\omega)|^2 = \frac{1}{1 + \omega^{2n}}$$

The term *maximally flat* refers to the fact that the derivatives of this function up to nth derivative are equal to zero at $\omega = 0$ and at $\omega = \infty$. For this reason the Butterworth function is also called the *maximally flat function*. Figure 2.2 shows the response of this function for different values of the parameter n. As the order

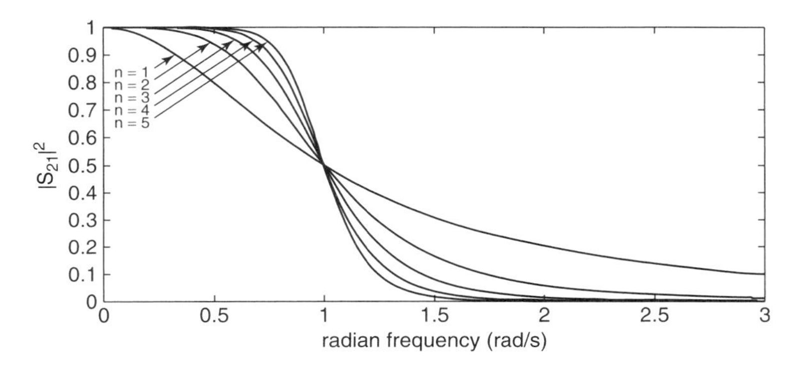

Figure 2.2 Low-pass Butterworth responses.

n of the Butterworth function increases, the better the function approximates the ideal low-pass magnitude response. Compared to the ideal low-pass response, the magnitude does not drop suddenly to zero after the cutoff frequency. One should then define three frequency bands: the passband, the transition region, and the stopband. A practical passband is not defined as the frequency band where the magnitude is equal to 1 but as the frequency band where the magnitude is greater than a given value. The stopband of a practical filter is the frequency band where the magnitude is smaller than a given value. The *transition region* is the frequency band between the passband and the stopband. The transition band should be as small as possible. How fast the magnitude of the function can drop from the passband to the stopband, or how small the transition region is, is characterized by what is called the *selectivity* of the filter. Figure 2.2 shows that the selectivity of the Butterworth function improves as the order n is increased. The passband edge for the Butterworth function is taken as the half power point (e.g., $|S_{21}|^2 = 0.5$) and occurs at the frequency $\omega = 1$ rad/s.

2.3.2 Chebyshev Function

The Chebyshev function [2.2] provides an equiripple behavior in the passband and a maximally flat response in the stopband. The Chebyshev function is given by

$$|S_{21}(j\omega)|^2 = \frac{1}{1 + \varepsilon^2 T_n^2(\omega)}$$

where $T_n(\omega)$ are Chebyshev polynomials of the first kind and can be generated using the recurrence formula

$$T_{n+1}(\omega) = 2\omega T_n(\omega) - T_{n-1}(\omega) \qquad \text{where} \quad T_0(\omega) = 1 \quad \text{and} \quad T_1(\omega) = \omega$$

The Chebyshev polynomials have the property that

$$T_n(\omega) = \begin{cases} \cos(n \cos^{-1} \omega) & 0 \leq |\omega| \leq 1 \\ \cosh(n \cosh^{-1} \omega) & 1 < |\omega| \end{cases}$$

The derivatives of the function up to nth derivative are equal to zero at $\omega = \infty$.

Figure 2.3 provides plots of the Chebyshev function for even and odd orders. When the order is odd, the function is equal to 1 at zero, and when the order is even, the function is equal to $1/(1 + \varepsilon^2)$ at zero. The function then ripples between 1 and $1/(1 + \varepsilon^2)$ until it reaches the frequency $\omega = 1$ rad/s. At that frequency, all the functions are at $1/(1 + \varepsilon^2)$ for the last time. Past $\omega = 1$ rad/s, the functions go to zero monotonically. The passband edge of the Chebyshev function is taken as the last crossing point at $1/(1 + \varepsilon^2)$ rather than the half-power point and is equal to $\omega = 1$ rad/s. As the order n of the Chebyshev function increases and as ε is kept small, the better the function approximates the ideal low-pass magnitude response. However, for the same order n, the Chebyshev function has better selectivity than that of the Butterworth function.

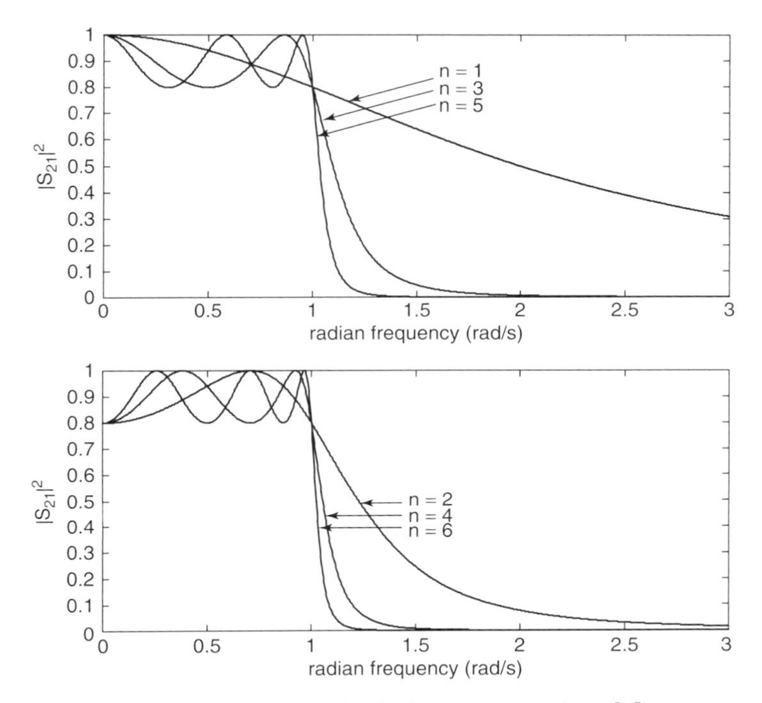

Figure 2.3 Low-pass Chebyshev responses ($\varepsilon = 0.5$).

2.3.3 Elliptic Function

The elliptic filters (or Cauer filters or Zolotarev filters) give equiripple responses [2.3] in both the passband and the stopband. The elliptic function is given as

$$|S_{21}(j\omega)|^2 = \frac{1}{1 + \varepsilon^2 F_n^2(\omega)}$$

The function $F_n(\omega)$ is a rational function of ω. The function is derived from the complete elliptic integral of the first kind and the Jacobian elliptic functions as shown below.

n Even	n Odd
$F_n(\omega) = C_e \prod_{r=1}^{n/2} \frac{\omega^2 - \omega_r^2}{\omega^2 - (\omega_s/\omega_r)^2}$	$F_n(\omega) = C_o \prod_{r=1}^{(n-1)/2} \frac{\omega^2 - \omega_r^2}{\omega^2 - (\omega_s/\omega_r)^2}$
$C_e = \prod_{r=1}^{n/2} \frac{1 - (\omega_s/\omega_r)^2}{1 - \omega_r^2}$	$C_o = \prod_{r=1}^{(n-1)/2} \frac{1 - (\omega_s/\omega_r)^2}{1 - \omega_r^2}$
$\omega_r = \mathrm{sn}\left(\frac{(2r-1)K(1/\omega_s^2)}{n}, \frac{1}{\omega_s^2}\right)$	$\omega_r = \mathrm{sn}\left(\frac{(2r)K(1/\omega_s^2)}{n}, \frac{1}{\omega_s^2}\right)$

In the expressions above, $K(m)$ is the complete elliptic integral of the first kind, defined as

$$K(m) = \int_0^{\frac{\pi}{2}} \frac{d\theta}{\sqrt{1 - m\sin^2\theta}}$$

and $\mathrm{sn}(u, m)$ is the elliptic sine provided by the Jacobian elliptic functions such that

$$\mathrm{sn}(u, m) = \sin\phi \qquad \text{where} \quad u(m, \phi) = \int_0^{\phi} \frac{d\theta}{\sqrt{1 - m\sin^2\theta}}$$

Note that $K(m)$ can be computed using the MATLAB function `ellipke(m)`, and the Jacobian elliptic sine $\mathrm{sn}(u, m)$ is given by `ellipj(u, m)`.

Figure 2.4(a) and (b) are examples of the elliptic function. The functions are defined primarily by the poles and zeros. For a third-order filter, the zero is equal to $\omega_{z1} = 0.9205$ and the pole is equal to $\omega_{p1} = \omega_s/\omega_{z1} = 1.3037$. For a fourth-order filter, the zeros are $\omega_{z1} = 0.4810$ and $\omega_{z2} = 0.9570$ and the poles are $\omega_{p1} = \omega_s/\omega_{z1} = 2.4949$ and $\omega_{p2} = \omega_s/\omega_{z2} = 1.2540$.

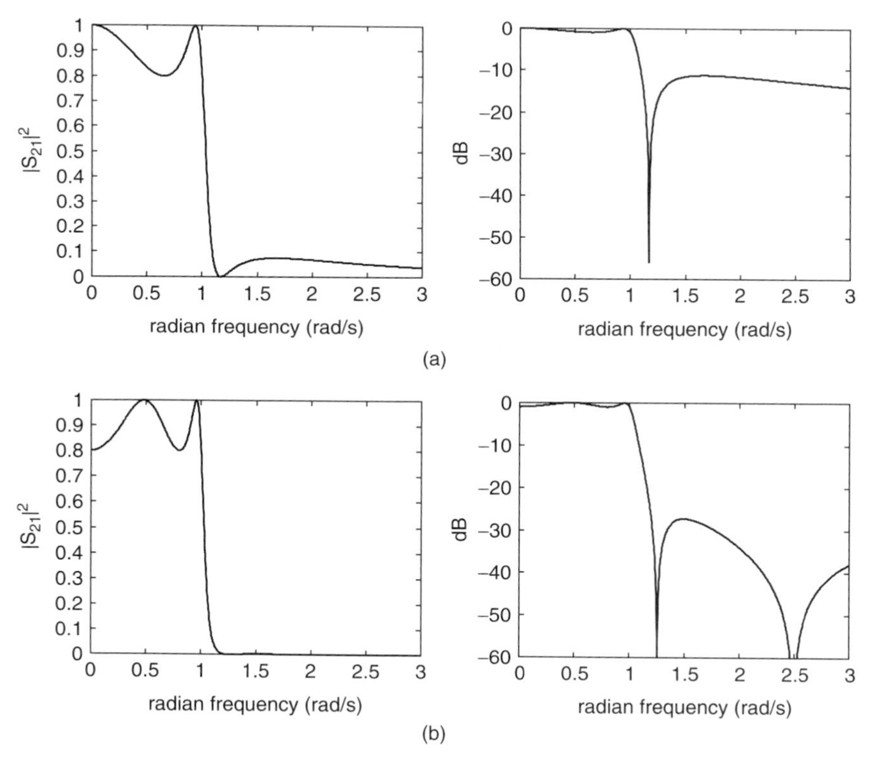

Figure 2.4 Low-pass elliptic magnitude responses: (a) $\varepsilon = 0.5$, $\omega_s = 1.2$, $n = 3$; (b) $\varepsilon = 0.5$, $\omega_s = 1.2$, $n = 4$.

The elliptic function provides much greater selectivity than the Chebyshev function and has often been considered as the ultimate approximation to the ideal low-pass filter in terms of magnitude response. However, it is difficult to design microwave filters with so many transmission zeros. Therefore, the generalized Chebyshev or pseudoelliptic function described next is preferred.

2.3.4 Generalized Chebyshev (Pseudoelliptic) Function

The generalized Chebyshev function provides an equiripple behavior in the passband and has a limited number of transmission zeros in the stopband. Sharpe [2.4] was one of the first to realize a generalized Chebyshev function as an electric ladder, and Cameron [2.5] gives a useful recursive technique to compute prototype elements of generalized Chebyshev filters.

The generalized Chebyshev function is expressed as

$$|S_{21}(\omega)|^2 = \frac{1}{1 + \varepsilon^2 F_n^2(\omega)}$$

where $F_n(\omega)$ is a function such that

$$F_n(\omega) = \begin{cases} \cos\left[(n - n_z)\cos^{-1}\omega + \sum_{k=1}^{n_z}\cos^{-1}\dfrac{1 - \omega\omega_k}{\omega - \omega_k}\right] & 0 \leq |\omega| \leq 1 \\ \cosh\left[(n - n_z)\cosh^{-1}\omega + \sum_{k=1}^{n_z}\cosh^{-1}\dfrac{1 - \omega\omega_k}{\omega - \omega_k}\right] & 1 < |\omega| \end{cases}$$

In these expressions, n_z is the number of purely imaginary zeros $(j\omega_1)$, $(j\omega_2)$, ..., $(j\omega_{n_z})$ that provide transmission zeros in the stopband.

When $n_z = 0$, the function $F_n(\omega)$ reduces to the traditional Chebyshev polynomial:

$$F_n(\omega) = T_n(\omega) = \begin{cases} \cos\left(n\cos^{-1}\omega\right) & 0 \leq |\omega| \leq \\ \cosh\left(n\cosh^{-1}\omega\right) & 1 < |\omega| \end{cases}$$

One should note that depending on the choice and number of frequencies ω_k, the response can be symmetrical or asymmetrical. Figure 2.5(a) shows a symmetrical response. In this case, $n = 4$ and $\varepsilon = 0.5$ and two zeros were specified at -1.2 and $+1.2$ rad/s. Figure 2.5(b) shows an asymmetrical response. In this case, $n = 4$ and $\varepsilon = 0.5$ and three zeros were specified at $-1.2 + 1.2$, and at 2.5 rad/s. The asymmetrical case requires special attention since due to the nonsymmetry around zero, the low-pass transfer function is complex valued, and traditional synthesis techniques provide circuits with complex-valued elements. As shown in Chapters 11 to 13, the design of microwave filters with asymmetrical generalized Chebyshev responses is a challenging problem.

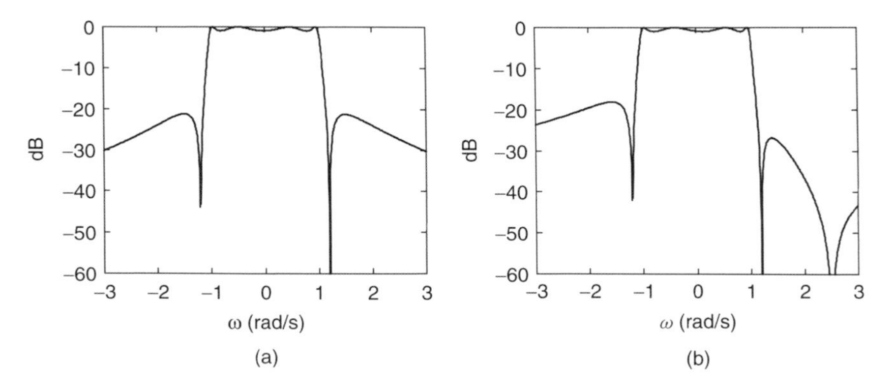

Figure 2.5 Low-pass generalized Chebyshev responses: (a) symmetrical; (b) asymmetrical.

2.4 FUNCTIONS APPROXIMATING THE IDEAL LOW-PASS PHASE RESPONSE

In this section, two transfer functions that provide an approximation to the linear-phase characteristic of the ideal low-pass filter are presented. The Bessel function [2.1] provides a maximally flat approximation while Rhodes equidistant linear-phase function [2.2] provides an equiripple approximation. Both functions only contain a denominator polynomial of order n.

When finding a design technique for linear-phase filters, the group delay function is often used. In the case of interest, the phase function can be expressed as

$$\phi(\omega) = -\tan^{-1}\frac{O(j\omega)}{E(j\omega)}$$

where $O(s)$ is the odd and $E(s)$ is the even polynomial that make the denominator polynomial. The group delay function is defined as the derivative of the phase function

$$T(\omega) = -\frac{\partial\phi(\omega)}{\partial\omega}$$

2.4.1 Bessel Function

The Bessel function provides a group delay function that presents maximally flat behavior at $\omega = 0$. The transfer function is given by

$$S_{21}(s) = \frac{1}{Q_n(s)}$$

where $Q_n(s)$ is a Bessel polynomial and can be generated [2.6] using the recurrence formula

$$Q_{n+1}(s) = Q_n(s) + \frac{s^2}{4n^2-1}Q_{n-1}(s) \qquad \text{with } Q_1(s) = 1 + s \quad \text{and} \quad Q_0(s) = 1$$

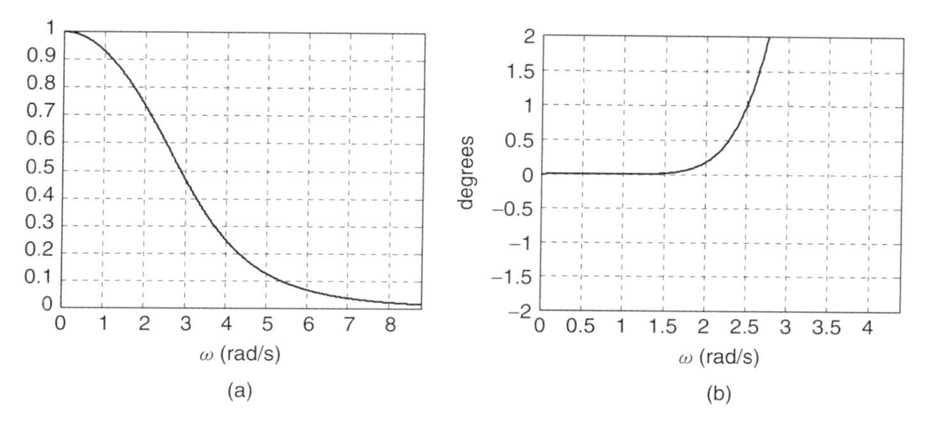

Figure 2.6 (a) Magnitude and (b) phase error responses for the Bessel function ($n = 4$).

Figure 2.6 shows the magnitude response and the phase error of the Bessel function. The phase error is defined as $\varepsilon(\omega) = \phi(\omega) - \omega$ and is also a good indicator of the linearity of the phase. It can be seen that the phase error is negligible around $\omega = 0$.

For magnitude-oriented functions the −3-dB point often serves as the reference value. For linear-phase filters there are no accepted phase error or group delay reference values. For example, the filter above has a phase error of less than 1° up to 2.5 rad/s. One could also use a 5° value. In addition, one should note that the −3-dB frequency is about 2 rad/s. A fifth-order Bessel function will have a −3-dB frequency of about 2.5 rad/s and a phase error of less than 1° up to 3.25 rad/s. The next function provides better means for controlling the frequency band where the phase is linear.

2.4.2 Rhodes Equidistant Linear-Phase Function

The equidistant linear-phase function provides a group delay function that interpolates the linear-phase at n equidistant frequencies [2.2]. The transfer function is given by

$$S_{21}(s) = \frac{1}{A_n(s)}$$

The polynomial $A_n(s)$ can be generated using the recurrence formula

$$A_{n+1}(s, \alpha) = A_n(s, \alpha) + \left(\frac{\tan\alpha}{\alpha}\right)^2 \frac{s^2 + (n\alpha)^2}{4n^2 - 1} A_{n-1}(s, \alpha)$$

with

$$A_1(s, \alpha) = 1 + \frac{\tan\alpha}{\alpha} s \quad \text{and} \quad A_0(s, \alpha) = 1 \quad \text{and} \quad \alpha < \frac{\pi}{2}$$

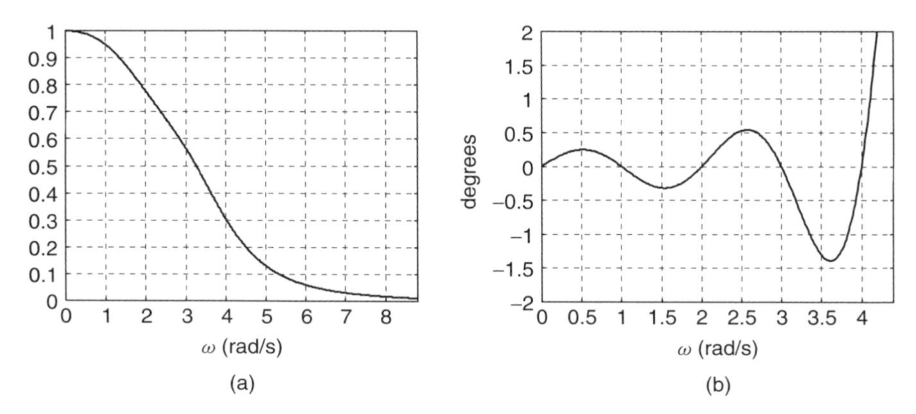

Figure 2.7 (a) Magnitude and (b) phase error responses for the ELP function ($\alpha = 1$ and $n = 4$).

The phase of this polynomial is exactly linear at a set of equidistant frequencies [2.2]:

$$\varepsilon(\omega) = \phi(\omega) - \omega = 0 \qquad \text{at } \omega = k\alpha \quad \text{for } k = 0, 1, 2, \ldots, n$$

Figure 2.7 shows the magnitude response and phase error of the equidistant linear-phase function. The phase error is zero at 0, α, 2α, 3α, and 4α where $\alpha = 1$. One can change the parameter α to change the frequency band where the phase is linear. For example, with $\alpha = 0.5$, the last interpolation point will occur at $n\alpha = 4 \times 0.5 = 2$ rad/s rather than the 4 rad/s shown in Figure 2.7. One notes that when taking $\alpha = 0$, the equidistant linear-phase function reduces to the Bessel function [e.g., $A_n(s, 0) = Q_n(s)$].

2.5 LOW-PASS LUMPED LADDER PROTOTYPES

2.5.1 General Synthesis Technique

Once the function $|S_{21}(j\omega)|^2$ has been defined, the first step in the synthesis of a filter will be to define a potential transfer function $S_{21}(s)$. This is done by letting $S_{21}(s)$ be the ratio of two polynomials and expressing them in terms of their roots:

$$S_{21}(s) = \frac{f(s)}{g(s)} = \frac{\prod_{j=1}^{m}(s - z_j)}{\prod_{k=1}^{n}(s - p_k)}$$

Note that

$$|S_{21}(j\omega)|^2 = S_{21}(j\omega)S_{21}^*(j\omega) = S_{21}(s)S_{21}(-s)|_{s=j\omega}$$

in which case

$$f(s)f(-s) = \prod_{j=1}^{m}(s - z_j)\prod_{j=1}^{m}(-s - z_j) = \pm\prod_{j=1}^{m}(s - z_j)\prod_{j=1}^{m}(s + z_j)$$

$$g(s)g(-s) = \prod_{k=1}^{n}(s - p_k)\prod_{k=1}^{n}(-s - p_k) = \pm\prod_{k=1}^{n}(s - p_k)\prod_{k=1}^{n}(s + p_k)$$

If $p_0 = a + jb$ is a complex pole of $S_{21}(s)S_{21}(-s)$, the poles p_0^*, $-p_0$, and $-p_0^*$ should also be present. If $p_0 = r$ is a real pole of $S_{21}(s)S_{21}(-s)$, the pole $-p_0$ should also be present. If $p_0 = j\omega_0$ is a purely imaginary pole of $S_{21}(s)S_{21}(-s)$, the pole p_0^* should also be present. In other words, one expects the poles (or zeros) of $S_{21}(s)S_{21}(-s)$ to come in groups of four or two:

$$\begin{array}{ccc} a + jb & & \\ a - jb & r & j\omega_0 \\ -a - jb & -r & -j\omega_0 \\ -a + jb & & \end{array}$$

For stability considerations, it is important to select the poles that have a real part that is negative (left-half-plane poles) when defining $S_{21}(s)$.

Example of a Third-Order Butterworth Function The starting function is given by

$$|S_{21}(j\omega)|^2 = \frac{1}{1 + \omega^6}$$

For $n = 3$, the poles are given by solving

$$\omega^6 + 1 = 0$$

For example, using the function `roots` in MATLAB, the poles in ω of $|S_{21}(j\omega)|^2$ are

```
-0.8660 + 0.5000i
-0.8660 - 0.5000i
 0.0000 + 1.0000i
 0.0000 - 1.0000i
 0.8660 + 0.5000i
 0.8660 - 0.5000i
```

There is a complex pole that comes in a group of four, and a purely imaginary pole that comes in a group of two. The corresponding poles of $S_{21}(s)S_{21}(-s)$ are given by multiplying the poles of $|S_{21}(j\omega)|^2$ by j. The poles in s of $S_{21}(s)S_{21}(-s)$ are therefore

```
-0.5000 - 0.8660i
 0.5000 - 0.8660i
-1.0000 + 0.0000i
 1.0000 + 0.0000i
-0.5000 + 0.8660i
 0.5000 + 0.8660i
```

For the stability of $S_{21}(s)$, the poles $-0.5 \pm j0.866$ and -1 are selected. This gives the transfer function

$$S_{21}(s) = \frac{1}{s^3 + 2s^2 + 2s + 1}$$

The second step in the synthesis is to define $S_{11}(s)$. For a lossless realization, $S_{21}(s)$ and $S_{11}(s)$ satisfy the condition

$$S_{21}(s)S_{21}(-s) + S_{11}(s)S_{11}(-s) = 1$$

This gives, for $S_{11}(s)S_{11}(-s)$,

$$S_{11}(s)S_{11}(-s) = 1 - S_{21}(s)S_{21}(-s) = \frac{g(s)g(-s) - f(s)f(-s)}{g(s)g(-s)}$$

The poles of $S_{11}(s)S_{11}(-s)$ are the same as those of $S_{21}(s)S_{21}(-s)$. Therefore, the denominator polynomial of $S_{11}(s)$ is taken as the denominator polynomial of $S_{21}(s)$. If $S_{11}(s)$ is expressed as the ratio of two polynomials,

$$S_{11}(s) = \frac{h(s)}{g(s)}$$

the next step is to define the polynomial $h(s)$ such that

$$h(s)h(-s) = g(s)g(-s) - f(s)f(-s)$$

The polynomial $g(s)g(-s) - f(s)f(-s)$ should have roots that appear in groups of four or groups of two as was the case in defining $S_{21}(s)$ so that $h(s)$ can be defined.

Example of a Third-Order Butterworth Function In this case, $f(s) = 1$ and $g(s) = s^3 + 2s^2 + 2s + 1$. This leads to

$$g(s)g(-s) - f(s)f(-s) = (s^3 + 2s^2 + 2s + 1)(-s^3 + 2s^2 - 2s + 1) \\ -(1)(1) = -s^6$$

The roots in this case are all equal to zero and $h(s)$ is given simply as $h(s) = \pm s^3$. This gives for $S_{11}(s)$,

$$S_{11}(s) = \frac{\pm s^3}{s^3 + 2s^2 + 2s + 1}$$

The third step in the synthesis is to define the input impedance $Z_{\text{in}}(s)$. From Chapter 1 the input impedance is given by

$$Z_{\text{in}}(s) = \frac{1 + S_{11}(s)}{1 - S_{11}(s)} = \frac{g(s) + h(s)}{g(s) - h(s)}$$

Example of a Third-Order Butterworth Function For the case $h(s) = +s^3$,

$$Z_{\text{in}}(s) = \frac{g(s) + h(s)}{g(s) - h(s)} = \frac{s^3 + 2s^2 + 2s + 1 + s^3}{s^3 + 2s^2 + 2s + 1 - s^3} = \frac{2s^3 + 2s^2 + 2s + 1}{2s^2 + 2s + 1}$$

The final step is to synthesize the input impedance $Z_{\text{in}}(s)$ as an electrical network. There are several different electrical networks that can create a given input impedance. For example, in the case of the third-order Butterworth function, the input impedance can be realized as the lumped ladder shown in Figure 2.8. The input impedance of this circuit is given by

$$Z_{\text{in}}(s) = Z_1(s) = L_1 s + \frac{1}{Y_1(s)}$$

$$Y_1(s) = C_1(s) + \frac{1}{Z_2(s)}$$

$$Z_2(s) = L_2 s + \frac{1}{Y_2(s)}$$

$$Y_2(s) = \frac{1}{R_0}$$

In the general case, the elements of the ladder are found by performing successive polynomial divisions:

$$Z_{\text{in}}(s) = g_1 s + \cfrac{1}{g_2 s + \cfrac{1}{g_3 s + \cfrac{\ddots}{g_n s + (1/R)\ (\text{or } R)}}}$$

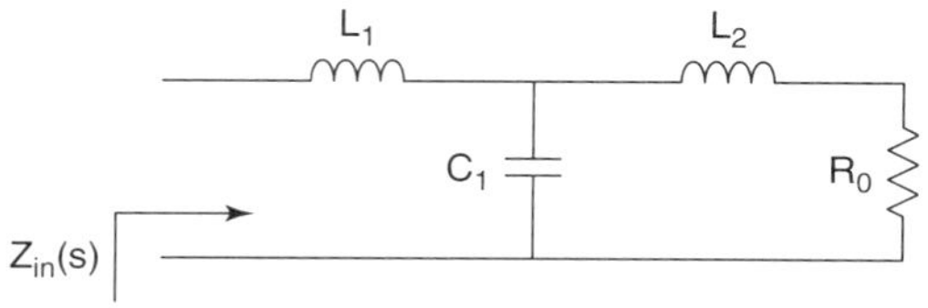

Figure 2.8 Synthesis of the third-order Butterworth function as a lumped LC ladder.

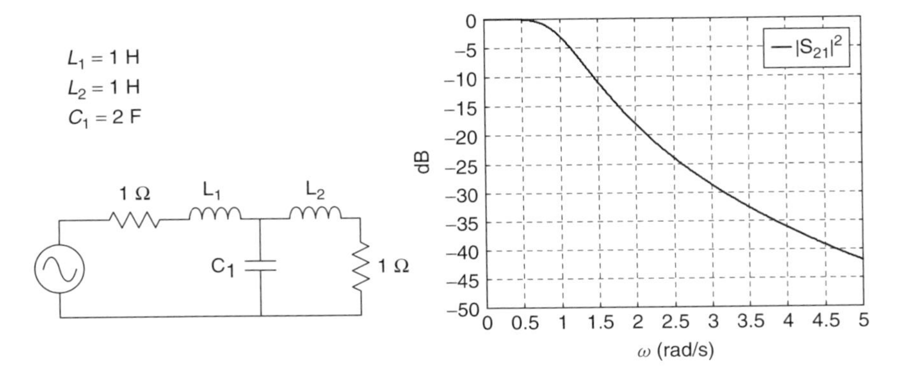

Figure 2.9 Third-order Butterworth function as a lumped LC ladder.

Example of a Third-Order Butterworth Function In this case the input impedance was given by

$$Z_{\text{in}}(s) = \frac{2s^3 + 2s^2 + 2s + 1}{2s^2 + 2s + 1}$$

Performing the successive polynomial divisions, we find that

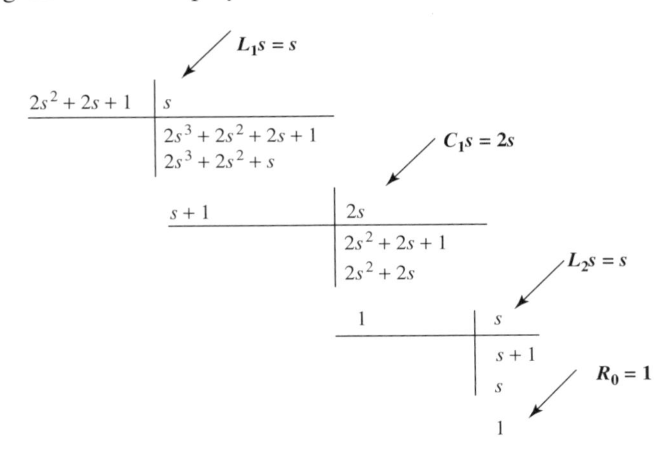

The lumped LC ladder that provides a third-order low-pass Butterworth response is shown in Figure 2.9 along with the simulation of the system using the $ABCD$ matrices of the simple elements and 1-Ω terminations for computing the S parameters.

2.5.2 Normalized Low-Pass Ladders

Several of the functions of Section 2.4 can be synthesized as low-pass LC prototype ladders (Figure 2.10). Note that these prototype circuits must be defined

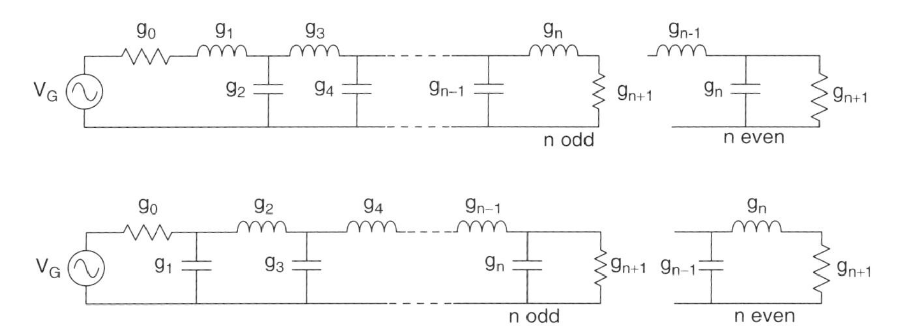

Figure 2.10 Low-pass *LC* prototype ladders.

with the terminations g_0 and g_{n+1} that were used in computing the elements $g_1, g_2, \ldots, g_n$.

Butterworth Low-Pass Ladder Prototype General closed-form expressions for the transfer functions, poles and zeros, and prototype elements can be provided for the Butterworth and Chebyshev functions. The low-pass Butterworth transfer function is defined by its poles:

$$S_{21}(s) = \frac{1}{\prod_{k=1}^{n}(s - p_k)}$$

where the poles are given by

$$p_k = -\sin\theta_k + j\cos\theta_k \qquad \text{with } \theta_k = \frac{2k-1}{2n}\pi \quad \text{for } k = 1, 2, \ldots, n$$

For a lossless system,

$$S_{11}(s) = \frac{\pm s^{2n}}{\prod_{k=1}^{n}(s - p_k)}$$

The elements of the prototype of Figure 2.10 for a low-pass Butterworth response normalized to a cutoff frequency of 1 rad/s and using terminations $g_0 = g_{n+1} = 1\ \Omega$ are given by

$$g_k = 2\sin\left(\frac{2k-1}{2n}\pi\right) \qquad \text{for } k = 1, 2, \ldots, n$$

where g_k represents the value of an inductance or the value of a capacitance, depending on the ladder model used.

Chebyshev Low-Pass Ladder Prototypes The low-pass Chebyshev transfer function is defined by its poles and a scaling factor K:

$$S_{21}(s) = \frac{K}{\prod_{k=1}^{n}(s - p_k)}$$

where the poles are given by

$$p_k = -\eta \sin\theta_k - j\sqrt{1+\eta^2}\cos\theta_k \qquad \text{with } \theta_k = \frac{2k-1}{2n}\pi \quad \text{for } k = 1, 2, \ldots, n$$

and where

$$\eta = \sinh\left(\frac{1}{n}\sinh^{-1}\frac{1}{\varepsilon}\right) \quad \text{and} \quad K = \prod_{k=1}^{n}\left(\sqrt{\eta^2 + \sin^2\frac{k\pi}{n}}\right)$$

For a lossless system,

$$S_{11}(s) = \frac{\pm\prod_{k=1}^{n}(s + j\cos\theta_k)}{\prod_{k=1}^{n}(s - p_k)}$$

The elements of the prototype of Figure 2.10 for a low-pass Chebyshev response normalized to a cutoff frequency of 1 rad/s and using terminations $g_0 = 1\ \Omega$ and g_{n+1} defined by the even or odd nature of the order n are given by

$$g_k g_{k+1} = \frac{4\sin\{[(2k-1)/2n]\pi\}\sin\{[(2k+1)/2n]\pi\}}{\eta^2 + \sin^2(k\pi/n)} \qquad k = 1, 2, \ldots, n-1$$

where

$$\eta = \sinh\left(\frac{1}{n}\sinh^{-1}\frac{1}{\varepsilon}\right) \quad \text{and} \quad g_1 = \frac{2}{\eta}\sin\frac{\pi}{2n}$$

When n is odd, the source and load resistance are $g_0 = g_{n+1} = 1\ \Omega$. However, when n is even, $|S_{21}(j\omega)|^2$ is equal to $1/(1+\varepsilon^2)$ at the origin. Therefore, the load resistor should be such that

$$|S_{21}(0)|^2 = \frac{4g_{n+1}}{(g_{n+1}+1)^2} = \frac{1}{1+\varepsilon^2}$$

The resistance values that satisfy this condition are given by

$$g_{n+1} = \begin{cases} (\varepsilon + \sqrt{1+\varepsilon^2})^2 & S_{11}(0) \geq 0 \\ \dfrac{1}{(\varepsilon + \sqrt{1+\varepsilon^2})^2} & S_{11}(0) \leq 0 \end{cases}$$

Other Functions Contrary to the Butterworth and Chebyshev functions, no closed-form expressions are available for the prototype elements of the Bessel and equidistant linear-phase transfer functions. The elements of the ladder must be computed using the general synthesis method. A low-pass equidistant linear-phase function synthesized as a lumped LC ladder and simulated phase error is shown in Figure 2.11. In the case of the elliptic and generalized Chebyshev functions,

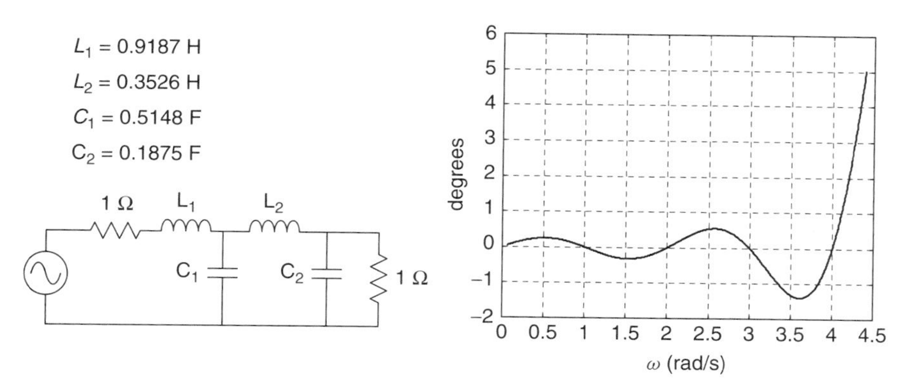

Figure 2.11 Equidistant linear-phase function as an LC ladder ($n = 4$, $\alpha = 1$).

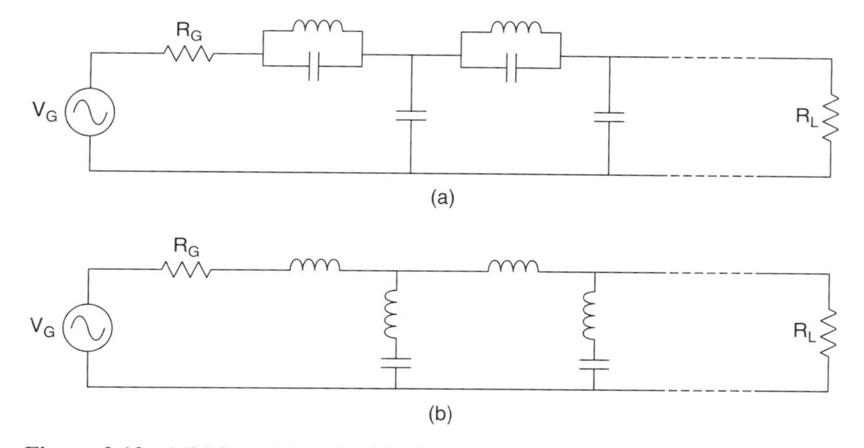

Figure 2.12 Midshunt (a) and midseries (b) realizations of the elliptic function.

the structure of the previous LC ladder prototypes is not adequate for producing transmission zeros. In the case of the elliptic function, it can generally be realized using the ladder structures shown in Figure 2.12 [2.7]. The generalized Chebyshev function will be synthesized as cross-coupled or in-line networks based on impedance or admittance inverters. Computing the transfer function is provided in Chapter 11, and defining the prototype elements is given in Chapters 11, 12, and 13.

2.6 IMPEDANCE AND FREQUENCY SCALING

2.6.1 Impedance Scaling

A practical filter requires terminations greater than 1 Ω. In this case the elements of the low-pass prototype must be scaled to accommodate the desired termination values. If the new termination conditions are

$$R_G = n_0 g_0$$
$$R_L = n_0 g_{n+1}$$

all the inductors and capacitors should be scaled and the new inductor and capacitor values L' and C' given in terms of the prototype inductor and capacitor values L and C by

$$L' = n_0 L$$
$$C' = \frac{1}{n_0} C$$

2.6.2 Frequency Scaling

From the normalized to 1 rad/s low-pass ladder prototypes, it is possible to define ladder prototypes that preserve the magnitude responses for low-pass, high-pass, bandpass, and bandstop characteristics. Unfortunately, the commonly used techniques, except for the low pass-to-low pass transformation, do not preserve the phase characteristics. Furthermore, transformations of low pass to low pass and low pass to bandpass are the most useful in microwave filter design and are presented below.

Low Pass-to-Low Pass In this case, the transfer function $S_{21}(s)$ is scaled according to the transformation

$$s \rightarrow \frac{s}{\omega_0}$$

where ω_0 is the desired cutoff. This amounts to scaling all the inductors and capacitors in the LC ladder prototypes in the following fashion:

$$L \quad \text{(inductor)} \longrightarrow \text{(inductor)} \quad \frac{L}{\omega_0}$$
$$C \quad \text{(capacitor)} \longrightarrow \text{(capacitor)} \quad \frac{C}{\omega_0}$$

Low Pass to Bandpass In this case, the transfer function $S_{21}(s)$ is scaled according to the transformation

$$s \rightarrow \alpha\left(\frac{s}{\omega_0} + \frac{\omega_0}{s}\right) \qquad \text{with } \alpha = \frac{\omega_0}{\omega_2 - \omega_1} \quad \text{and} \quad \omega_0 = \sqrt{\omega_1\,\omega_2}$$

where ω_1 and ω_2 are the desired passband edges of the bandpass filter. This amounts to replacing an inductor by a series inductor/capacitor combination and

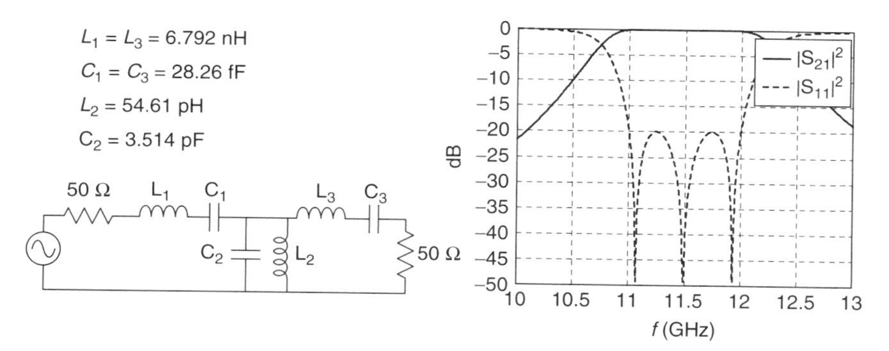

Figure 2.13 Third-order bandpass Chebyshev filter.

a capacitor by a parallel inductor/capacitor combination in the following fashion:

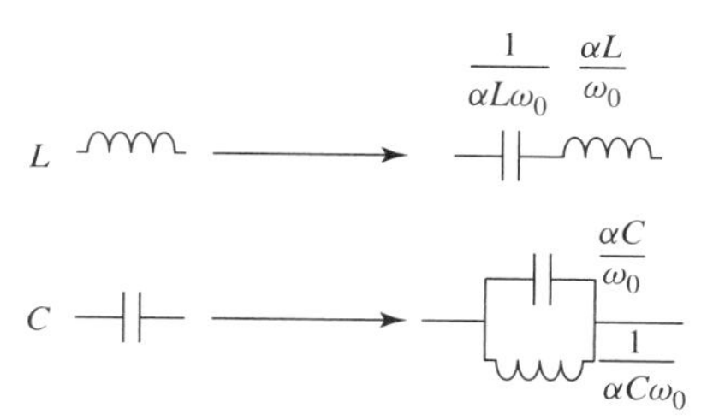

2.7 *LC* FILTER EXAMPLE

We wish to design a bandpass filter with passband edges of 11 and 12 GHz. The filter should be terminated with 50-Ω resistors. The filter should have a Chebyshev response of order 3 with a return loss of −20 dB. The bandpass prototype circuit and the forward and return loss coefficients of the filter are given in Figure 2.13. It can be seen that the values needed for the lumped inductors and capacitors are impossible to manufacture. Therefore, the lumped *LC* ladder prototypes are not used to design microwave filters directly. Instead, they are used as a reference when modeling a microwave structure. However, the prototype circuits based on impedance and admittance inverters presented next are more flexible and more appropriate when modeling a microwave structure.

2.8 IMPEDANCE AND ADMITTANCE INVERTER LADDERS

2.8.1 Low-Pass Prototypes

When designing microwave filters [2.8], it is often preferable to use ladder prototypes based on impedance or admittance inverters as shown in Figure 2.14.

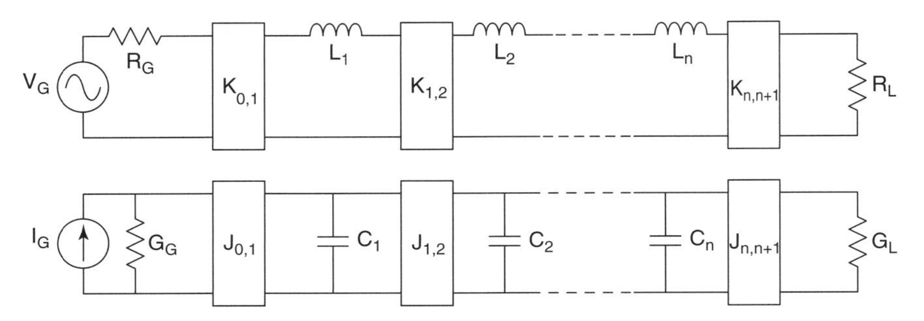

Figure 2.14 Low-pass ladder prototypes using impedance/admittance inverters.

The element values using impedance inverters are given by

$$K_{k,k+1} = \sqrt{\frac{L_k L_{k+1}}{g_k g_{k+1}}} \qquad \text{for } k = 1, 2, \ldots, n-1$$

$$K_{0,1} = \sqrt{\frac{R_G L_1}{g_0 g_1}} \quad \text{and} \quad K_{n,n+1} = \sqrt{\frac{L_n R_L}{g_n g_{n+1}}}$$

where $g_0, g_1, \ldots, g_n, g_{n+1}$ are the LC ladder prototype element values of Figure 2.10. The inductor values are totally arbitrary. One should note that there is flexibility for the terminations of the circuits in Figure 2.14. For example, generally, $g_1, \ldots, g_n$ in Figure 2.10 are defined for $g_0 = g_{n+1} = 1$. Here one could chose $R_G = R_L = 50\ \Omega$ and the effect appears only on $K_{0,1}$ and $K_{n,n+1}$.

The element values using admittance inverters are given by

$$J_{k,k+1} = \sqrt{\frac{C_k C_{k+1}}{g_k g_{k+1}}} \qquad \text{for } k = 1, 2, \ldots, n-1$$

$$J_{0,1} = \sqrt{\frac{G_G C_1}{g_0 g_1}} \quad \text{and} \quad J_{n,n+1} = \sqrt{\frac{C_n G_L}{g_n g_{n+1}}}$$

where $g_0, g_1, \ldots, g_n, g_{n+1}$ are the LC ladder prototype element values of Figure 2.10. The capacitor values are totally arbitrary. Note that the issue of the termination $g_{n+1} \neq 1$ in the case of even-order Chebyshev filters can be avoided by taking $K_{0,1} = K_{n,n+1}$ or $J_{0,1} = J_{n,n+1}$. In this case, the circuits of Figure 2.14 can provide an even-order Chebyshev response when the terminations are both equal.

2.8.2 Scaling Flexibility

In the case of the low-pass ladders using impedance or admittance inverters, we note the following property [2.9]:

$$S_{21}(s)S_{21}(-s) = 2R_G \frac{Z_{\text{in}}(s) + Z_{\text{in}}(-s)}{[R_G + Z_{\text{in}}(s)][R_G + Z_{\text{in}}(s)]}$$

$$S_{21}(s)S_{21}(-s) = 2n_0 R_G \frac{n_0 Z_{\text{in}}(s) + n_0 Z_{\text{in}}(-s)}{[n_0 R_G + n_0 Z_{\text{in}}(s)][n_0 R_G + n_0 Z_{\text{in}}(s)]}$$

This means that changing R_G to $n_0 R_G$ and $Z_{\text{in}}(s)$ to $n_0 Z_{\text{in}}(s)$ does not modify the response. In the case of an impedance inverter realization, the input impedance $n_0 Z_{\text{in}}(s)$ is given by the fraction expansion

$$n_0 Z_{\text{in}}(s) = \cfrac{n_0 n_1 K_{01}^2}{n_1 L_1 s + \cfrac{n_1 n_2 K_{12}^2}{n_2 L_2 s + \cfrac{n_2 n_3 K_{23}^2}{\ddots \quad n_n L_n s + \cfrac{n_n n_{n+1} K_{n,n+1}^2}{n_{n+1} R_L}}}}$$

This property allows for great flexibility to modify the low-pass ladder to accommodate the natural behavior of a microwave structure. For example, suppose that in Figure 2.15 the first ladder was designed with normalized terminations and normalized inductor values. However, the microwave structure might not be able to provide for this particular combination. The property above shows that one can change the values of the inductors to better comply with the range of values obtainable from the microwave structure. The change in inductor value is translated to changes in the inverter values, as shown in the second ladder in Figure 2.15.

For example, one could choose new values for the resistors and inductors and then scale the impedance inverters appropriately:

$$n_0 = \frac{R_G(\text{new})}{R_G(\text{old})} \qquad n_{n+1} = \frac{R_L(\text{new})}{R_L(\text{old})} \quad \text{and} \quad n_k = \frac{L(\text{new})}{L(\text{old})} \quad \text{then} \quad K_{k,k+1}(\text{new}) = \sqrt{n_k n_{k+1}}\; K_{k,k+1}(\text{old})$$

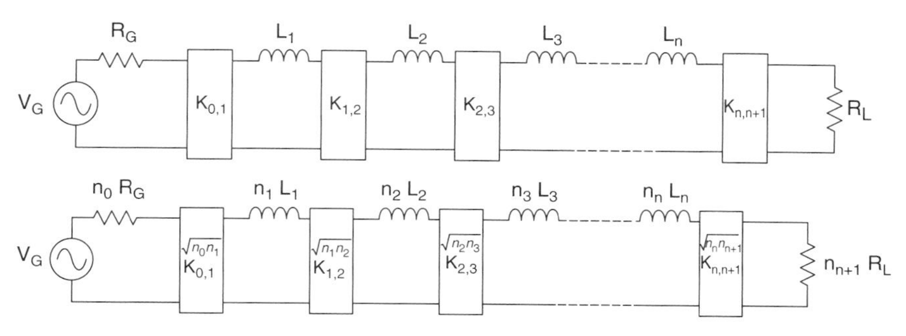

Figure 2.15 Impedance scaling options for impedance inverter ladders.

If one is mainly interested in finding the ladder prototype terminated on 50 Ω rather than on 1 Ω, one takes $n_0 = n_{n+1} = 50$ and then chooses $n_k = 1$ for $k = 1, 2, \ldots, n$. This places the scaling only on the first and last impedance inverters.

2.8.3 Bandpass Ladders

The bandpass impedance or admittance inverter prototypes are shown in Figure 2.16. For the impedance inverter case, the series inductor L has been replaced by a series $L_a + C_a$ combination similarly to the low pass-to-bandpass transformation of Section 2.6. For the admittance inverter case, the parallel capacitor C has been replaced by a parallel $L_a || C_a$ combination. In addition, the effects of the center frequency ω_0 and the parameter α are also seen on the value of the inverters.

Impedance Inverter	Admittance Inverter
$K_{k,k+1} = \dfrac{\omega_0}{\alpha}\sqrt{\dfrac{L_{ak}L_{ak+1}}{g_k g_{k+1}}}$ for $k = 1, 2, \ldots, n-1$	$J_{k,k+1} = \dfrac{\omega_0}{\alpha}\sqrt{\dfrac{C_{ak}C_{ak+1}}{g_k g_{k+1}}}$ for $k = 1, 2, \ldots, n-1$
$K_{0,1} = \sqrt{\dfrac{\omega_0}{\alpha}\dfrac{R_G L_{a1}}{g_0 g_1}}$ and $K_{n,n+1} = \sqrt{\dfrac{\omega_0}{\alpha}\dfrac{L_{an} R_L}{g_n g_{n+1}}}$	$J_{0,1} = \sqrt{\dfrac{\omega_0}{\alpha}\dfrac{G_G C_{a1}}{g_0 g_1}}$ and $J_{n,n+1} = \sqrt{\dfrac{\omega_0}{\alpha}\dfrac{C_{an} G_L}{g_n g_{n+1}}}$

where $\omega_0 = \sqrt{\omega_1\omega_2}$, $\alpha = \omega_0/(\omega_2 - \omega_1)$, and $g_0, g_1, \ldots, g_n, g_{n+1}$ are given from the low-pass LC ladder prototype elements. In addition, one can either use arbitrary inductors and capacitors as long as they satisfy $L_a C_a \omega_0^2 = 1$. This is done automatically when applying the low pass-to-bandpass transformation on the arbitrary inductors L or arbitrary capacitors C of the low-pass ladder.

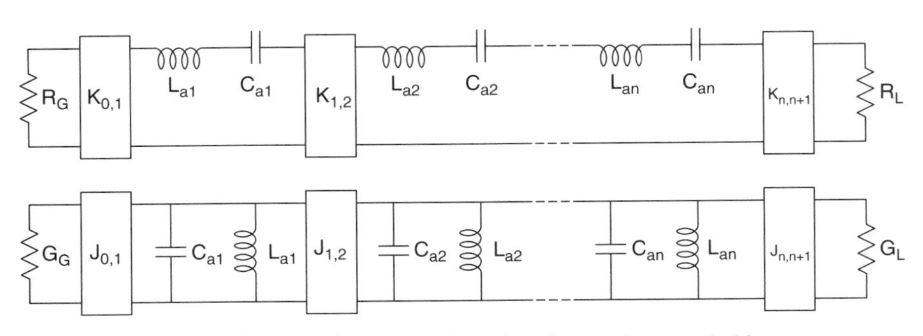

Figure 2.16 Bandpass impedance/admittance inverter ladders.

2.8.4 Filter Examples

The first example provides the impedance inverter and inductor values used for a low-pass ladder normalized to 1 rad/s. The inductor is selected arbitrarily as 10 mH. The filter should have a Chebyshev response of order 4 with a return loss of −20 dB.

K	L
$K_{01} = 0.103515$	$L_1 = 10$ mH
$K_{12} = 0.009106$	$L_2 = 10$ mH
$K_{23} = 0.006999$	$L_3 = 10$ mH
$K_{34} = 0.009106$	$L_4 = 10$ mH
$K_{45} = 0.114441$	

For an even-order Chebyshev function, the last inverter is different from the first and the ladder should be terminated on $R_G = 1\ \Omega$ and $R_L = 1.22\ \Omega$. Note that R_G was taken as g_0 and R_L as g_{n+1}. One would get the same response by taking $K_{01} = K_{45} = 0.103515$ and using $R_G = R_L = 1\ \Omega$. The forward and return coefficients are plotted in Figure 2.17. One observes the symmetrical behavior of this filter around zero. One can also check the expected 1-rad/s normalized cutoff frequency and the −20-dB return loss.

The second example provides the admittance inverter and capacitor/inductor values used for a 11- to 12-GHz bandpass ladder. The ladder is based on the low-pass ladder where all capacitors C were equal to 1 μF. The filter should have a Chebyshev response of order 4 with a return loss of −20 dB. In this case we use $G_G = G_L = 1/50\ \Omega$ and $J_{01} = J_{45}$:

J	C_a	L_a
$J_{01} = 1.46 \times 10^{-4}$	$C_{a1} = 0.159$ fF	$L_{a1} = 1.21$ nH
$J_{12} = 9.11 \times 10^{-7}$	$C_{a2} = 0.159$ fF	$L_{a2} = 1.21$ nH
$J_{23} = 7.00 \times 10^{-7}$	$C_{a3} = 0.159$ fF	$L_{a3} = 1.21$ nH
$J_{34} = 9.11 \times 10^{-7}$	$C_{a4} = 0.159$ fF	$L_{a4} = 1.21$ nH
$J_{45} = 1.46 \times 10^{-4}$		

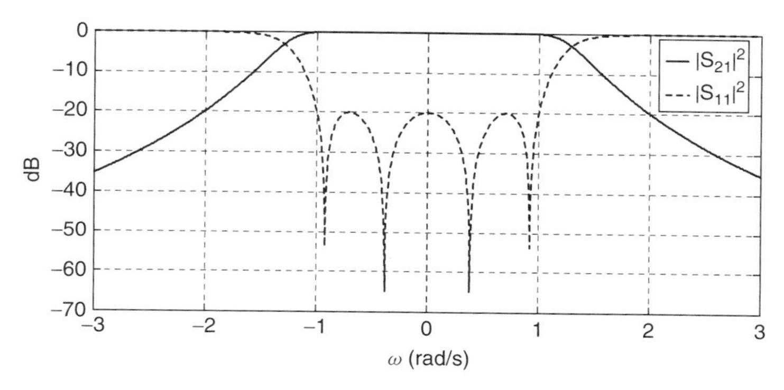

Figure 2.17 Forward and return coefficients of the low-pass example.

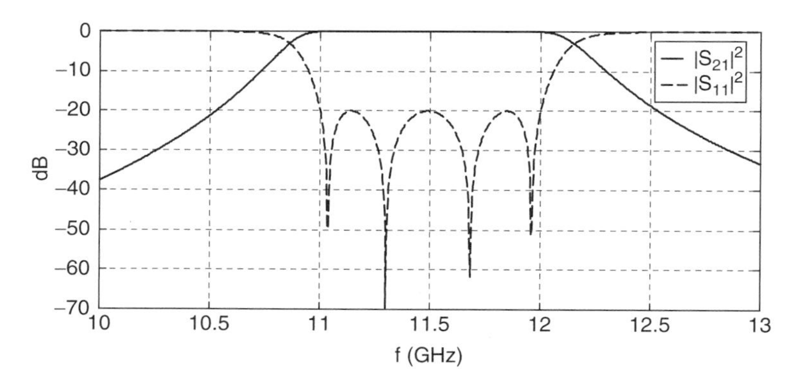

Figure 2.18 Forward and return coefficients of the bandpass example.

The forward and return coefficients are plotted in Figure 2.18. One observes the correct passband edges and return loss.

2.9 CONCLUSIONS

In this chapter we surveyed the different approximations used in filter synthesis. The classical amplitude approximations using Butterworth, Chebyshev, or elliptic functions were reviewed and the generalized Chebyshev or pseudoelliptic function was introduced. These are more appropriate for microwave filters than the elliptic response that requires a large number of transmission zeros. Low-pass phase approximations are given by the Bessel and Rhodes equidistant linear-phase functions. The general synthesis technique for realizing these functions under the form of *LC* ladders was also recalled. Impedance and frequency transformation techniques were also provided since formulas and tables of ladder prototypes are often normalized to 1 Ω and 1 rad/s. The chapter ended with a review of impedance/admittance inverter ladder realizations of the most common responses. The flexibility of these ladders was introduced and gives an indication as to why they are preferred when designing microwave filters.

REFERENCES

[2.1] H. Baher, *Synthesis of Electrical Networks*, Wiley, New York 1984.

[2.2] J. D. Rhodes, *Theory of Electrical Filters*, Wiley, New York, 1976.

[2.3] G. L. Matthaei, L. Young, and E. M. T. Jones, *Microwave Filters, Impedance-Matching Networks, and Coupling Structures*, McGraw-Hill, New York, 1964.

[2.4] C. B. Sharpe, A general Tchebycheff rational function, *Proc. IRE*, pp. 454–457, Feb. 1954.

[2.5] R. J. Cameron, Fast generation of Chebyschev filter prototypes with asymmetrically-prescribed transmission zeros, *ESA J.*, vol. 6, pp. 83–95, 1982.

[2.6] W. E. Thomson, Delay networks having a maximally-flat frequency characteristics *Proc. IEE*, vol. 96, pp. 487–490, Nov. 1949.

[2.7] P. Amstutz, Elliptic approximation and elliptic filter design on small computers, *IEEE Trans. Circuits Syst.*, vol. 25, no. 12, 1978.

[2.8] J. S. Hong, and M. J. Lancaster, *Microstrip Filters for RF/Microwave Applications*, Wiley, New York, 2001.

[2.9] P. Jarry, *Microwave Filters and Couplers*, University of Bordeaux, Bordeaux, France, 2003.

3

WAVEGUIDES AND TRANSMISSION LINES

3.1 INTRODUCTION

In this chapter we review the field equations, modes, and cavity properties for two of the most common types of waveguides. We also describe transmission lines such as the stripline and the suspended substrate stripline. We conclude with a brief review of distributed circuits.

3.2 RECTANGULAR WAVEGUIDES AND CAVITIES

3.2.1 Rectangular Waveguides

Rectangular waveguides are the easiest structure to construct. They provide low losses and good power handling. Unfortunately, they are often bulky and difficult

Advanced Design Techniques and Realizations of Microwave and RF Filters,
By Pierre Jarry and Jacques Beneat

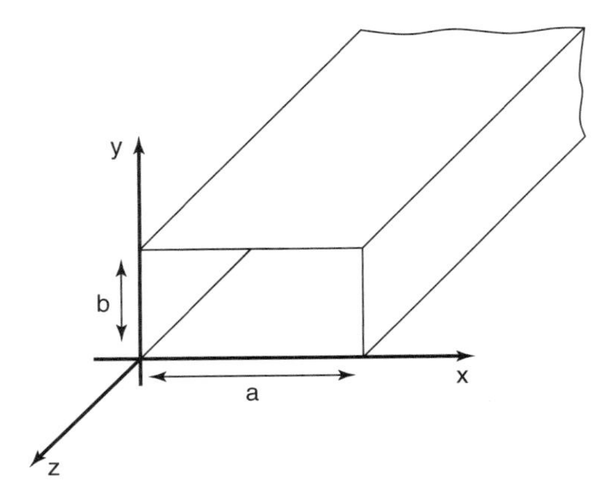

Figure 3.1 Rectangular waveguide.

to integrate with active components. A typical rectangular waveguide is shown in Figure 3.1. The waveguides properties are derived from solving the Maxwell equations with the given boundary conditions. The waveguide is made of a single conductor, and as a result, only transverse electric (TE) and transverse magnetic (TM) wave propagation is possible. However, there can be an infinity of waves denoted as $TE_{n,m}$ and $TM_{n,m}$ that satisfy the Maxwell equations. These wave solutions are also referred to as *modes*. In the case of $TE_{n,m}$ modes, the electric field in the z direction is zero ($E_z = 0$), while in the case of $TM_{n,m}$ modes, the magnetic field in the z direction is zero ($H_z = 0$). The equations for these electric and magnetic fields, along with useful parameters for rectangular waveguides, are summarized in Table 3.1.

Since, theoretically, many modes can exist in a rectangular waveguide, it is important to check how one could control their number. For example, in Table 3.1, $\beta_{n,m}$ represents the propagation function of a mode n,m. In this expression $k_0 = \omega/c$, where c is the speed of light and ω is the frequency of operation. For a given $k_{c,n,m}$, as the frequency of operation ω drops, $\beta_{n,m}$ can become purely imaginary and the term $e^{-j\beta_{mn}z}$ will attenuate the amplitude of the wave rapidly as it travels in the z direction. The lowest frequency of operation for a given mode is therefore

$$\frac{\omega}{c} = k_{c,n,m} \quad \text{or} \quad f_{c,n,m} = \frac{c}{2\pi} k_{c,n,m}$$

For a rectangular waveguide, the TE_{10} mode has the smallest cutoff frequency and is referred to as the *fundamental* or *dominant mode*. This means that a TE_{10} mode can still propagate without attenuation in the waveguide when all other modes have vanished if the frequency of operation is above the cutoff frequency of the TE_{10} mode and below the cutoff frequency of all other modes. This is why the TE_{10} mode is often used when designing monomode microwave systems based on rectangular waveguides.

TABLE 3.1. Mode Properties for a Rectangular Waveguide

	TE	TM
H_z	$\cos\frac{n\pi x}{a}\cos\frac{m\pi y}{b}e^{-j\beta_{mn}z}$	0
E_z	0	$\sin\frac{n\pi x}{a}\sin\frac{m\pi y}{b}e^{-j\beta_{nm}z}$
E_x	$Z_{h,n,m}H_y$	$\frac{-j\beta_{nm}m\pi}{ak_{c,n,m}^2}\cos\frac{n\pi x}{a}\sin\frac{m\pi y}{b}e^{-j\beta_{nm}z}$
E_y	$-Z_{h,n,m}H_x$	$\frac{-j\beta_{nm}n\pi}{bk_{c,n,m}^2}\sin\frac{n\pi x}{a}\cos\frac{m\pi y}{b}e^{-j\beta_{nm}z}$
H_x	$\frac{j\beta_{nm}n\pi}{ak_{c,n,m}^2}\sin\frac{n\pi x}{a}\cos\frac{m\pi y}{b}e^{-j\beta_{nm}z}$	$-\frac{E_y}{Z_{e,n,m}}$
H_y	$\frac{j\beta_{nm}m\pi}{bk_{c,n,m}^2}\cos\frac{n\pi x}{a}\sin\frac{m\pi y}{b}e^{-j\beta_{nm}z}$	$\frac{E_x}{Z_{e,n,m}}$
$Z_{h,n,m}$	$\frac{k_o}{\beta_{n,m}}Z_c$	
$Z_{e,n,m}$		$\frac{\beta_{n,m}}{k_o}Z_c$
$k_{c,n,m}$	$\sqrt{\left(\frac{n\pi}{a}\right)^2+\left(\frac{m\pi}{b}\right)^2}$	
$\beta_{n,m}$	$\sqrt{k_0^2-k_{c,n,m}^2}$	
$f_{c,n,m}$	$\frac{c}{2\pi}k_{c,n,m}$	
$\lambda_{c,n,m}$	$\frac{2\pi}{k_{c,n,m}}$	

In Table 3.1, the fields are multiplied by the term $e^{-j\beta_{mn}z}$, and under normal circumstances, $\beta_{n,m}$ is purely real. This means that the wave will propagate without any loss over large values of z. This occurs when the waveguide is a perfect conductor. In reality, the waveguide is not a perfect conductor and there are losses that occur as the wave travels in the z direction. The term $e^{-j\beta_{mn}z}$ should then be replaced by a more general term $e^{-(\alpha+j\beta_{mn})z}$, where α is called the attenuation constant.

The attenuation constant α_{TE} for a $\mathrm{TE}_{n,m}$ mode is given by

$$\alpha_{\mathrm{TE}} =$$

$$\frac{2R_m}{bZ_0\sqrt{1-k_{c,n,m}^2/k_0^2}}\left[\left(1+\frac{b}{a}\right)\frac{k_{c,n,m}^2}{k_0^2}+\frac{b}{a}\left(\frac{\varepsilon_{0m}}{2}-\frac{k_{c,n,m}^2}{k_0^2}\right)\frac{n^2ab+m^2a^2}{n^2b^2+m^2a^2}\right]$$

where $R_m = \sqrt{\omega\mu_0/2\sigma}$ is the resistive part of the surface impedance and $\varepsilon_{0m} = 1$ for $m = 0$ and 2 for $m \geq 1$. The attenuation constant α_{TM} for a $\mathrm{TM}_{n,m}$ mode is given by

$$\alpha_{\mathrm{TM}} = \frac{2R_m}{bZ_0\sqrt{1 - k_{c,n,m}^2/k_0^2}} \frac{n^2b^2 + m^2a^2}{n^2b^2a + m^2a^2}$$

It is often too complex to include the effects of the attenuation constants in expressions when defining a method for designing microwave filters. Additional results may be found in a book by Collin [3.1]: in particular, expressions for surface impedance and power loss.

3.2.2 Rectangular Cavities

Figure 3.2 shows a rectangular cavity of length d. It may be considered as a section of a rectangular waveguide terminated in a short at both ends. A cavity can be used in microwave filters to create a resonating cell similar to lumped *RLC* resonating cells. The main goal is to define the resonating frequency of the resonating cell. For a $\mathrm{TE}_{n,m}$ or $\mathrm{TM}_{n,m}$ mode, the propagation constant is given by

$$\beta_{n,m} = \sqrt{k_0^2 - \left(\frac{n\pi}{a}\right)^2 - \left(\frac{m\pi}{b}\right)^2}$$

The electric tangential field should be equal to 0 at $z = 0$ and $z = d$. This provides a condition on the propagation constant such that

$$\beta_{n,m} = \frac{l\pi}{d} \qquad l = 1,\ 2,\ 3,\ \ldots$$

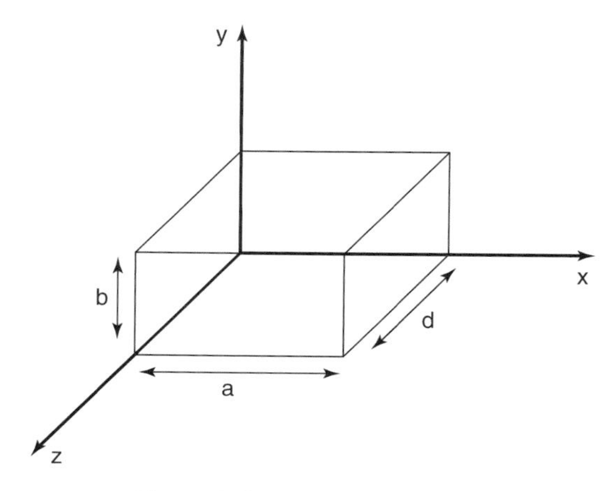

Figure 3.2 Rectangular cavity.

This can only occur for discrete values of k_0 such that

$$k_0 = k_{n,m,l} = \sqrt{\left(\frac{n\pi}{a}\right)^2 + \left(\frac{m\pi}{b}\right)^2 + \left(\frac{l\pi}{d}\right)^2}$$

These particular values of k_0 give the resonant frequencies of the cavity

$$f_{n,m,l} = \frac{c}{2\pi} k_{n,m,l} = c\sqrt{\left(\frac{n}{2a}\right)^2 + \left(\frac{m}{2b}\right)^2 + \left(\frac{l}{2d}\right)^2}$$

The TE_{10} mode can provide the lowest resonant frequency when $l = 1$. The lowest resonant mode is therefore the TE_{101} resonant mode.

One should note that it is possible to obtain the same resonant frequency, depending on the mode, the cross sections of the waveguide, and the length of the cavity. In this book the resonant modes for rectangular cavities TE_{102}, TE_{301}, TE_{100}, TM_{120}, TM_{210}, ... will be of particular interest.

3.3 CIRCULAR WAVEGUIDES AND CAVITIES

3.3.1 Circular Waveguides

Figure 3.3 shows a circular waveguide with a circular cross section of radius a. For a circular waveguide, cylindrical coordinates are most appropriate for the analysis [3.2]. The properties of the $TE_{n,m}$ and $TM_{n,m}$ modes in a circular waveguide are given in Table 3.2. The parameter $x_{n,m}$ is the mth root of the nth Bessel function $[J_n(x) = 0]$, and the parameter $x'_{n,m}$ is the mth root of the derivative of the nth Bessel function $[J'_n(x) = 0]$.

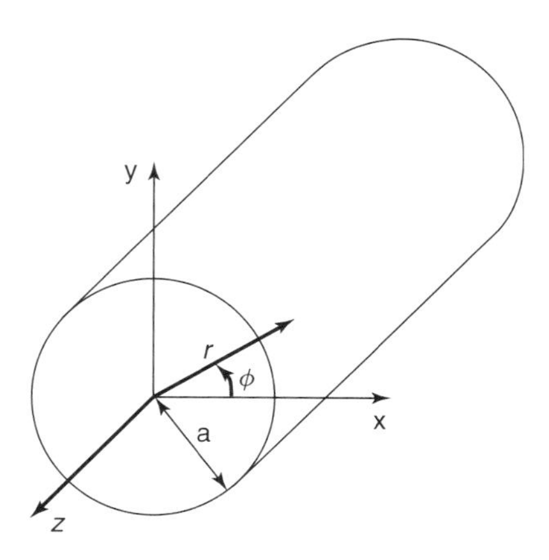

Figure 3.3 Circular waveguide.

TABLE 3.2. Mode Properties for a Circular Waveguide

	TE	TM
H_z	$J_n\left(\frac{x'_{nm}r}{a}\right)e^{-j\beta_{nm}z}\begin{cases}\cos n\phi\\ \sin n\phi\end{cases}$	0
E_z	0	$J_n\left(\frac{x_{nm}r}{a}\right)e^{-j\beta_{nm}z}\begin{cases}\cos n\phi\\ \sin n\phi\end{cases}$
H_r	$-\frac{j\beta_{nm}x'_{nm}}{ak^2_{c,nm}}J'_n\left(\frac{x'_{nm}r}{a}\right)e^{-j\beta_{nm}z}\begin{cases}\cos n\phi\\ \sin n\phi\end{cases}$	$-\frac{E_\phi}{Z_{e,nm}}$
H_ϕ	$-\frac{jn\beta_{nm}}{rk^2_{c,nm}}J_n\left(\frac{x'_{nm}r}{a}\right)e^{-j\beta_{nm}z}\begin{cases}-\sin n\phi\\ \cos n\phi\end{cases}$	$\frac{E_r}{Z_{e,nm}}$
E_r	$Z_{h,nm}H_\phi$	$-\frac{j\beta_{nm}x_{nm}}{ak^2_{c,nm}}J'_n\left(\frac{x_{nm}r}{a}\right)e^{-j\beta_{nm}z}\begin{cases}\cos n\phi\\ \sin n\phi\end{cases}$
E_ϕ	$-Z_{h,nm}H_r$	$-\frac{jn\beta_{nm}}{rk^2_{c,nm}}J_n\left(\frac{x_{nm}r}{a}\right)e^{-j\beta_{nm}z}\begin{cases}-\sin n\phi\\ \cos n\phi\end{cases}$
$Z_{h,n,m}$	$\frac{k_o}{\beta_{n,m}}Z_c$	
$Z_{e,n,m}$		$\frac{\beta_{n,m}}{k_o}Z_c$
$k_{c,n,m}$	$\frac{x'_{n,m}}{a}$	$\frac{x_{n,m}}{a}$
$\beta_{n,m}$	$\sqrt{k_0^2-k^2_{c,n,m}}$	
$f_{c,n,m}$	$\frac{c}{2\pi}k_{c,n,m}$	
$\lambda_{c,n,m}$	$\frac{2\pi}{k_{c,n,m}}$	

Table 3.3 lists values for $x'_{n,m}$ and $x_{n,m}$ for the first few $\mathrm{TE}_{n,m}$ and $\mathrm{TM}_{n,m}$ modes. Note that $x'_{0m}=x_{1m}$ since $dJ_0(x)/dx=-J_1(x)$ and the TE_{0m} and TM_{1m} modes are said to be degenerate. From Table 3.3, the smallest value, 1.841, occurs for the TE_{11} mode. The TE_{11} mode therefore has the lowest cutoff frequency and is the fundamental or dominant mode in a circular waveguide. The attenuation constants α_{TE} and α_{TM} for TE and TM modes in a circular waveguide are given by

$$\alpha_{\mathrm{TE}}=\frac{R_m}{aZ_0\sqrt{1-k^2_{c,n,m}/k_0^2}}\left(\frac{k^2_{c,n,m}}{k_0^2}+\frac{n^2}{x'^2_{nm}-n^2}\right)$$

$$\alpha_{\mathrm{TM}}=\frac{R_m}{aZ_0\sqrt{1-k^2_{c,n,m}/k_0^2}}$$

TABLE 3.3. Roots of Bessel Functions

	Values of $x'_{n,m}$ for $TE_{n,m}$ Modes			Values of $x_{n,m}$ for $TM_{n,m}$ Modes		
n	$x'_{n,1}$	$x'_{n,2}$	$x'_{n,3}$	$x_{n,1}$	$x_{n,2}$	$x_{n,3}$
0	3.832	7.016	10.174	2.405	5.520	8.654
1	1.841	5.331	8.536	3.832	7.016	10.174
2	3.054	6.706	9.970	5.135	8.417	11.620

One should note that even if the TE_{11} mode is the fundamental mode, TE_{0m} modes are often preferred, due to their attenuation, which decreases with frequency.

3.3.2 Cylindrical Cavities

A cylindrical cavity is a section of a circular waveguide of length d and radius a, with short-circuiting plates at each end, as shown in Figure 3.4. In the case of a circular cavity the resonant frequencies are not the same for the TE and TM modes. The resonating frequency of a $TE_{n,m,l}$ resonant mode is given by

$$f_{n,m,l} = \frac{c}{2\pi}\sqrt{\left(\frac{x'_{n,m}}{a}\right)^2 + \left(\frac{l\pi}{d}\right)^2}$$

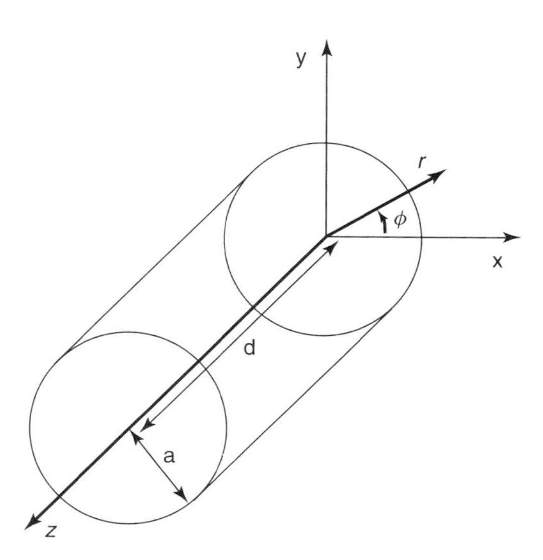

Figure 3.4 Cylindrical cavity.

and the resonating frequency of a $TM_{n,m,l}$ resonant mode is given by

$$f_{n,m,l} = \frac{c}{2\pi}\sqrt{\left(\frac{x_{n,m}}{a}\right)^2 + \left(\frac{l\pi}{d}\right)^2}$$

When deciding on a given cavity, it is important to check their unloaded Q factors. The Q factor gives a sense of how much energy is lost in a structure. Clearly, one seeks structures that can provide high Q factors.

The quality factor for a $TE_{n,m,l}$ resonating mode of a circular cavity is given by

$$2\pi Q\frac{\delta_S}{\lambda_0} = \frac{[1-(n/x'_{n,m})^2]\sqrt{(x'_{n,m})^2+(l\pi a/d)^2}}{(x'_{n,m})^2+(2a/d)(l\pi a/d)^2+(1-2a/d)(nl\pi a/x'_{n,m}\,d)^2}$$

In the case of a $TM_{n,m,l}$ resonant mode, the quality factor is given by

$$2\pi Q\frac{\delta_S}{\lambda_0} = \begin{cases} \dfrac{x_{n,m}}{1+a/d} & l = 0 \\ \dfrac{\sqrt{x_{n,m}^2+(l\pi a/d)^2}}{1+2a/d} & l > 0 \end{cases}$$

It can be shown that the quality factor of the TE_{011} resonant mode is greater than that of the TE_{111} resonant mode. This is another reason why TE_{01} modes are usually preferred over the fundamental TE_{11} mode. Also, in general, $TM_{n,m,l}$ resonant modes are not often used with circular waveguides. In this book, the resonant modes for circular cavities that are of interest are

$$TE_{011},\ TE_{113},\ \ldots$$

3.4 EVANESCENT MODES

Evanescent modes occur when the propagation constant $\beta_{n,m}$ is imaginary, resulting in modes that vanish after some distance into the waveguide, due to the $e^{-j\beta_{nm}z}$ term. This occurs when the quantity $k_0^2 - k_{c,n,m}^2$ is negative [3.3]. This property was already in use to eliminate higher-order modes in the waveguide so that only the fundamental mode was propagating in the system. This has the effect of placing a limitation on the reduction in size of the waveguide for a given frequency of operation since the cutoff frequencies are inverse proportional to the cross-sectional dimensions of the waveguide. In the case of evanescent-mode waveguide filters, the fundamental mode will also be considered as an evanescent mode, at least in parts of the structure. For a given frequency of operation, this has the effect of relaxing the requirements regarding cross-sectional dimensions. On the other side, as will be seen in some detail in Chapter 6, the evanescent nature of the modes results in both characteristic impedances that are no longer real and other unusual properties.

3.5 PLANAR TRANSMISSION LINES

Planar transmission lines (i.e., microstrip, stripline, suspended substrate stripline, etc.) provide useful transmission media for microwave filters. The structure of microstrip and stripline is shown in Figure 3.5. The microstrip consists of a thin conducting strip of width W placed on top of a substrate of relative permittivity ε_r. The substrate itself lies on top of a conducting ground plane. The stripline is made of a conducting strip of width W placed in the center of a substrate of relative permittivity ε_r. The substrate is placed between two conducting ground planes.

Even if microstrip lines are commonly used in microwave circuits, they have several disadvantages. The characteristic impedances of the transmission line is often small (10 to 100 Ω). They also have significant transmission losses, and only low-quality factors Q are possible. Striplines can provide higher characteristic impedances (20 to 150 Ω) and have smaller losses. An exact analysis of the homogeneous stripline was first described by Cohn [3.4,3.5]. In practice, the dimensions of the stripline are chosen so as to operate in a TEM mode while keeping losses to a minimum.

The characteristic impedance of stripline can be computed using the equation [3.6]

$$Z_0 = \begin{cases} \dfrac{30\pi}{\sqrt{\varepsilon_r}} \dfrac{1}{(W/b) + 0.441} & \dfrac{W}{b} > 0.35 \\ \dfrac{30\pi}{\sqrt{\varepsilon_r}} \dfrac{1}{(W/b) - (0.35 - W/b)^2 + 0.441} & \dfrac{W}{b} < 0.35 \end{cases}$$

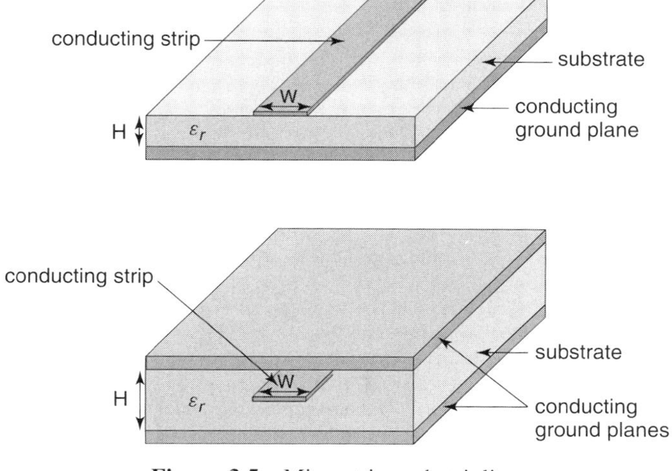

Figure 3.5 Microstrip and stripline.

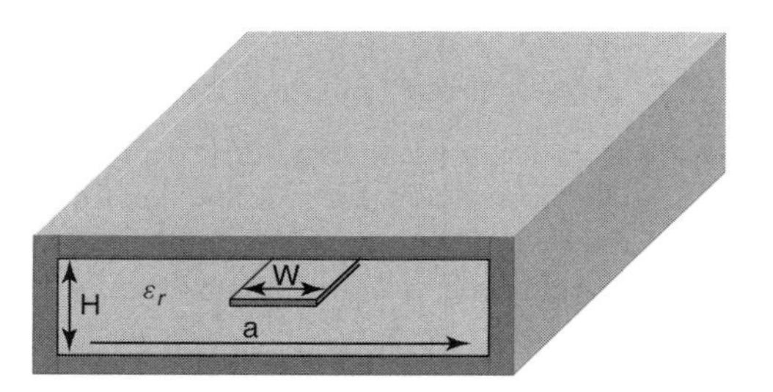

Figure 3.6 Enclosed stripline.

An enclosed stripline is shown in Figure 3.6. Besides the top and bottom conducting ground planes, the substrate is closed by conductors on both the left and right sides. The characteristic impedance for the enclosed stripline is given by [3.6]

$$Z_0 = \frac{\sqrt{\varepsilon_r}}{c}\frac{1}{C} \quad \text{where } C = \frac{W}{\displaystyle\sum_{\substack{n=1\\ \text{odd}}}^{\infty} \frac{4a \sin(n\pi W/2a)\sinh(n\pi H/4a)}{(n\pi)^2 \varepsilon_0 \varepsilon_r \cosh(n\pi H/2a)}} \quad \text{F/m}$$

As the relative dielectric permittivity increases, the characteristic impedance decreases. The highest characteristic impedances would be obtained when the relative permittivity is that of air (e.g., 1). A structure that makes use of this property is the suspended substrate stripline (SSS) shown in Figure 3.7. The conducting strip is placed on top of a thin substrate of thickness h. The substrate and the conducting strip are enclosed in a housing conductor.

It is useful to define the concept of effective relative permittivity ε_{eff}. Figure 3.8 shows the suspended substrate stripline on the left and an approximation to the

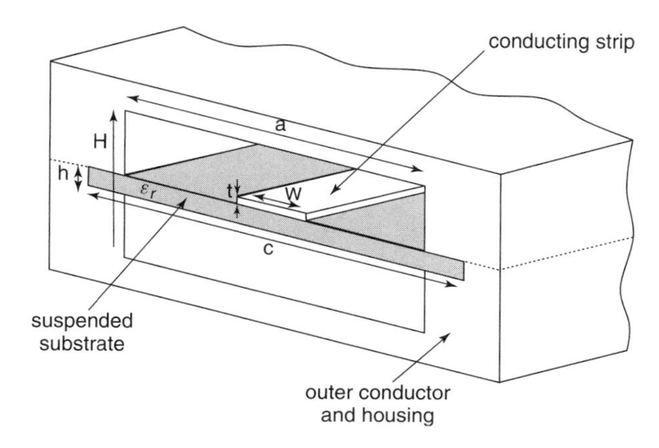

Figure 3.7 Suspended substrate stripline.

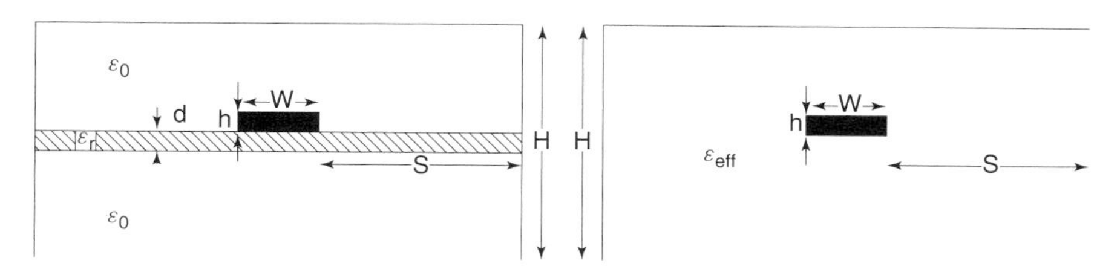

Figure 3.8 Concept of effective relative permittivity ε_{eff}.

behavior of the structure on the right. Since the surroundings of the conducting strip consist mainly of air, the effective relative permittivity of the structure will be close to 1. The approximation consists of replacing the suspended substrate stripline by a regular stripline of relative permittivity ε_{eff}.

The effective relative permittivity in the case of the suspended substrate stripline is provided by [3.7]

$$\varepsilon_{\text{eff}} = 1 + \frac{2(\varepsilon_r - 1)(h/S)}{4[W/(H-h)] + \dfrac{8}{\pi}\log\{1 + \coth[\pi S/(H-h)]\}} \quad \text{where } S = \frac{a - W}{2}$$

The characteristic impedance of the suspended substrate stripline can then be approximated by using the equation for the enclosed stripline with the effective relative permittivity ε_{eff} instead of the relative permittivity of the substrate ε_r.

There are several numerical methods for better estimating the characteristic impedance of striplines and SSS. A variational technique has been described by Tong and Jarry [3.8] and compared to the moment method. Figure 3.9 shows the

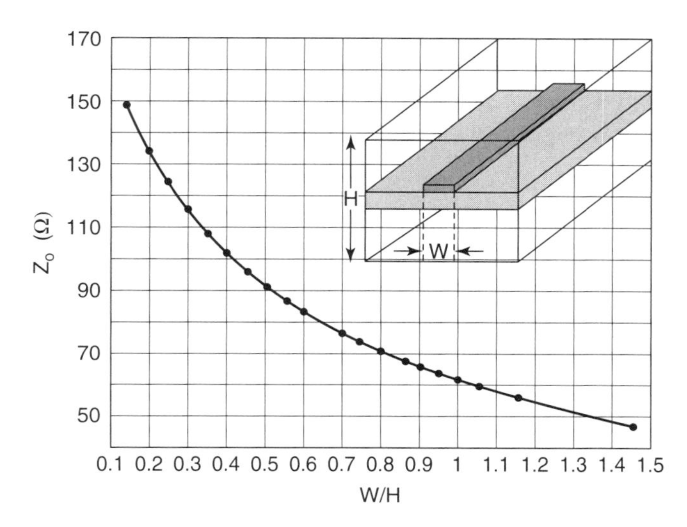

Figure 3.9 Characteristic impedance of a suspended substrate stripline.

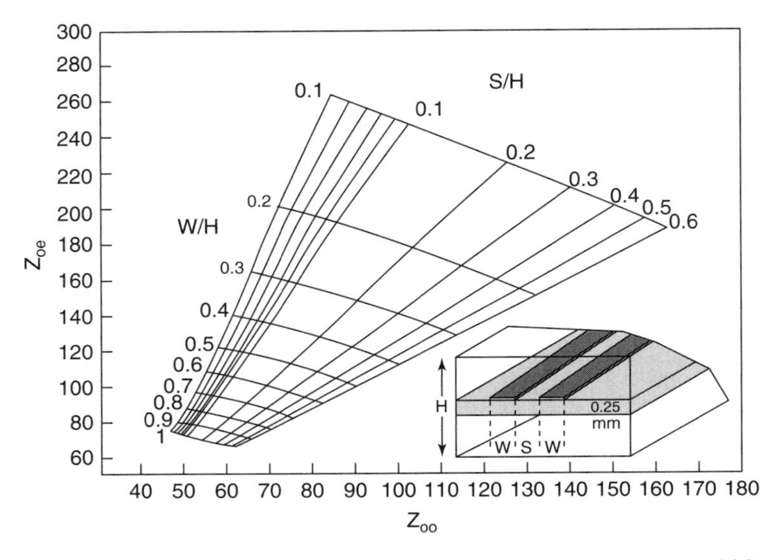

Figure 3.10 Even- and odd-mode impedances of a coupled line in SSS.

characteristic impedance of a single suspended stripline using this method. As will be seen in capacitive-gap coupled line filters, it will also be important to characterize the coupling between two lines. Figure 3.10 provides an example of the even- and odd-mode impedances of a coupled SSS structure.

3.6 DISTRIBUTED CIRCUITS

There exists a modern distributed-circuit theory based on the work of Paul Richards. This theory is based on systems made of transmission lines of equal length. The results of lumped circuit theory can be applied to distributed circuits through the use of the Richards transform [3.9]:

$$t = \tanh\left(\frac{l}{v}s\right) = \tanh \tau_0 s = j \tan\left(\frac{\pi}{2}\frac{\omega}{\omega_o}\right)$$

where v is the wave propagation speed, τ_0 is the delay of the transmission line of length l_0, and s and t are complex variables such that

$$s = \sigma + j\omega$$
$$t = \Sigma + j\Omega$$

The domain of the variable t is referred to as the *distributed domain*. The main element in distributed circuits is a transmission line of length l_0. The *ABCD* matrix of a transmission line of characteristic impedance Z_c or characteristic admittance $Y_c = 1/Z_c$ is expressed in the distributed domain as

$$\begin{pmatrix} A & B \\ C & D \end{pmatrix} = \frac{1}{\sqrt{1-t^2}} \begin{pmatrix} 1 & tZ_c \\ tY_c & 1 \end{pmatrix}$$

This most basic element is referred to as the *unit element*. When a unit element is connected to a load Z_L, the input impedance Z_{in} is given in the distributed domain by

$$Z_{\text{in}}(t) = Z_c \frac{Z_L + Z_c t}{Z_c + Z_L t}$$

In this book we do not use the theory of distributed circuits; instead, the unit elements will be modeled to represent an element of our prototype circuits, such as the impedance inverter.

A unit element of electric length θ and characteristic impedance K is shown in Figure 3.11. This unit element is defined entirely by its $ABCD$ matrix:

$$\begin{pmatrix} A & B \\ C & D \end{pmatrix} = \begin{pmatrix} \cos\theta & jK\sin\theta \\ j\,(1/K)\sin\theta & \cos\theta \end{pmatrix}$$

where $\theta = \beta l$ and $\beta = \omega/c$. This matrix can be written as

$$\begin{pmatrix} A & B \\ C & D \end{pmatrix} = \begin{pmatrix} \cos(\theta/2) & jK\sin(\theta/2) \\ (j/K)\sin(\theta/2) & \cos(\theta/2) \end{pmatrix}^2$$

Since this matrix corresponds to a symmetric network, we can define the half network and its reverse. These are given by

$$\begin{pmatrix} A_1 & B_1 \\ C_1 & D_1 \end{pmatrix} = \begin{pmatrix} \cos(\theta/2) & jK\sin(\theta/2) \\ (j/K)\sin(\theta/2) & \cos(\theta/2) \end{pmatrix} \quad \text{and}$$

$$\begin{pmatrix} D_1 & B_1 \\ C_1 & A_1 \end{pmatrix} = \begin{pmatrix} \cos(\theta/2) & jK\sin(\theta/2) \\ (j/K)\sin(\theta/2) & \cos(\theta/2) \end{pmatrix}$$

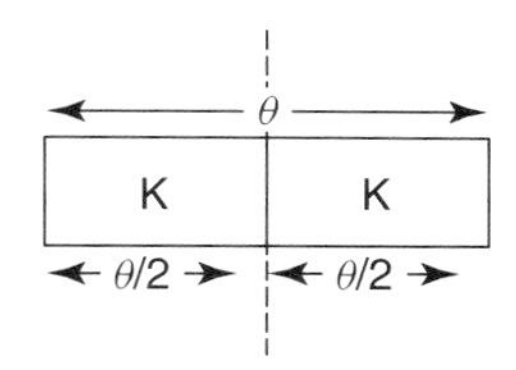

Figure 3.11 Representation of a unit element of electrical length θ.

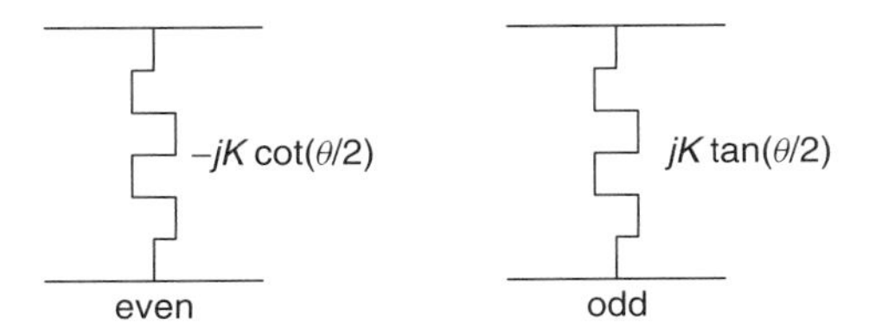

Figure 3.12 Even and odd impedances of a unit element.

According to the bisection theorem, the even and odd impedances of a transmission line of length θ are given by

$$Z_e = \frac{A_1}{C_1} = -jK \text{ cotan } \frac{\theta}{2}$$

$$Z_o = \frac{B_1}{D_1} = jK \tan \frac{\theta}{2}$$

The unit element can then be represented as even and odd impedances, as shown in Figure 3.12. A particular case occurs when the electric length is $\theta = \pi / 2$. In this case the $ABCD$ matrix reduces to

$$\begin{pmatrix} A & B \\ C & D \end{pmatrix} = \begin{pmatrix} 0 & jK \\ j(1/K) & 0 \end{pmatrix}$$

which is that of an ideal impedance inverter. One can also check that the even- and odd-mode impedances in this case are $Z_e = -jK$ and $Z_o = jK$. Therefore, it is possible to use a unit element to create an impedance inverter.

Another way to obtain impedance inverters is to use two transmission lines of length $\phi/2$ separated by a parallel reactance X (an inductor or capacitor) as shown in Figure 3.13. The system provides the following equations from which the transmission line length and reactance can be computed to provide the desired impedance inverter:

$$X = \omega L \quad \text{or} \quad X = -\frac{1}{\omega C}$$

$$K = Z_o \tan \left| \frac{\phi}{2} \right|$$

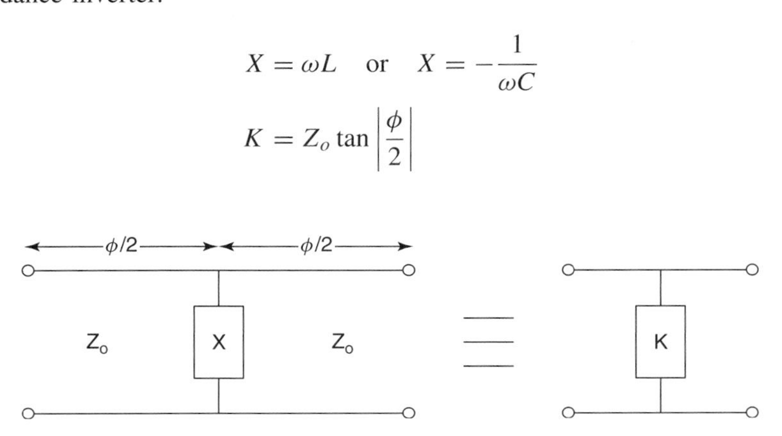

Figure 3.13 Impedance inverter based on a transmission line and a reactance.

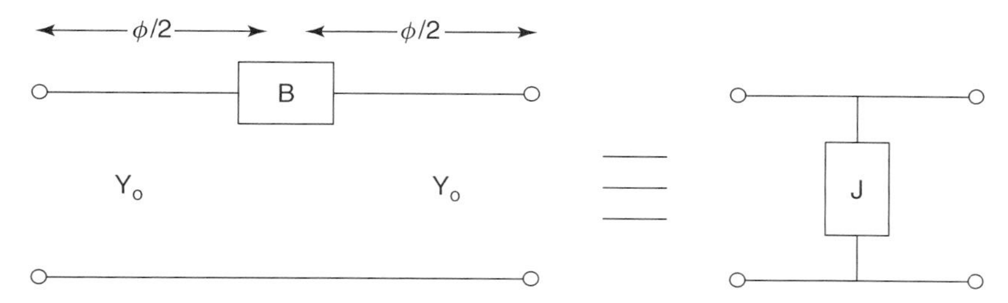

Figure 3.14 Admittance inverter based on a transmission line and a reactance.

$$\phi = -\tan^{-1}\frac{2X}{Z_o}$$

$$\left|\frac{X}{Z_o}\right| = \frac{K/Z_o}{1-(K/Z_o)^2}$$

The dual case of an admittance inverter is obtained using the two transmission lines of length $\phi/2$ and a series reactance B as shown in Figure 3.14.

The system provides the following equations, from which the transmission line length and reactance can be computed to provide the desired admittance inverter:

$$B = \omega C \quad \text{or} \quad B = -\frac{1}{\omega L}$$

$$J = Y_o \tan\left|\frac{\phi}{2}\right|$$

$$\phi = -\tan^{-1}\frac{2B}{Y_o}$$

$$\left|\frac{B}{Y_o}\right| = \frac{J/Y_o}{1-(J/Y_o)^2}$$

One should note that these inverters are dependent of the frequency ω at which the reactance X or B is computed. They are also dependent on the frequency ω through the angle θ. These dependences must be minimized wherever possible. Additional results concerning the use of redundant unit elements are provided in Appendix 2.

3.7 CONCLUSIONS

In this chapter we reviewed several key concepts and equations related to the microwave structures used in this book. The material was limited to the most fundamental elements of waveguides (rectangular and circular), evanescent modes, planar transmission lines, and distributed circuits.

REFERENCES

[3.1] R. E. Collin, *Foundations for Microwave Engineering*, McGraw-Hill, New York, 1966.

[3.2] L. Lewin, *Theory of Waveguides*, Butterworth, London, 1975.

[3.3] R. E. Collin, *Field Theory of Guided Waves*, 2nd ed., McGraw-Hill, New York, 1992.

[3.4] S. B. Cohn, Problems in strip transmission lines, *IRE Trans. Microwave Theory Tech.*, vol. 3, pp. 119–126, Mar. 1955.

[3.5] S. B. Cohn and G. D. Osterhudes, A more accurate model in the TE_{10} type waveguide mode in suspended substrate, *IEEE Trans. Microwave Theory Tech.*, vol. 30, pp. 293–294, Mar. 1982.

[3.6] D. M. Pozar, *Microwave Engineering*, 2nd ed., Wiley, New York, 1998.

[3.7] J. D. Rhodes, Suspended substrates provide alternative to coax, *Microwave Syst. News*, vol. 9, pp. 134–143, Aug. 1979.

[3.8] D. Lo Hine Tong and P. Jarry, Microwave filters on suspended substrate, *Ann. Telecommun.*, vol. 46, no. 3–4, pp. 216–222, 1991.

[3.9] P. I. Richards, Resistor transmission line networks, *Proc. IRE*, vol. 30, pp. 217–220, Feb. 1948

Selected Bibliography

Alseyab, S. A., A novel class of generalized Chebyshev low-pass prototype for suspended substrate stripline filters, *IEEE Trans. Microwave Theory Tech.*, vol. 30, no. 9, 1982.

Argollo, D. F., H. Abdalla, and A. J. M. Soares, Method of lines applied to broadside suspended stripline coupler design, *IEEE Trans. Magn.*, vol. 31, no. 3, 1995.

Chiang, Y. C., C. K. C. Tzuang, and S. Su, Design of a three-dimensional gap-coupled suspended substrate stripline bandpass filter, *IEE Proc.-H*, vol. 139, no. 4, 1992.

Hislop, A. and D. Rubin, Suspended substrate Ka-band multiplexer, *Microwave J*. June 1981.

Kumar, R., Design model for broadside-coupled suspended substrate stripline for microwave and millimeter-wave applications, *Microwave Opt. Technol. Lett.*, vol. 2, no. 2, 2004.

Maloratsky, L. G., Reviewing the basics of suspended striplines, *Microwave J.*, Oct. 2002.

Menzel, W. A novel miniature suspended stripline filter, presented at the 33rd European Microwave Conference, Munich, Germany, 2003.

Menzel, W. and W. Schwab, Compact bandpass filters with improved stop-band characteristics using planar multilayer structures, *IEEE Trans. Microwave Theory Tech.*, MTT-S Digest, 1992.

Mobbs, C. I., and J. D. Rhodes, A generalized Chebyshev suspended substrate stripline bandpass filter, *IEEE Trans. Microwave Theory Tech.* vol. 31, no. 5, 1983.

Rhodes, J. D., Suspended substrate filters and multiplexers, presented at the 16th European Microwave Conference, Dublin, Ireland, Sept. 8–12, 1986.

Tomar, R., Y. Antar, and P. Bhartia, CAD of suspended substrate microstrips: an overview, *Int. J. RF Microwave Comput Aided Eng.*, vol. 15, Jan. 2005.

4

CATEGORIZATION OF MICROWAVE FILTERS

4.1 INTRODUCTION

In this chapter we provide a categorization of the microwave filters that are presented in the book. The first category of filters is referred to as *minimum-phase filters*. The low-pass function of these filters consists of a numerator polynomial of order zero. Therefore, the phase characteristic is provided primarily by the denominator polynomial. Another consequence is that these filters cannot provide zeros of transmission,

The second category of filters will be referred to as *non-minimum-phase* filters. In this case the low-pass numerator polynomial can have an order greater than zero. The non-minimum-phase term comes from the additional phase coming

Advanced Design Techniques and Realizations of Microwave and RF Filters,
By Pierre Jarry and Jacques Beneat

from the numerator polynomial. Another consequence is that the numerator polynomial can provide zeros of transmission. Among the non-minimum-phase filters, there are two subcategories. The first regroups filters referred to as *non-minimum-phase symmetrical filters*. The term *symmetrical* refers to the symmetry of the low-pass magnitude response around zero. From Fourier analysis this means that the impulse response will be real. In contrast, the second subcategory regroups filters referred to as *non-minimum-phase asymmetrical filters*. For these filters, the low-pass magnitude response is no longer symmetrical around zero. The low-pass transfer functions have complex parameters and are not realizable. However, bandpass filters can be realized using special techniques.

The majority of the filters presented in this book are concerned with magnitude responses. However, a few microwave filters that exhibit a bandpass linear-phase characteristic are also presented. Finally, a new approach to microwave filter design based on the use of the genetic algorithm is introduced.

4.2 MINIMUM-PHASE MICROWAVE FILTERS

4.2.1 General Design Steps

The minimum-phase bandpass filters are based on frequency transform techniques applied to a low-pass realization of a minimum-phase function such as the Butterworth or Chebyshev functions [4.1,4.2]. A typical minimum-phase function is provided by the Chebyshev function

$$|S_{21}(j\omega)|^2 = S_{21}(s)S_{21}(-s)|_{s=j\omega} = \frac{1}{1+\varepsilon^2 T_n^2(\omega)}$$

where $T_n(\omega)$ are Chebyshev polynomials so that the numerator is of order zero.

Figure 4.1 shows a low-pass realization of the Chebyshev function using a ladder made of ideal admittance inverters and parallel capacitors. This realization provides an attenuation that is equiripple in the passband, as shown in Figure 4.2. A bandpass realization of the minimum-phase Chebyshev function consists of applying the low pass-to-bandpass frequency transform

$$s \to \alpha\omega_c \left(\frac{s}{\omega_0} + \frac{\omega_0}{s}\right) \quad \text{with} \quad \alpha = \frac{\omega_0}{\omega_2 - \omega_1} \quad \text{and} \quad \omega_0 = \sqrt{\omega_1 \omega_2}$$

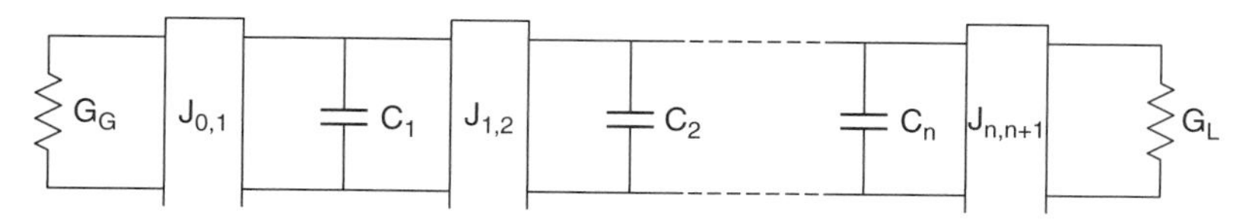

Figure 4.1 Low-pass realization of a minimum-phase function.

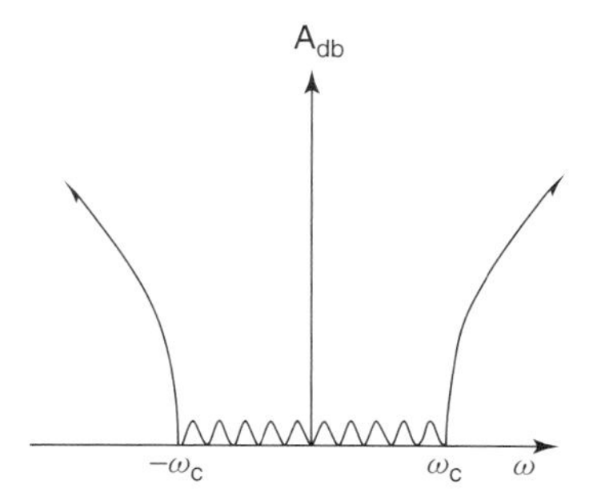

Figure 4.2 Low-pass attenuation characteristic of a minimum-phase function.

This transform preserves the attenuation characteristic as shown in Figure 4.3 by transforming the frequency range $[-\omega_C, +\omega_C]$ to the frequency range $[\omega_1, \omega_2]$, which is centered at ω_0.

In general, one starts with the low-pass prototype with $\omega_c = 1$ rad/s and then computes ω_0 and α from the specifications of the passband edges ω_1 and ω_2 of the bandpass filter. For the bandpass realization, the low-pass parallel capacitor is transformed into a parallel $L_a C_a$ resonating cell as shown in Figure 4.4. The termination admittances as well as the impedance inverters are ideally independent of frequency. The new capacitors and inductors are related to the low-pass capacitors by $C_{ak} = (\alpha/\omega_0)C_k$ and $L_{ak} = 1/\alpha\omega_0 C_k$. The resonating frequency of each parallel cell is equal to the center frequency of the bandpass filter $\omega_0 = 1/\sqrt{L_a C_a}$.

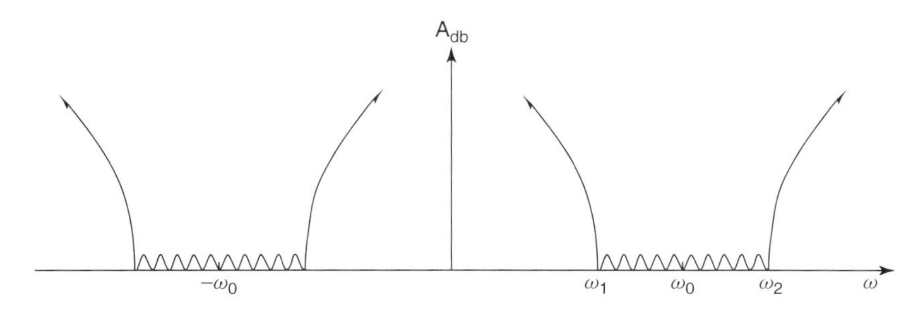

Figure 4.3 Bandpass attenuation characteristic of a minimum-phase function.

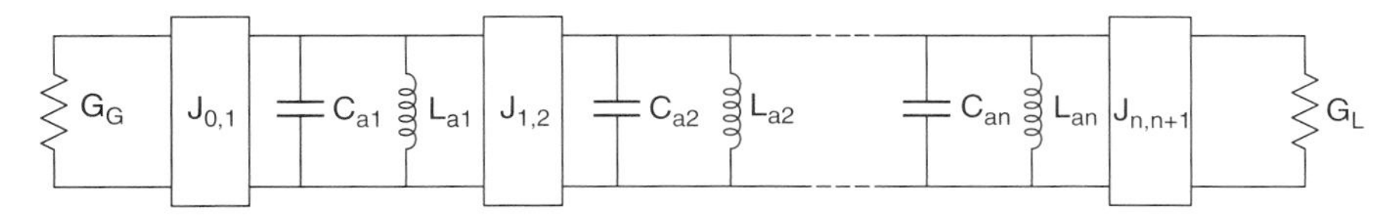

Figure 4.4 Bandpass realization of a minimum-phase function.

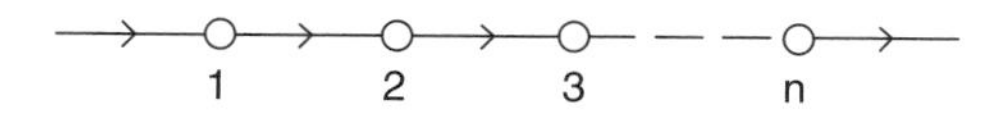

Figure 4.5 Graph representation of a minimum-phase filter.

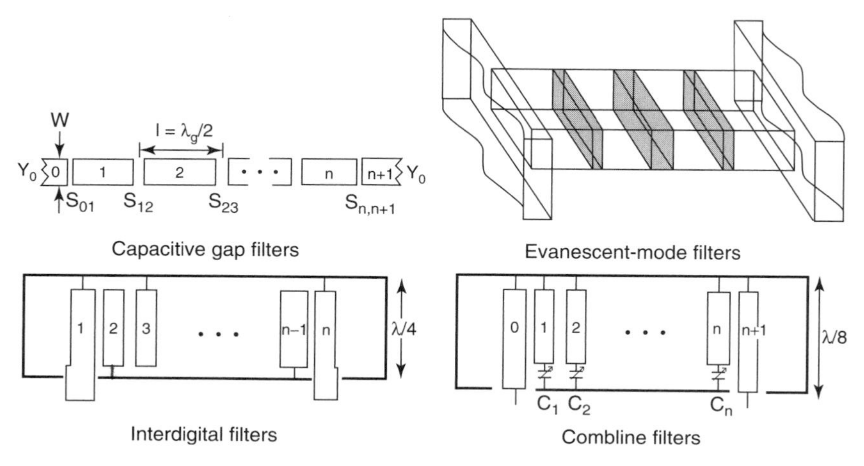

Figure 4.6 Minimum-phase microwave filters.

A graph representation of this type of filter is shown in Figure 4.5. In this graph the nodes $1, 2, \ldots, n$ represent the capacitances and inductances (C_{ak}, L_{ak}), while the branches between the nodes represent the ideal inverters $J_{i,i+1}$.

4.2.2 Minimum-Phase Filter Examples

There exist a multitude of minimum-phase microwave filters. In this book we provide four microwave filter design examples using suspended substrate stripline and evanescent-mode waveguide structures. The capacitive-gap (Chapter 5) and evanescent-mode waveguide (Chapter 6) filters will be based on lumped elements, while the interdigital (Chapter 7) and combline (Chapter 8) filters will be based on distributed elements. These filters are shown in Figure 4.6.

Minimum-phase filters are often the easiest to design. However, for significant rejection, one must use high-order filters. This, in turn, increases the complexity, cost, and size of the realizations. The non-minimum-phase filters presented next might be more difficult to design, but they will provide greater rejection without increasing the order of the filter.

4.3 NON-MINIMUM-PHASE SYMMETRICAL RESPONSE MICROWAVE FILTERS

Elliptic filters could be thought of as the optimum filters for providing small return loss in the passband and large rejections in the stopband. Unfortunately, they are difficult to realize using common microwave structures, due to the large number of zeros of transmission they require. A compromise was proposed by Atia and Williams [4.3], where quasielliptic responses are presented. These responses require only a limited number of transmission zeros (in general, up to 4) for improving the rejection while preserving a Chebyshev equiripple behavior in the passband. Cameron [4.4] was the first to describe a recursive technique to calculate the transfer function $S_{21}(s)$ of symmetrical and asymmetrical quasielliptic filters.

4.3.1 General Design Steps

Non-minimum-phase filters are such that

$$|S_{21}(j\omega)|^2 = S_{21}(s)S_{21}(-s)|_{s=j\omega} = \frac{1}{1+\varepsilon^2 D_n^2(\omega)}$$

where in the case of quasielliptic filters, $D_n(\omega)$ is given as a rational function:

$$D_n(\omega) = \frac{R_n(j\omega)}{P_{n_z}(j\omega)}$$

where $R_n(j\omega) = R_n(s)|_{s=j\omega}$ and $P_{n_z}(j\omega) = P_{n_z}(s)|_{s=j\omega}$, n is the degree of the polynomial $R_n(s)$ and also the order of the filter, and n_z is the degree of the polynomial $P_{n_z}(s)$ and represents the number of transmission zeros. In the case of symmetrical filters, the zeros should be chosen such that the low-pass amplitude characteristic is symmetrical around zero.

For example, Figure 4.7 shows the typical attenuation characteristic of a low-pass non-minimal-phase symmetrical filter. It can be seen that the attenuation characteristic is symmetrical around zero. This means in particular that a zero of transmission occurring at a frequency $+\omega_z$ will also occur at a frequency $-\omega_z$. A low-pass non-minimum-phase symmetrical filter can be realized using the

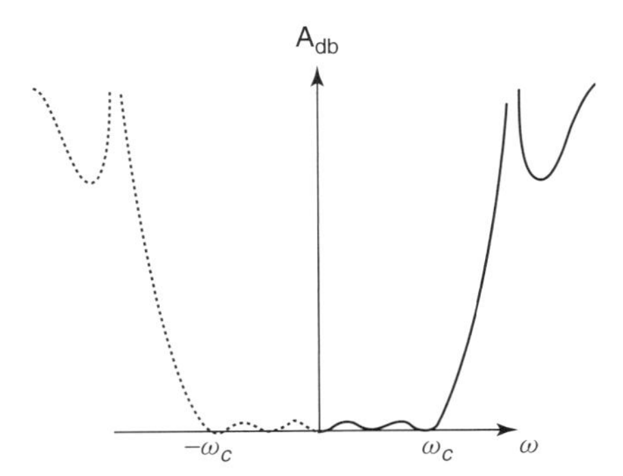

Figure 4.7 Low-pass attenuation characteristic of a non-minimum-phase symmetrical function.

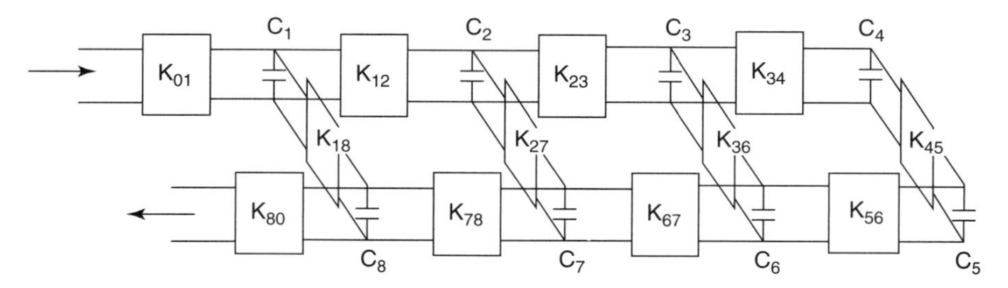

Figure 4.8 Low-pass realization of a non-minimum-phase symmetrical function.

structure of Figure 4.8. In this realization, all the elements have real (as opposed to complex) values. A bandpass realization of a non-minimum-phase symmetrical function consists of applying the same low pass-to-bandpass frequency transform as for the case of minimum-phase filters:

$$s \rightarrow \alpha\omega_c\left(\frac{s}{\omega_0}+\frac{\omega_0}{s}\right) \quad \text{with} \quad \alpha=\frac{\omega_0}{\omega_2-\omega_1} \quad \text{and} \quad \omega_0=\sqrt{\omega_1\omega_2}$$

The bandpass attenuation characteristic is shown in Figure 4.9.

For the bandpass realization, the low-pass parallel capacitor is transformed into a parallel $L'_k C'_k$ resonating cell as shown in Figure 4.10. The termination admittances as well as the impedance inverters are ideally independent of frequency. The new capacitors and inductors are related to low-pass capacitors by $C'_k = (\alpha/\omega_0)C_k$ and $L'_k = 1/\alpha\omega_0 C_k$. The resonating frequency of each parallel cell is equal to the center frequency of the bandpass filter $\omega_0 = 1/\sqrt{L'C'}$. A graph representation of this type of filter is shown in Figure 4.11.

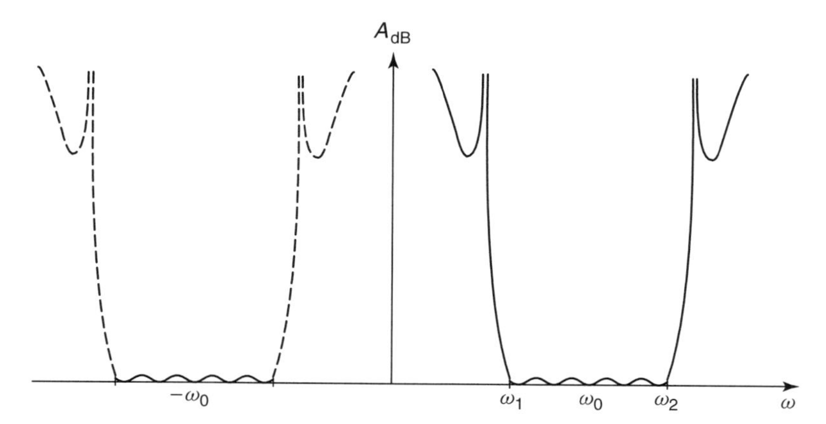

Figure 4.9 Bandpass attenuation characteristic of a non-minimum-phase symmetrical function.

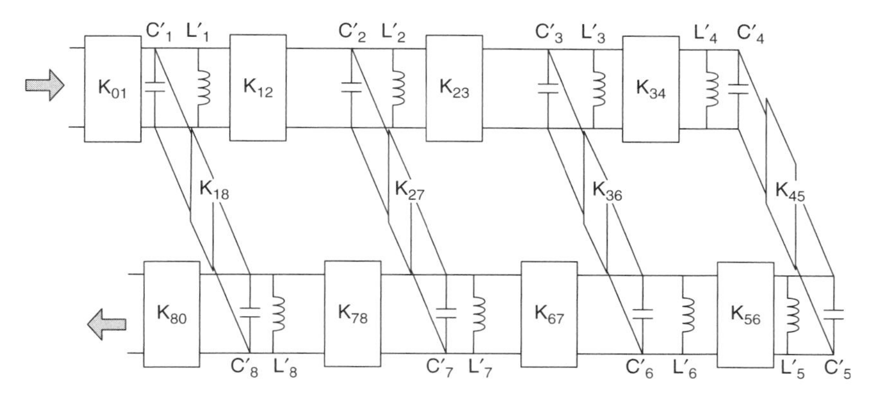

Figure 4.10 Bandpass realization of a non-minimum-phase symmetrical function.

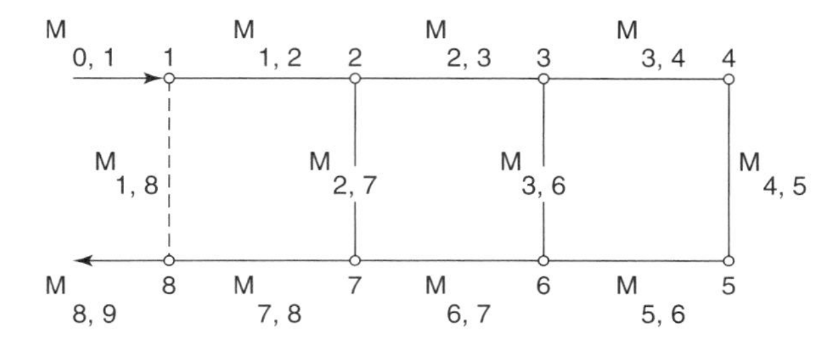

Figure 4.11 Graph representation of a non-minimum-phase symmetrical filter.

4.3.2 Non-Minimum-Phase Symmetrical Response Filter Examples

From the graph representation it is possible to define a design method for non-minimum-phase TE_{011} filters [4.5], generalized interdigital filters [4.6], and generalized direct-coupled cavity filters [4.7]. Non-minimum-phase dual-mode filters with parallel irises require additional transformations of the graph [4.8]. In the case of symmetrical responses, the bisection theorem can be used and the filters can be designed based on the even-mode equivalent circuit of the structure. These filters are shown in Figure 4.12.

4.3.3 Microwave Linear-Phase Filters

Generally, the low pass-to-bandpass frequency transformation does not preserve the linearity of the phase. The transformation tends to preserve a good linearity but only in the vicinity of the center frequency of the passband. There is no systematic design method for bandpass linear-phase filters. One technique consists of approximating the linear phase by decomposing the system into its odd part $Y_o(s)$ and even part $Y_e(s)$ when the structure has a geometrical symmetry (e.g., in Figure 4.8 one would have $K_{01} = K_{80}$, $K_{12} = K_{78}, \ldots, K_{34} = K_{56}$, and

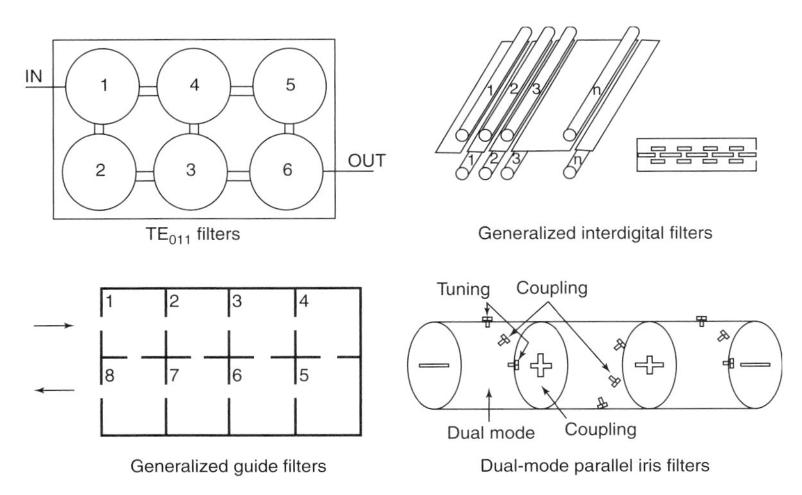

Figure 4.12 Non-minimum-phase symmetrical microwave filters.

$C_1 = C_8, \ldots, C_4 = C_5$). From the bisection theorem [4.9], the forward transmission function can be expressed as

$$S_{21}(s) = \frac{Y_e - Y_o}{(1 + Y_e)(1 + Y_o)} = \frac{Y(s) - Y^*(s)}{[1 + Y(s)][1 + Y^*(s)]}$$

where $Y_e = Y(s)$ and $Y_o = Y^*(s)$. From these equations it is possible to define conditions for the phase and amplitude of the filter. In Chapter 9 the technique is described for generalized interdigital filters. It is also shown that it is possible to approximate the ideal group delay characteristic for non-minimum-phase TE_{011} filters (Chapter 10).

In summary, non-minimum-phase symmetrical filters will have a low-pass realization with real elements from which a bandpass realization can be defined using traditional frequency transform techniques. In addition, the symmetrical response can at times be associated with the geometrical symmetry of the structure to reduce the complexity of the design technique. For the non-minimum-phase asymmetrical filters presented next, new techniques are needed and the structures will not in general present a geometrical symmetry.

4.4 NON-MINIMUM-PHASE ASYMMETRICAL RESPONSE MICROWAVE FILTERS

4.4.1 General Design Steps

The general function of non-minimum-phase asymmetrical filters is the same as that of non-minimum-phase symmetrical filters:

$$|S_{21}(j\omega)|^2 = S_{21}(s)S_{21}(-s)|_{s=j\omega} = \frac{1}{1 + \varepsilon^2 D_n^2(\omega)}$$

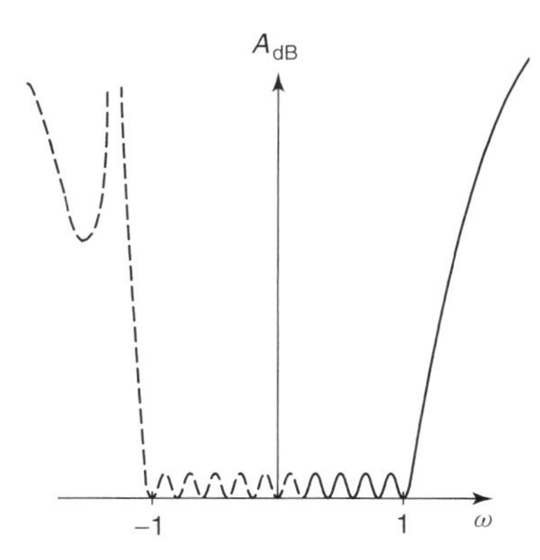

Figure 4.13 Low-pass attenuation characteristic of a non-minimum-phase nonsymmetrical function.

where $D_n(\omega)$ is defined in the same manner as for symmetrical filters. In the case of asymmetrical filters, however, the zeros are chosen such that the low-pass amplitude characteristic is not symmetrical around zero. The asymmetrical behavior of the low-pass attenuation characteristic is shown in Figure 4.13. From Fourier analysis, such a function will not have a real impulse response.

The general technique for designing bandpass non-minimum-phase asymmetrical microwave filters starts by defining an imaginary low-pass realization as shown in Figure 4.14. These low-pass prototypes can be thought of as an imaginary realization since the elements (C_i, jB_i) or (L_i, jX_i) do not correspond to any known physical elements [4.10–4.12]. In fact, it is not possible to realize a

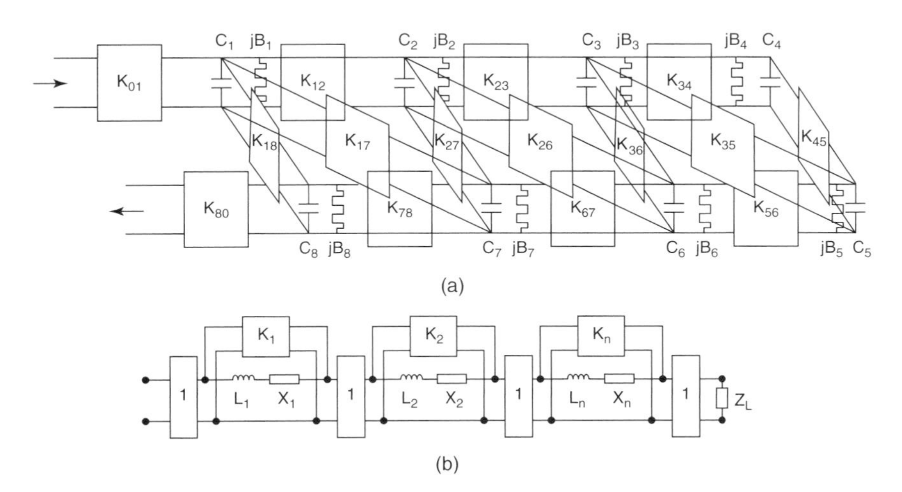

Figure 4.14 Imaginary low-pass realizations of a non-minimum-phase asymmetrical function: (a) cross-coupled network; (b) in-line network.

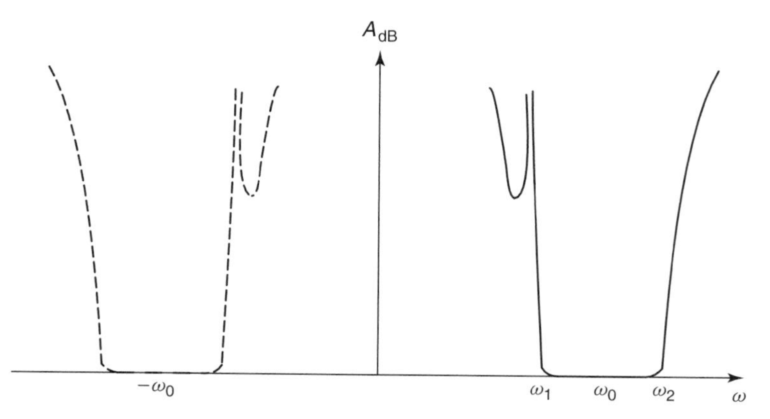

Figure 4.15 Bandpass attenuation characteristic of a non-minimum-phase asymmetrical function.

low-pass asymmetrical filter. It is, however, possible to realize a bandpass filter based on this imaginary low-pass non-minimum-phase asymmetrical filter.

The attenuation characteristic of a bandpass asymmetrical filter is shown in Figure 4.15. One notes that the attenuation characteristic of the bandpass filter is now symmetrical around zero. Therefore, it should be realizable. The bandpass attenuation characteristic can be obtained from the low-pass characteristic by applying the traditional frequency transformation when $\omega_C = 1$:

$$s \to \alpha\left(\frac{s}{\omega_0} + \frac{\omega_0}{s}\right) \quad \text{with} \quad \alpha = \frac{\omega_0}{\omega_2 - \omega_1} \quad \text{and} \quad \omega_0 = \sqrt{\omega_1\omega_2}$$

However, for the realization to be composed of real-valued elements, the following narrowband approximations are needed:

$$\omega \to \alpha\left(\frac{\omega}{\omega_0} - \frac{\omega_0}{\omega}\right) \approx \frac{2\alpha}{\omega_0}(\omega - \omega_0) \quad \text{when} \quad \omega \text{ is close to } \omega_0$$

The transformation from low pass to bandpass is then given by

$$s \to \frac{2}{\Delta\omega}(s - j\omega_0) \quad \text{with} \quad \Delta\omega = \omega_2 - \omega_1$$

This new transformation corresponds approximately to a frequency translation $s \to s - s_0$, where $s_0 = j\omega_0$. It is possible when applying this transform on the imaginary low-pass realization to transform the parallel C_k and imaginary jB_k cell into a real-valued parallel capacitor and inductor C'_k and L'_k as shown in Figure 4.16. Applying this new transform on the capacitor C_k and imaginary element jB_k, we find that

$$2\left(\frac{\omega - \omega_0}{\Delta\omega}\right) C_k + B_k = \omega C'_k - \frac{1}{\omega L'_k}$$

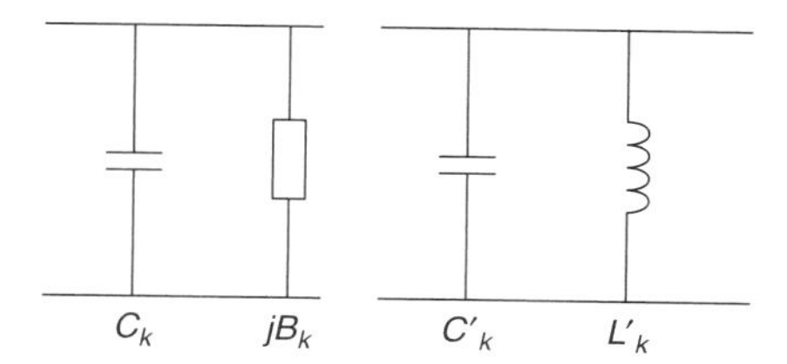

Figure 4.16 Effect of the new frequency transform technique.

At this stage it is difficult to express C'_k and L'_k in terms of C_k and B_k. A second equation is needed. The second equation consists of matching the derivatives of the expressions above with respect to ω:

$$\frac{2}{\Delta\omega}C_k = C'_k + \frac{1}{\omega^2 L'_k}$$

These two equations should be satisfied for all frequencies and in particular for $\omega = \omega_0$. In this case we have

$$\begin{aligned} B_k &= \omega_0 C'_k - \frac{1}{\omega_0 L'_k} \qquad & C'_k &= \frac{1}{\omega_0}\left(\frac{\omega_0}{\Delta\omega}C_k + \frac{B_k}{2}\right) \\ \frac{2}{\Delta\omega}C_k &= C'_k + \frac{1}{\omega_0^2 L'_k} \qquad & \frac{1}{L'_k} &= \omega_0\left(\frac{\omega_0}{\Delta\omega}C_k - \frac{B_k}{2}\right) \end{aligned}$$

Therefore, a bandpass non-minimum-phase asymmetrical filter can be realized using the real-valued structure of Figure 4.17. One should note that the resonant frequency of $L'_k C'_k$ resonators is given by

$$\omega_k = \frac{1}{\sqrt{L'_k C'_k}} \cong \omega_0 - \frac{B_k}{C_k}\frac{\Delta\omega}{2}$$

This frequency is no longer exactly equal to the center frequency ω_0, but to a frequency shifted slightly from the center frequency by the value $(B_k/C_k)(\Delta\omega/2)$. The graph representations of this type of filter are shown in Figure 4.18.

4.4.2 Non-Minimum-Phase Asymmetrical Response Filter Examples

Using the graph representation, it is possible to provide a design technique for non-minimum-phase asymmetrical capacitive-gap-coupled line (CGCL) filters [4.13], and generalized shifted rectangular guide filters [4.14]. Dual-mode filters such as the circular with nonparallel iris [4.15] and rectangular in-line filters [4.16] require additional transformation of the graph. These filters are shown in Figure 4.19.

In summary, non-minimum-phase asymmetrical low-pass filters are not realizable. Only an imaginary structure can be proposed, due to the nonsymmetry of

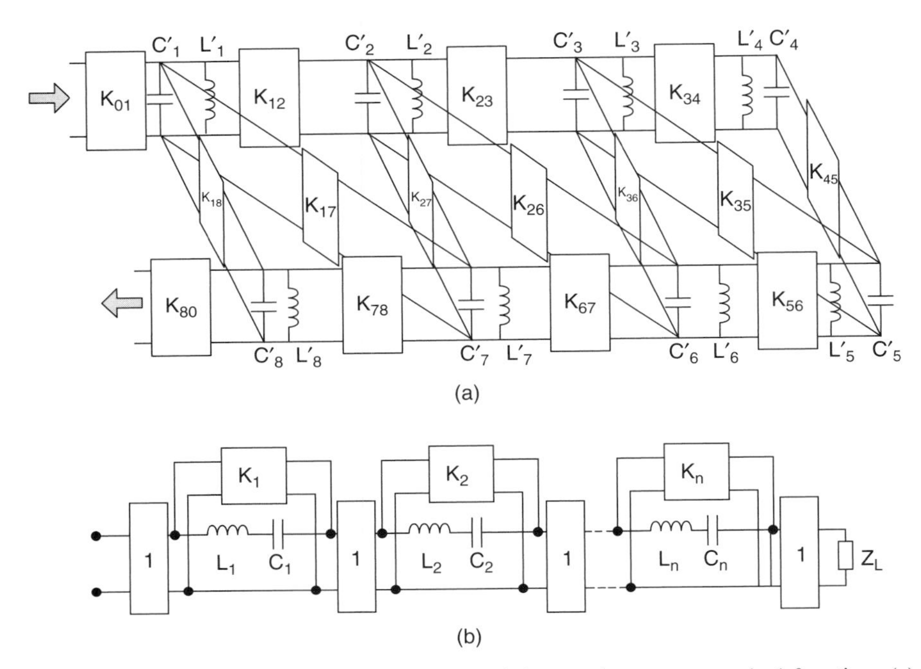

Figure 4.17 Bandpass realizations of a non-minimum-phase asymmetrical function: (a) cross-coupled network; (b) in-line network.

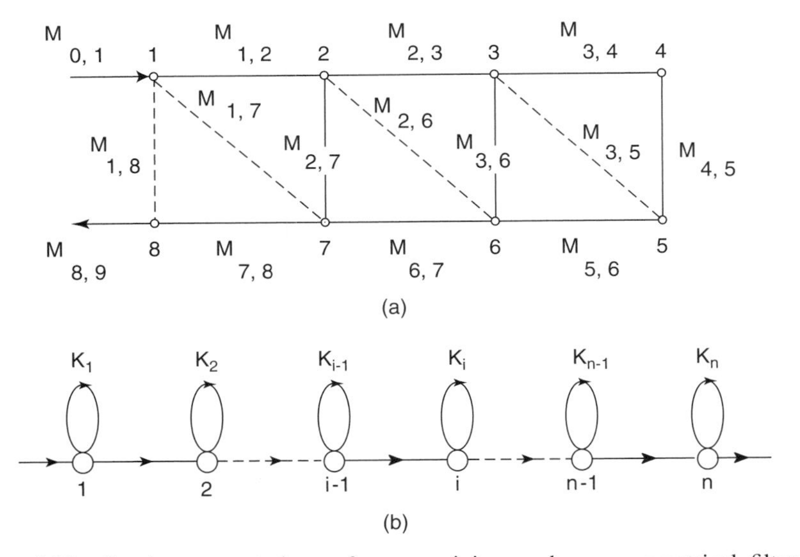

Figure 4.18 Graph representations of a non-minimum-phase asymmetrical filter: (a) cross-coupled network; (b) in-line network.

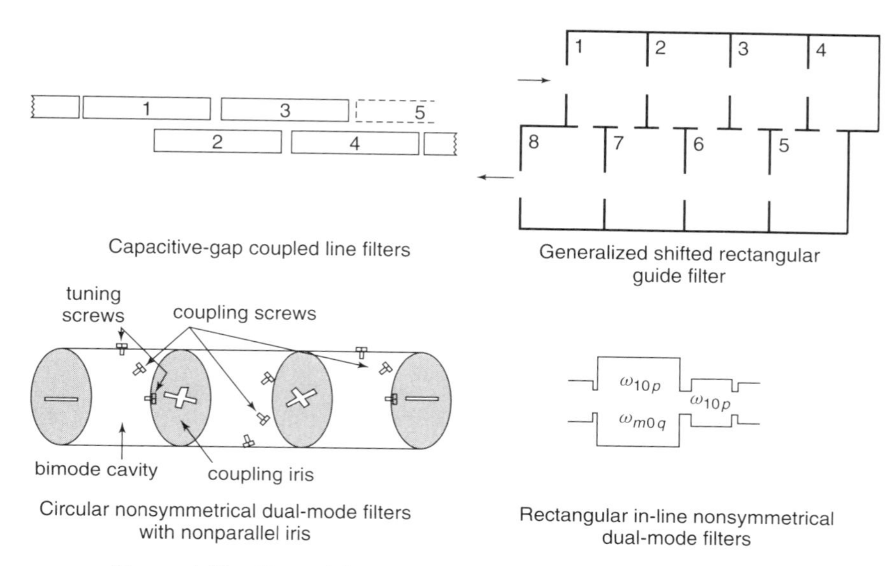

Figure 4.19 Non-minimum-phase asymmetrical microwave filters.

the low-pass response around zero. The traditional low pass-to-bandpass transformation must be modified so that a bandpass realization with real-valued elements can be obtained. Since there is usually no geometrical symmetry in these cases, the bisection theorem cannot be used to reduce the complexity of the design technique.

4.4.3 Multimode Microwave Filters by Optimization

The design of multimode filters is based on a generalization of the technique for dual-mode filters [4.17]. However, these filters will be designed using the electromagnetic equations directly rather than an equivalent circuit of the structure. In this new method, basic building blocks are defined that can be used to generate a multitude of microwave structures, as shown in Figure 4.20.

The design technique for generating microwave filters with prescribed responses will be accomplished using a powerful optimization procedure known as the *genetic algorithm*. These microwave filters tend to be very compact and require only simple manufacturing techniques. Due to the complete electromagnetic characterization of these structures, the use of manual tuning screws can be eliminated.

4.5 CONCLUSIONS

The categorization of microwave filters presented in this chapter will be used for the remainder of the book. Microwave filters can be regrouped into two

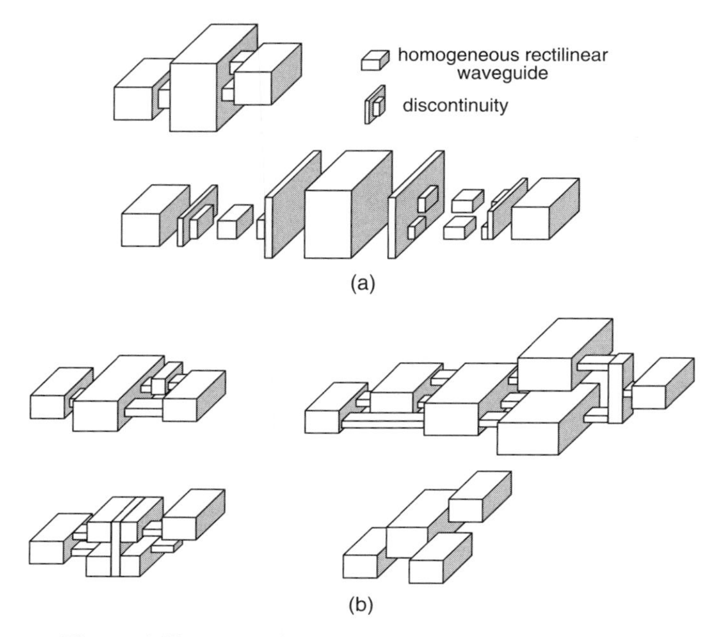

Figure 4.20 (a) Basic elements; (b) analyzable structures.

broad categories: minimum-phase filters and non-minimum-phase filters. Among non-minimum-phase filters, an additional categorization is used to separate filters with low-pass characteristics that are symmetrical around zero and filters with low-pass responses that are asymmetrical around zero. Due to the complex-valued elements of the low-pass filter, the design of asymmetrical filters is difficult. These categories were based primarily on the amplitude characteristics of the filters. At times, phase linearity can be important, depending on the application. As seen in Chapter 2, there are a few low-pass functions that approximate the linear phase. Unfortunately, the traditional low pass-to-bandpass frequency transformation technique does not preserve the phase characteristic, and new approaches are needed. Finally, with the advance of computer technology it is currently possible to implement and run optimization programs in a timely manner. This book provides an introduction to these emerging techniques as they are applied to the design of microwave filters.

REFERENCES

[4.1] H. Baher, *Synthesis of Electrical Networks*, Wiley, New York, 1984.

[4.2] J. D. Rhodes, *Theory of Electrical Filters*, Wiley, New York, 1976.

[4.3] E. A. Atia and A. E. Williams, New type of waveguide band-pass filters for satellite transponders, *Comsat Tech. Rev.*, vol. 1, pp. 21–43, 1971.

[4.4] R. J. Cameron, A novel realisation for microwave bandpass filters, *ESA J.*, vol. 3, pp. 281–287, 1979.

[4.5] J. D. Rhodes and I. H. Zabalawi, Synthesis of symmetrical dual mode in-line prototype networks, Int. J. Circuit Theory Appl., vol. 8, no. 2, pp. 145–160, 1980.

[4.6] J. D. Rhodes, The theory of generalized interdigital linear phase networks, *IEEE Trans. Circuit Theory*, vol. 16, no. 3, pp. 280–288, 1969.

[4.7] J. D. Rhodes, The generalized direct-coupled cavity linear-phase filter, *IEEE Trans. Microwave Theory Tech.*, vol. 18, pp. 308–313, June 1970.

[4.8] E. A. Atia, A. E. Williams, and R. W. Newcomb, Narrow-band multiple coupled cavities synthesis, *IEEE Trans. Circuits Syst.*, vol. 21, no. 5, pp. 649–655, 1974.

[4.9] H. Abdalla, Jr., and P. Jarry, Microwave linear phase filters, *Proc. 14th European Microwave Conference*, EMC'84, Leuven, Belgium, pp. 358–363, Sept. 1984.

[4.10] R. J. Cameron, Fast generation of Chebyshev filter prototypes with asymmetrically prescribed transmission zeros, *ESA J.*, vol. 6, pp. 83–95, 1982.

[4.11] S. Amari, Synthesis of cross-coupled resonator filters using an analytical gradient-based optimization technique, *IEEE Trans. Microwave Theory Tech.*, vol. 48, no. 9, pp. 1559–1564, 2000.

[4.12] S. A. Mohammed and L. A. Find, Construction of lowpass filter transfer functions having prescribed phase and amplitude characteristics, *IEEE Colloq. Dig., Microwave Filters*, pp. 8–15, Jan. 1982.

[4.13] E. Hanna, P. Jarry, E. Kerherve, and J. M. Pham, A design approach for capacitive gap parallel-coupled line filters and fractal filters with the suspended substrate technology, *Research Signpost, Special Volume on Microwave Filters and Amplifiers*, ed. P. Jarry, pp. 49–71, 2005.

[4.14] R. J. Cameron, Dual-mode realization for asymmetric filter characteristics, *ESA J.*, vol. 6, pp. 339–356, 1982.

[4.15] C. Guichaoua and P. Jarry, Asymmetric frequency dual-mode satellite filters: simple realizations using slot irises and input/output realignment, *Proc. 8th B. Franklin Symposium on Advances in Antenna and Microwave Technology*, IEEE AP/MTT-S, University of Pennsylvania, Philadelphia, PA, pp. 32–36, Mar. 1990.

[4.16] P. Jarry, M. Guglielmi, J. M. Pham, E. Kerhervé, O. Roquebrun, and D. Schmitt, Synthesis of dual-mode in-line asymmetric microwave rectangular filters with higher modes, *Int. J. RF Microwave Comput. Aided Eng.*, pp. 241–248, Feb. 2005.

[4.17] N. Boutheiller, P. Jarry, E. Kerherve, J. M. Pham, and S. Vigneron, Two zero four poles building block for microwave filters using all resonant modes at their cut-off frequency, *Int. J. RF Microwave Comput. Aided Eng.*, vol. 13, no. 6, pp. 429–437, 2003.

PART II

MINIMUM-PHASE FILTERS

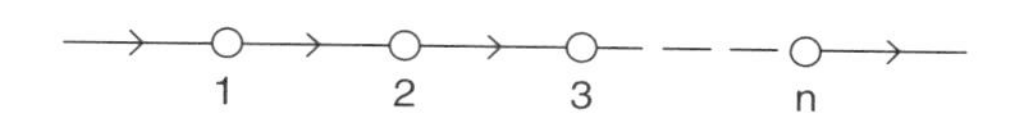

5

CAPACITIVE-GAP FILTERS FOR MILLIMETER WAVES

5.1 INTRODUCTION

The development of compact and low-cost filters is in demand for modern microwave systems. Suspended substrate stripline (SSS) structures satisfy these requirements and are also capable of providing low loss, good temperature stability, and the possibility of integrating with active components. In this chapter the design of capacitive-gap filters realized using suspended substrates and striplines is presented. The design technique provides the physical dimensions of the structure to achieve specified Chebyshev bandpass responses [5.1]. Additional experimental work shows that the design technique can be extended to filters operating in the millimeter range [5.2–5.4]. The chapter concludes with introductory

Advanced Design Techniques and Realizations of Microwave and RF Filters,
By Pierre Jarry and Jacques Beneat

material on a rigorous electromagnetic characterization of the structures that will be needed for improving the design techniques at millimeter wavelengths.

5.2 CAPACITIVE-GAP FILTERS

5.2.1 Capacitive-Gap Filter Structure

A capacitive-gap filter is depicted in Figure 5.1. It consists of several conducting strips of length L separated by gaps of length S. The width W of the strips is kept equal so that each strip presents a characteristic admittance Y_0. A transmission line of length L and of characteristic admittance Y_0 has an $ABCD$ matrix of the form

$$\begin{pmatrix} A & B \\ C & D \end{pmatrix} = \frac{1}{\sqrt{1+\tan^2\theta}} \begin{pmatrix} 1 & j\dfrac{\tan\theta}{Y_0} \\ jY_0\tan\theta & 1 \end{pmatrix}$$

where the angle $\theta = (2\pi/\lambda_g)L$.

A gap of length S between two lines of width W and characteristic admittance Y_0 has been characterized [5.5] and behaves as a Π network made of susceptances B_a and B_b, as shown in Figure 5.2, with

$$\frac{B_a}{Y_0} = \frac{-2H}{\lambda_g} \ln\left(\cosh\frac{\pi S}{2H}\right)$$

$$\frac{B_b}{Y_0} = \frac{H}{\lambda_g} \ln\left(\coth\frac{\pi S}{2H}\right)$$

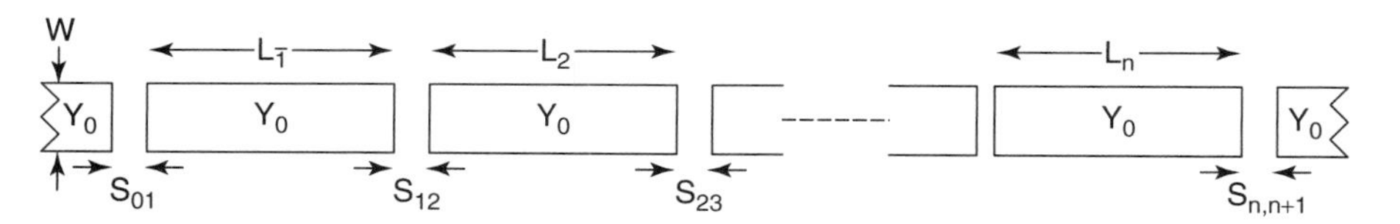

Figure 5.1 Capacitive gap filter.

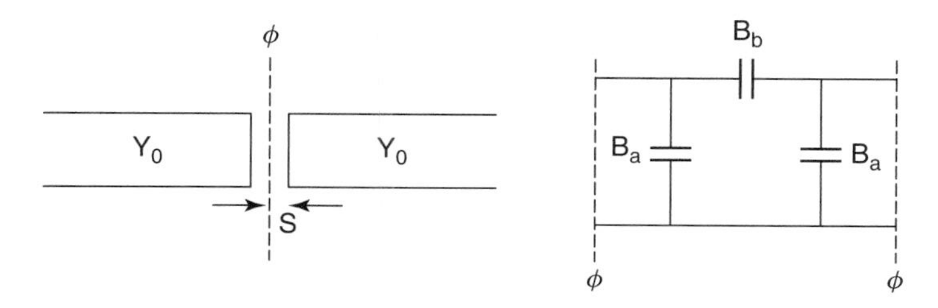

Figure 5.2 Equivalent circuit of a gap of length S.

where H is the distance between the two ground planes in a SSS. The $ABCD$ matrix of the gap is then given by

$$\begin{pmatrix} A & B \\ C & D \end{pmatrix} = \begin{pmatrix} 1 + \dfrac{B_a}{B_b} & \dfrac{1}{jB_b} \\ jB_a\left(2 + \dfrac{B_a}{B_b}\right) & 1 + \dfrac{B_a}{B_b} \end{pmatrix}$$

5.2.2 Design Procedures

The design technique is based on finding the line lengths L and the gap lengths S such that the structure can approximate the ideal bandpass admittance inverter prototype of Figure 5.3, where

$$J_{k,k+1} = \frac{\omega_0}{\alpha}\sqrt{\frac{C_{ak}C_{ak+1}}{g_k g_{k+1}}} \qquad \text{for } k = 1, 2, \ldots, n-1$$

$$J_{0,1} = \sqrt{\frac{\omega_0}{\alpha}\frac{G_G C_{a1}}{g_0 g_1}} \quad \text{and} \quad J_{n,n+1} = \sqrt{\frac{\omega_0}{\alpha}\frac{C_{an} G_L}{g_n g_{n+1}}}$$

with $\omega_0 = \sqrt{\omega_1\omega_2}$, $\alpha = \omega_0/(\omega_2 - \omega_1)$, and the g_k's are given from the low-pass LC ladder prototype elements of Chapter 2. There are several approaches to solving the problem. The solution proposed consists of associating lines of length L with the resonating L_aC_a cells of the prototype, and using the gaps of length S to produce the admittance inverters J. The association of lines of length L and the bandpass prototype of Figure 5.3 will be based on the approximation technique described below.

One notes that the susceptances B between the admittance inverters in Figure 5.3 are provided by a parallel combination of an inductor and a capacitor:

$$B(\omega) = C_a\omega - \frac{1}{L_a\omega}$$

The normalized derivative b of the susceptance $B(\omega)$ at the center frequency ω_0 is defined as

$$b = \frac{\omega_0}{2}\left.\frac{\partial B(\omega)}{\partial\omega}\right|_{\omega=\omega_0}$$

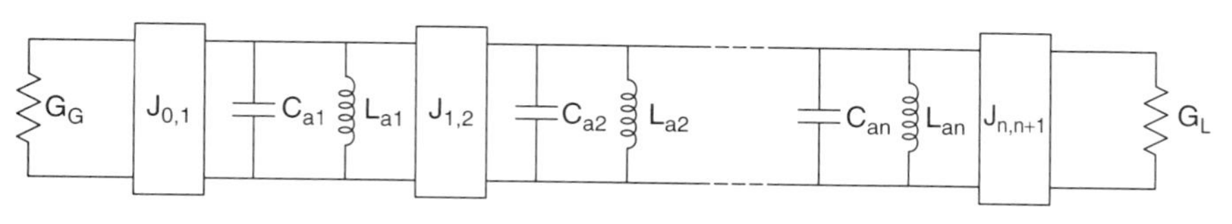

Figure 5.3 Admittance inverter bandpass prototype.

In the case of the bandpass prototype of Figure 5.3, this gives

$$b = \frac{\omega_0}{2}\left(C_a + \frac{1}{L_a}\frac{1}{\omega_0^2}\right) = \frac{1}{2}\left(C_a\omega_0 + \frac{1}{L_a\omega_0}\right)$$

Since $L_a C_a \omega_0^2 = 1$, the normalized derivative b reduces to

$$b = C_a\omega_0 \quad \text{or} \quad b = \frac{1}{L_a\omega_0}$$

The equations of the bandpass prototype of Figure 5.3 can then be reformulated in terms of the normalized derivative b:

$$J_{k,k+1} = \frac{1}{\alpha}\sqrt{\frac{b_k b_{k+1}}{g_k g_{k+1}}} \qquad \text{for} \quad k = 1, 2, \ldots, n-1$$

$$J_{0,1} = \sqrt{\frac{1}{\alpha}\frac{G_G b_1}{g_0 g_1}} \qquad \text{and} \qquad J_{n,n+1} = \sqrt{\frac{1}{\alpha}\frac{G_L b_n}{g_n g_{n+1}}}$$

where $b_k = C_{ak}\omega_0$.

One notes that the equations above provide the exact specified response of the bandpass prototype of Figure 5.3. The approximation will take place when the normalized derivative b of a line of length L is calculated and used instead of $b_k = C_{ak}\omega_0$ to compute the values of the admittance inverters.

The next stage consists of computing the normalized derivative b provided by a line of length L. The lengths L will be restricted to be close to half-wavelengths $(\lambda_g/2)$. When the length L is close to half a wavelength, the angle θ becomes close to π and the line of length $L \approx \lambda_g/2$ corresponds to a parallel admittance Y as shown in Figure 5.4.

The susceptance B for computing the normalized derivative b when $\theta \approx \pi$ is therefore

$$B(\theta) = Y_0 \tan\theta \approx Y_0(\theta - \pi)$$

and the normalized derivative is

$$b = \frac{\omega_0}{2}\left.\frac{\partial B(\omega)}{\partial\omega}\right|_{\omega=\omega_0} = \frac{\theta_0}{2}\left.\frac{\partial B(\theta)}{\partial\theta}\right|_{\theta=\theta_0} \qquad \text{where} \quad \theta_0 = \pi$$

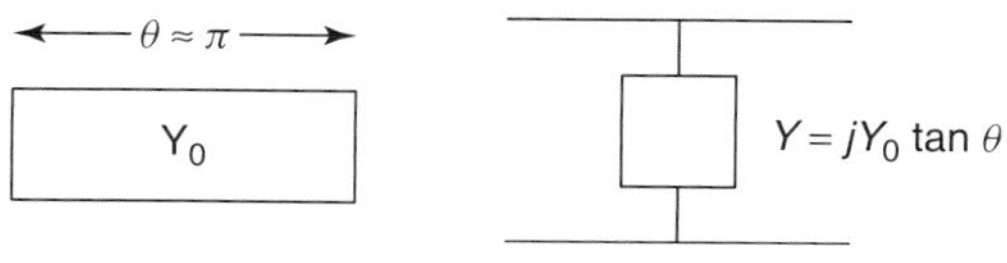

Figure 5.4 Representation of a $L \approx \lambda_g/2$ line.

This gives for the normalized derivative of the line of length $L \approx \lambda_g/2$,

$$b = Y_0 \frac{\pi}{2}$$

The prototype admittance inverters to be used for the structure will now be given by

$$J_{k,k+1} = \frac{\pi}{2} \frac{Y_0}{\alpha} \sqrt{\frac{1}{g_k g_{k+1}}} \quad \text{for } k = 1, 2, \ldots, n-1$$

$$J_{0,1} = \sqrt{\frac{1}{\alpha} \frac{G_G Y_0}{g_0 g_1} \frac{\pi}{2}} = Y_0 \sqrt{\frac{\pi}{2\alpha g_0 g_1}} \quad \text{and} \quad J_{n,n+1} = \sqrt{\frac{1}{\alpha} \frac{G_L Y_0}{g_n g_{n+1}} \frac{\pi}{2}} = Y_0 \sqrt{\frac{\pi}{2\alpha g_n g_{n+1}}}$$

where the terminations are taken as Y_0.

The remainder of the design technique consists of defining the length S of the gaps to produce the admittance inverters. It is clear that a gap alone (B_a and B_b) cannot produce an admittance inverter, especially not an admittance inverter with a specific value. The method proposed to solve this problem is to take a gap of length S surrounded by two lines of length $l/2$, as shown in Figure 5.5. It will be seen that this process can result in negative lengths $l/2$. The method consists of adding these potential negative lengths $l/2$ to the half-wavelength lines L.

The $ABCD$ matrix of the gap surrounded by the lines $l/2$ should be made equal to the $ABCD$ matrix of an admittance inverter J at the center frequency ω_0:

$$\left(\frac{1}{\sqrt{1+\tan^2(\phi/2)}}\right)^2 \begin{pmatrix} 1 & j\dfrac{\tan(\phi/2)}{Y_0} \\ jY_0 \tan\dfrac{\phi}{2} & 1 \end{pmatrix} \begin{pmatrix} 1+\dfrac{B_a}{B_b} & \dfrac{1}{jB_b} \\ jB_a\left(2+\dfrac{B_a}{B_b}\right) & 1+\dfrac{B_a}{B_b} \end{pmatrix}$$

$$\begin{pmatrix} 1 & j\dfrac{\tan(\phi/2)}{Y_0} \\ jY_0 \tan\dfrac{\phi}{2} & 1 \end{pmatrix} = \begin{pmatrix} 0 & \dfrac{j}{J} \\ jJ & 0 \end{pmatrix}$$

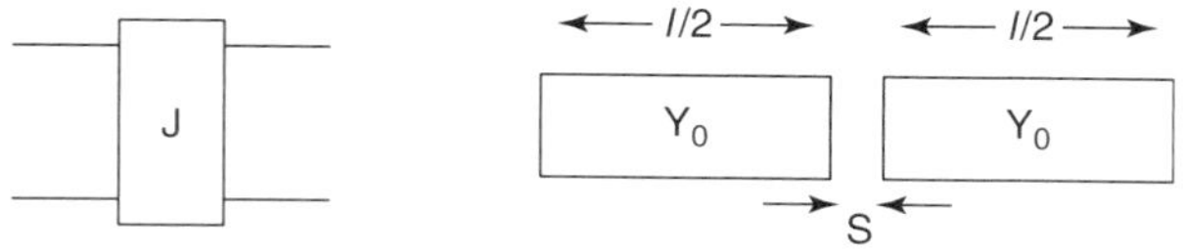

Figure 5.5 Inverter admittance using a gap surrounded by lines of length $l/2$.

where $\phi = (2\pi/\lambda_g)l$. This results in the following equations:

$$\left(1 - \tan^2\frac{\phi}{2}\right)\left(1 + \frac{B_a}{B_b}\right) - \tan\frac{\phi}{2}\frac{B_a}{Y_0}\left(2 + \frac{B_a}{B_b}\right) + \tan\frac{\phi}{2}\frac{Y_0}{B_b} = 0$$

$$2\tan\frac{\phi}{2}\left(1 + \frac{B_a}{B_b}\right) + \frac{B_a}{Y_0}\left(2 + \frac{B_a}{B_b}\right) + \tan^2\frac{\phi}{2}\frac{Y_0}{B_b} = \left(1 + \tan^2\frac{\phi}{2}\right)\frac{J}{Y_0}$$

$$2\tan\frac{\phi}{2}\left(1 + \frac{B_a}{B_b}\right) - \tan^2\frac{\phi}{2}\frac{B_a}{Y_0}\left(2 + \frac{B_a}{B_b}\right) - \frac{Y_0}{B_b} = \left(1 + \tan^2\frac{\phi}{2}\right)\frac{Y_0}{J}$$

The first equation provides a condition on the length ϕ so that the network will act as an admittance inverter

$$\tan\phi = -\frac{(B_a/Y_0) + (2B_b/Y_0 + B_a/Y_0)}{1 - (B_a/Y_0)(2B_b/Y_0 + B_a/Y_0)}$$

In other words, the angle ϕ should be such that

$$\phi = \phi_1 + \phi_2$$

where

$$\phi_1 = -\tan^{-1}\frac{B_a}{Y_0}$$

$$\phi_2 = -\tan^{-1}\left(\frac{2B_b}{Y_0} + \frac{B_a}{Y_0}\right)$$

The second equation provides a relation between the lengths ϕ_1 and ϕ_2 and the value of the admittance inverter:

$$\frac{J}{Y_0} = \frac{\tan(\phi/2) + (B_a/Y_0)}{1 - \tan(\phi/2)(B_a/Y_0)} = \frac{\tan(\phi/2) - \tan\phi_1}{1 + \tan(\phi/2)\tan\phi_1} = \tan\left(\frac{\phi}{2} - \phi_1\right) = \tan\frac{\phi_2 - \phi_1}{2}$$

At this stage, one should find the length ϕ and the length S of the gap that satisfies the equations and provides the desired admittance inverter value J. There are several approaches to this problem.

The proposed method is based on noting that B_a is small compared to B_b when S/H is small. This has the effect that the angle ϕ_1 is rather small and is smaller than ϕ_2. The equations can then be simplified and a noniterative design method can be proposed:

$$\frac{J}{Y_0} \approx \tan\frac{\phi_2}{2} \quad \text{and} \quad \phi_2 \approx -\tan^{-1}\frac{2B_b}{Y_0}$$

When the admittance inverter value J is given, the susceptance B_b of the gap can be defined using

$$\frac{B_b}{Y_0} = \frac{J/Y_0}{1 - (J/Y_0)^2}$$

The gap length S is computed simply by taking the inverse of the equation in B_b:

$$S = \frac{2H}{\pi} \coth^{-1}\left[\exp\left(\frac{B_b}{Y_0}\frac{\lambda_g}{H}\right)\right]$$

At this stage, the values of B_a and B_b are computed for the value of S found above. The complete angle ϕ is used to make the network that of an admittance inverter:

$$\phi = \phi_1 + \phi_2$$

where

$$\phi_1 = -\tan^{-1}\frac{B_a}{Y_0}$$

$$\phi_2 = -\tan^{-1}\left(\frac{2B_b}{Y_0} + \frac{B_a}{Y_0}\right)$$

Since ϕ_1 is in general small, the final value of the admittance inverter from $\tan[(\phi_2 - \phi_1)/2]$ will be close to the desired value.

5.2.3 Step-by-Step Design Example

The first step consists in choosing the dielectric material (ε_r), the metallic enclosure dimensions (H and a), the thickness of the substrate, and the width of the conducting strip (h and W), as shown in Figure 5.6. We are interested in designing a bandpass filter from 8125 to 8475 MHz with a Chebyshev response of order 5 with a return loss of -20 dB. We select $H = 4$ mm, $a = 7.65$ mm, and the dielectric material as Duroid with a relative permittivity $\varepsilon_r = 2.22$ and a thickness of $h = 0.127$ mm. To achieve a characteristic impedance $Z_c = 50\ \Omega$, $W = 5.47$ mm and $\varepsilon_{\text{eff}} = 1.024$. The wavelength to be used for simulations will

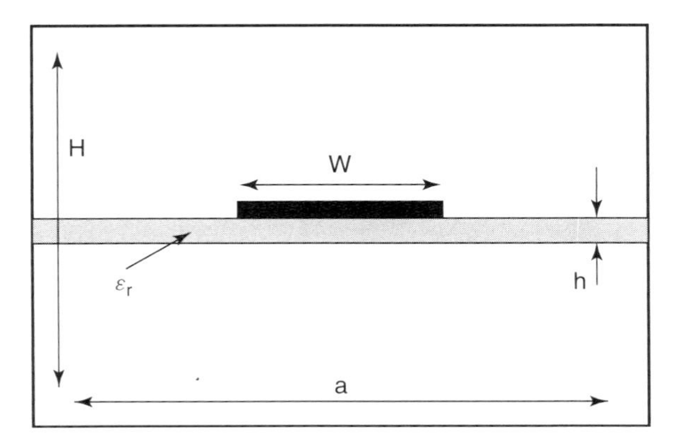

Figure 5.6 Suspended substrate stripline elements.

be $\lambda_g = (c/f)/\sqrt{\varepsilon_{\text{eff}}}$. For the design, only the wavelength $\lambda_{g0} = 35.7264$ mm at the center frequency $f_0 = 8298$ MHz is needed.

The rest of the technique consists of computing the admittance inverters based on the normalized slope. Then the susceptances B_b are computed, from which the gap lengths S can be deduced (Table 5.1). The angles ϕ_1 and ϕ_2 are computed from B_a and B_b obtained from the gap lengths S. Finally, the lengths $l/2$ are added to the perfect $\lambda_g/2$ lines to provide the final line lengths L (Table 5.2). This results in the filter shown in Figure 5.7.

The simulated response of the resulting filter is provided in Figure 5.8. Note that this simulation is approximate since it uses the *ABCD* matrices of the transmission lines and the equivalent circuit of the gap. The effects of the various

TABLE 5.1

Normalized Admittance Inverters, J/Y_0	Normalized Susceptances, B_b/Y_0	Gap Lengths, S (mm)
J0, 1 = 0.2609	B0, 1 = 0.2800	S0,1 = 0.209
J1, 2 = 0.0573	B1, 2 = 0.0575	S1,2 = 1.758
J2, 3 = 0.0421	B2, 3 = 0.0422	S2,3 = 2.140
J3, 4 = 0.0421	B3, 4 = 0.0422	S3,4 = 2.140
J4, 5 = 0.0573	B4, 5 = 0.0575	S4,5 = 1.758
J5, 6 = 0.2609	B5, 6 = 0.2800	S5,6 = 0.209

TABLE 5.2

Lengths Related to ϕ_1 (mm)	Lengths Related to ϕ_2 (mm)	Line Lengths L (mm)
li1 0, 1 = 0.004	li2 0, 1 = −2.899	L1 = 16.371
li1 1, 2 = 0.282	li2 1, 2 = −0.372	L2 = 17.983
li1 2, 3 = 0.404	li2 2, 3 = −0.075	L3 = 18.193
li1 3, 4 = 0.404	li2 3, 4 = −0.075	L4 = 17.983
li1 4, 5 = 0.282	li2 4, 5 = −0.372	L5 = 16.371
li1 5, 6 = 0.004	li2 5, 6 = −2.899	

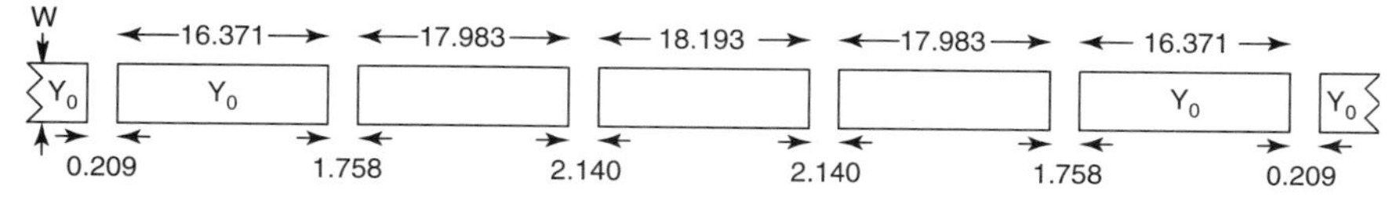

Figure 5.7 Line and gap lengths of the filter example (Chebyshev −20 dB, order 5).

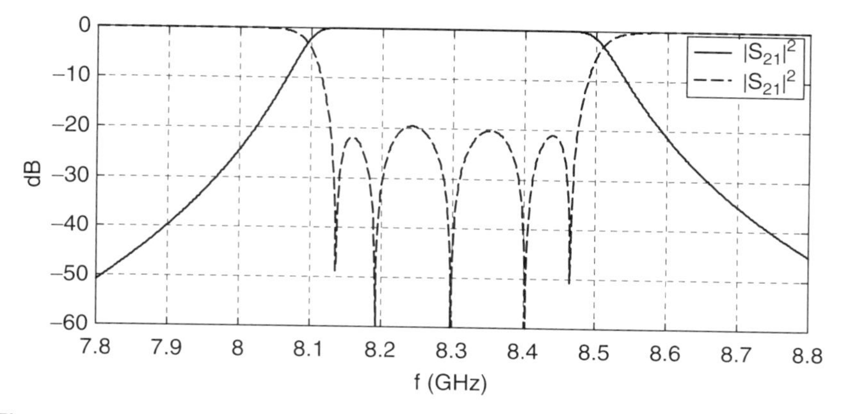

Figure 5.8 Simulated response of the filter example (Chebyshev −20 dB, order 5).

approximations (the normalized slope and the admittance inverter formed by B_a, B_b, and $l/2$) are seen to be rather minimal.

5.2.4 Filter Realizations

Two bandpass filters were designed according to specifications requested by telecommunications companies. The first is a narrowband filter centered at 8.3 GHz, and the second is a broadband filter centered at 16.5 GHz.

Narrowband Fourth-Order Chebyshev Filter at 8.3 GHz The filter was to have a center frequency of $f_0 = 8.3$ GHz with a passband width of 300 MHz. The filter should follow a Chebyshev response of order 4 with a return loss of −16.5 dB. The dielectric material was taken as Duroid that has a relative permittivity of $\varepsilon_r = 2.2$. The dimensions of the SSS were chosen as $H = 4$ mm, $h = 0.127$ mm, and $a = 7.65$ mm. For a characteristic impedance of 50 Ω, the effective permittivity and the strip width were $\varepsilon_{\text{eff}} = 1.024$ and $W = 5.47$ mm. The filter lengths were found to be

Gap Lengths (mm)	Line Lengths (mm)
$S_{01} = 0.22$	$L_1 = 15.14$
$S_{12} = 1.87$	$L_2 = 15.28$
$S_{23} = 2.16$	$L_3 = 15.28$
$S_{34} = 1.87$	$L_4 = 15.14$
$S_{45} = 0.22$	

The measured responses of the filter are shown in Figure 5.9. The filter did match the predictions well except for the insertion loss, which was about 1 dB.

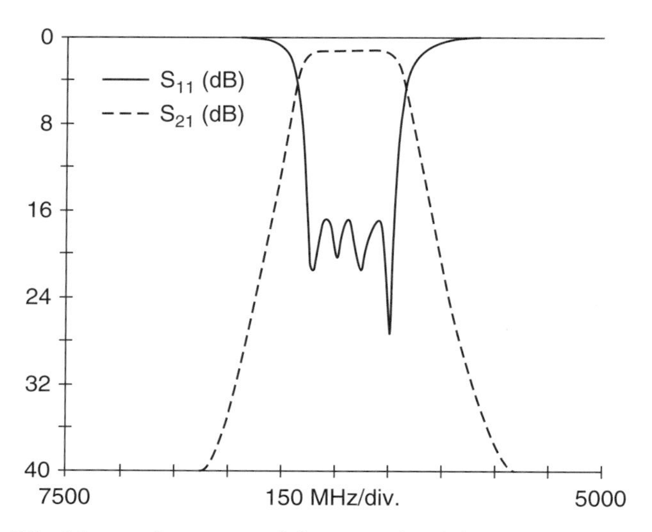

Figure 5.9 Measured response of the narrowband filter centered at 8.3 GHz.

Broadband Fifth-Order Chebyshev Filter at 16.5 GHz The filter was to have a center frequency of $f_0 = 16.5$ GHz with a passband width of 2.4 GHz. The filter should follow a Chebyshev response of order 5 with a return loss of -12 dB. The dielectric material was taken as Duroid that has a relative permittivity of $\varepsilon_r = 2.2$. The dimensions of the SSS were chosen as $H = 3.5$ mm, $h = 0.25$ mm, and $a = 8$ mm. For a characteristic impedance of 50 Ω, the effective permittivity and the strip width were $\varepsilon_{\text{eff}} = 1.024$ and $W = 5.47$ mm.

Gap Lengths (mm)	Line Lengths (mm)
$S_{01} = 0.19$	$L_1 = 6.95$
$S_{12} = 1.16$	$L_2 = 7.10$
$S_{23} = 1.42$	$L_3 = 7.05$
$S_{34} = 1.42$	$L_4 = 7.10$
$S_{45} = 1.16$	$L_5 = 6.95$
$S_{56} = 0.19$	

The measured responses of the filter are shown in Figure 5.10. The response shows the limitations of the technique, as the passband ripple is no longer equiripple. However, the return loss is below -12 dB throughout the passband, as intended. The insertion loss also increased to 1.8 dB. Figure 5.11 provides a picture of the realization of the capacitive-gap filter centered at 16.5 GHz.

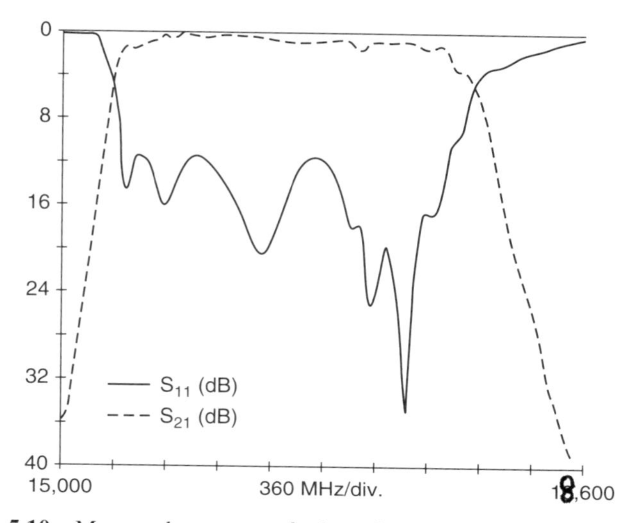

Figure 5.10 Measured response of a broadband filter centered at 16.5 GHz.

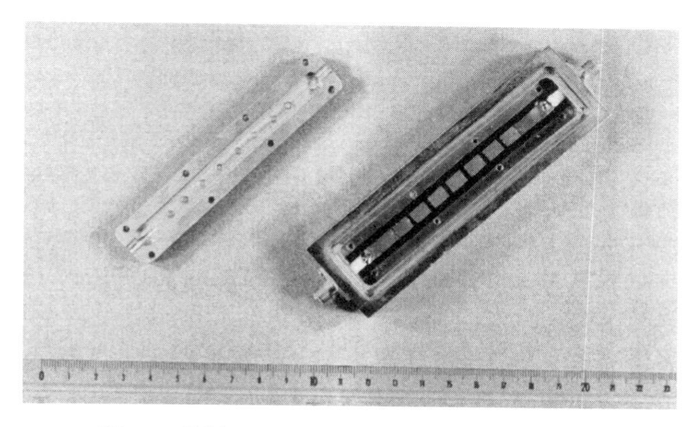

Figure 5.11 A 16.5-GHz capacitive-gap filter.

5.3 EXTENSION TO MILLIMETER WAVES

5.3.1 Millimeter-Wave Technology

The millimeter-wave region spans from 30 to 300 GHz and represents a vast spectrum resource. Figure 5.12 shows the average atmospheric absorption that occurs at these wavelengths. It can be seen that there are bands around 35, 94, 140, and 220 GHz where the attenuation is at its lowest. Each band can provide relative bandwidths of about 20%. These bands can be used for communication systems that require large bandwidths to support high data rates or large signal spectrums. Another advantage of these bands is the possibility for compact realizations of spatial systems (satellites) or military systems (missiles) due to the

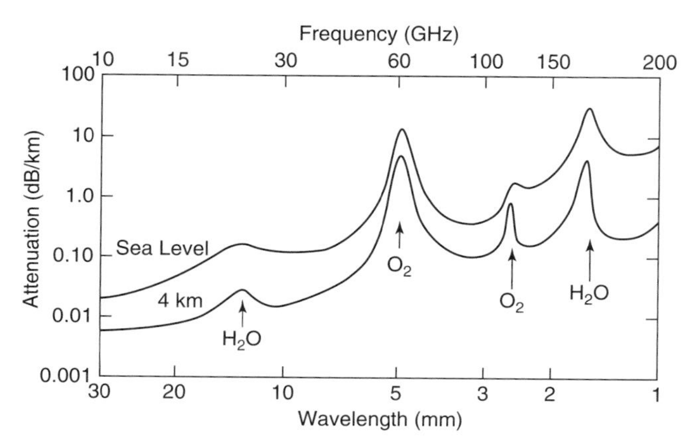

Figure 5.12 Average atmospheric absorption.

small wavelengths. This can also be seen as a problem since millimeter-wave circuits such as filters, mixers, and couplers will require very accurate tolerances and costly construction techniques. The suspended substrate stripline structure has an advantage over conventional microstrip structures, since in general it requires larger elements for the same functionality. In this section we show results that show the feasibility of capacitive-gap filters using suspended substrate structure at millimeter wavelengths.

A main issue when designing the system at millimeter waves are the connections from the suspended stripline structure to the rest of the system. In our case we studied the effects of connecting the SSS to rectangular waveguides. Figure 5.13 provides an illustration of the casing of the SSS and the two side holes for connecting the waveguides.

Experiments were conducted using a sweep frequency technique to define the direct losses from the combination waveguide–SSS–waveguide when the SSS was a 50-Ω continuous line. After some manual adjustments using a tuning screw, the insertion and return loss of this combination shown in Figure 5.14 were obtained. These results are very encouraging since they show that this technique can be used to produce systems with insertion losses of less than 1 dB and return losses on the order of 20 dB in a band from 26 to 40 GHz [5.6]. Building on this experimental work of a SSS fed by waveguides, millimeter-wave filters using suspended stripline structures were designed and tested [5.6]. An example is provided in the next section.

5.3.2 Fifth-Order Chebyshev Capacitive-Gap Filter at 35 GHz

The filter was to have a center frequency of $f_0 = 35$ GHz with a passband width of 2 GHz. The filter should follow a Chebyshev response of order 5 with a return loss of −18 dB. The dielectric material was taken as Duroid that has a relative permittivity of $\varepsilon_r = 2.2$.

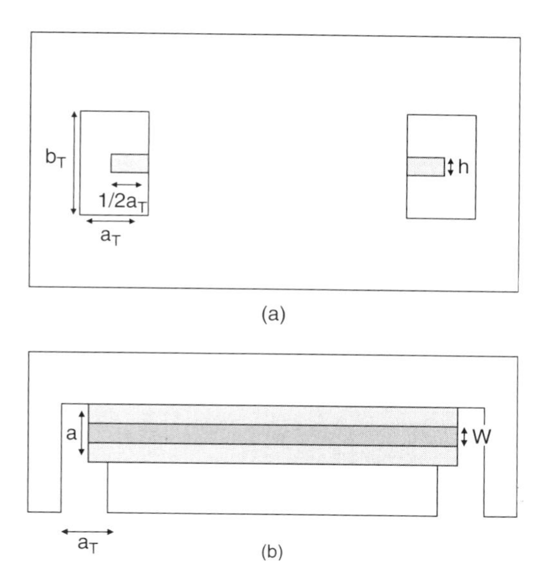

Figure 5.13 Connections to rectangular waveguides (a_T, b_T): (a) side view; (b) top view.

The dimensions of the SSS were chosen as $H = 1.55$ mm, $h = 0.25$ mm, $a = 3.55$ mm. For a characteristic impedance of 50 Ω, the effective permittivity and the strip width were $\varepsilon_{\text{eff}} = 1.156$ and $W = 2.009$ mm. The dimensions selected for the suspended stripline structure are in accordance with the prescribed performance characteristics of the millimeter-wave filter (Figure 5.14).

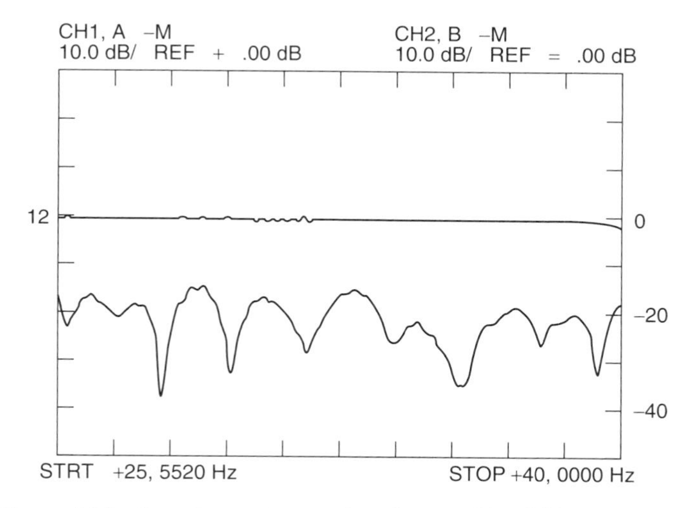

Figure 5.14 Insertion and return loss (waveguide–SSS–waveguide).

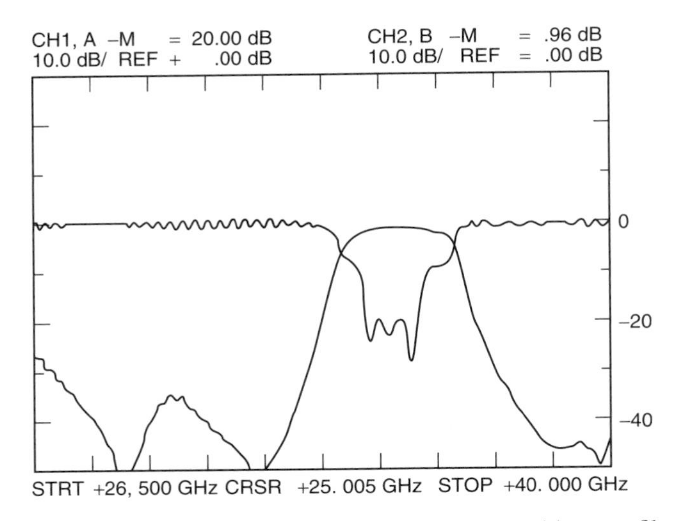

Figure 5.15 Measured responses of a 35-GHz capacitive-gap filter.

Gap Lengths (mm)	Line Lengths (mm)
$S_{01} = 0.134$	$L_1 = 3.068$
$S_{12} = 0.800$	$L_2 = 2.949$
$S_{23} = 0.948$	$L_3 = 2.898$
$S_{34} = 0.948$	$L_4 = 2.949$
$S_{45} = 0.800$	$L_5 = 3.068$
$S_{56} = 0.134$	

The measured responses of the filter are shown in Figure 5.15, and the system is shown in Figure 5.16. The response shows an insertion loss of about 1 dB at

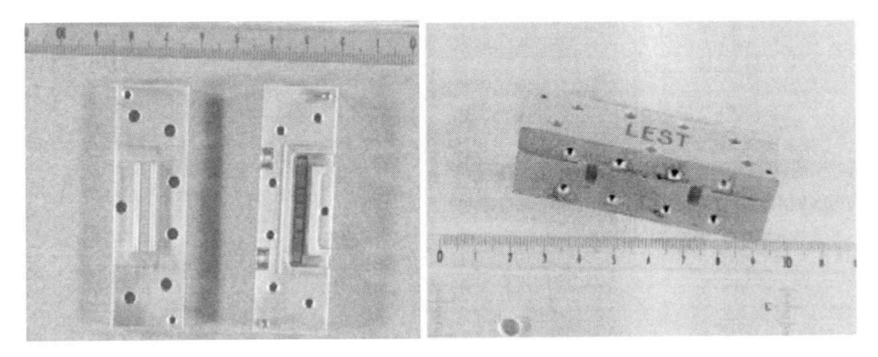

Figure 5.16 Millimeter-wave SSS capacitive-gap filter.

the center frequency. The return loss meets the −18-dB requirements in most of the passband. The five zeros of reflection typical of a fifth-order Chebyshev can be seen somewhat if we consider the dips at the passband edges.

5.4 ELECTROMAGNETIC CHARACTERIZATION OF SSS

As wavelengths decrease when increasing the frequencies of interest, as is the case at millimeter wavelengths, more rigorous approaches are needed for characterizing the structures. In this section we provide some introductory material on the electromagnetic study of microwave structures based on a spectral domain method [5.7–5.9]. The method is applicable for several classes of E-plane suspended bandpass millimeter filters. An E-plane suspended line is shown in Figure 5.17. It is made of a suspended stripline of width W and of infinite length placed on a substrate of permittivity ε_r and thickness h_2 placed in a casing of height $h_1 + h_2 + h_3$ and width b.

On the stripline, hybrid TE and TM modes are present. The electric potentials ϕ_i^e and the magnetic potentials ϕ_i^h are a solution of the Helmholtz equation

$$\nabla_t^2 \phi^{e,h}(x, y) + k_c^2 \phi^{e,h}(x, y) = 0$$

The spectral domain approach consists of transforming the (x, y) system to a (α_n, y) system, where α_n is understood as a spectral component. The potentials $\phi_i^{e,h}(x, y)$ are expressed as a Fourier series, as follows:

$$\phi_i^{e,h}(x, y) = \sum_{n=-\infty}^{n=+\infty} \phi_i^{e,h}(\alpha_n, y)\, e^{j\alpha_n x}$$

The parameter α_n is obtained by examining the behavior of the fields in the x direction. When the line is centered, the E modes are even and the H modes

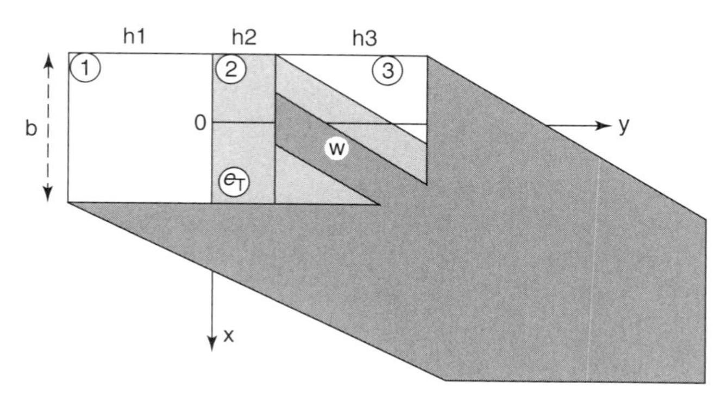

Figure 5.17 E-plane suspended stripline.

are odd [$E_z(x, y, z) = E_z(-x, y, z)$ and $H_z(x, y, z) = -H_z(-x, y, z)$] and the parameter α_n is given by

$$\alpha_n = \frac{(2n-1)\pi}{b}$$

It is also found that when the E modes are odd and the H modes are even, the parameter α_n is given by

$$\alpha_n = \frac{(2n)\pi}{b}$$

Applying boundary conditions and after some manipulations, the following algebraic equations are found relating the surface currents to the electric waves expressed in the spectral domain:

$$\begin{pmatrix} G_{11}(\alpha_n, \beta) & G_{12}(\alpha_n, \beta) \\ G_{21}(\alpha_n, \beta) & G_{22}(\alpha_n, \beta) \end{pmatrix} \begin{pmatrix} J_x(\alpha_n) \\ J_z(\alpha_n) \end{pmatrix} = \begin{pmatrix} E_x(\alpha_n) \\ E_z(\alpha_n) \end{pmatrix}$$

In this matrix equation, $E_x(\alpha_n)$, $E_z(\alpha_n)$, $J_x(\alpha_n)$, and $J_z(\alpha_n)$ are unknown functions. It is possible to eliminate $J_x(\alpha_n)$ and $J_z(\alpha_n)$ by using Galerkin's method and Parseval's theorem and solving for $E_x(\alpha_n)$ and $E_z(\alpha_n)$. A numerical solution of this matrix equation is obtained by introducing a known set of basis functions:

$$J_x(\alpha_n) = \sum_{i=1}^{i=I} a_i j_{x,i}(\alpha_n)$$

$$J_z(\alpha_n) = \sum_{k=1}^{k=K} b_k j_{z,k}(\alpha_n)$$

The characteristic equation for determining the propagation constant of the dominant and higher modes is obtained by setting the determinant of the coefficient matrix to zero.

$$\sum_{i=1}^{i=I} a_i \sum_{n=-\infty}^{n=+\infty} G_{11} j_{x,i} j_{x,m}^* + \sum_{k=1}^{k=K} b_k \sum_{n=-\infty}^{n=+\infty} G_{12} j_{z,k} j_{x,m}^* = 0 \quad m = 1, \ldots, I$$

$$\sum_{i=1}^{i=I} a_i \sum_{n=-\infty}^{n=+\infty} G_{21} j_{x,i} j_{x,m}^* + \sum_{k=1}^{k=K} b_k \sum_{n=-\infty}^{n=+\infty} G_{22} j_{z,k} j_{x,m}^* = 0 \quad m = 1, \ldots, K$$

The technique was applied to E-plane centered and displaced suspended lines. The results provided the value of the characteristic impedance at 30 GHz for several line widths, as shown in Figures 5.18 and 5.19.

The method can also be used to define more accurate equivalent-circuit elements, as in the case of a capacitive gap, as shown in Figure 5.20. The technique can be applied to finding the characteristic impedances and other characteristics of coupled lines, such as those needed in Chapter 11 for capacitive-gap coupled line filters.

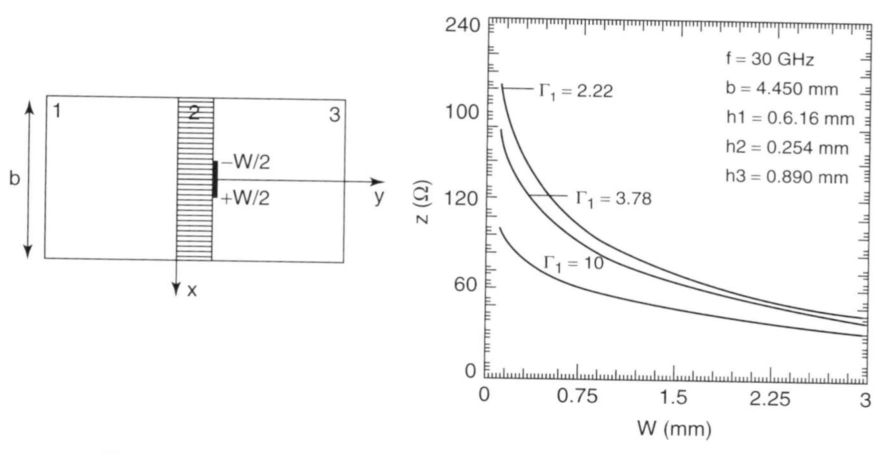

Figure 5.18 Characteristic impedance for a centerline at 30 GHz.

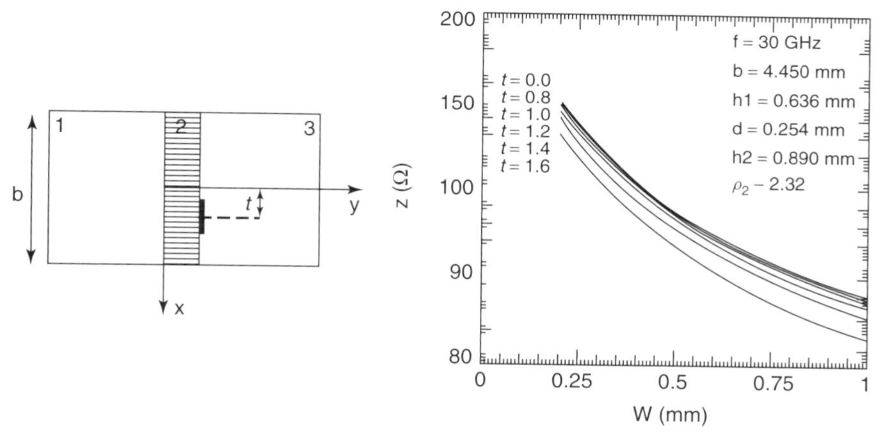

Figure 5.19 Characteristic impedance for a displaced line at 30 GHz.

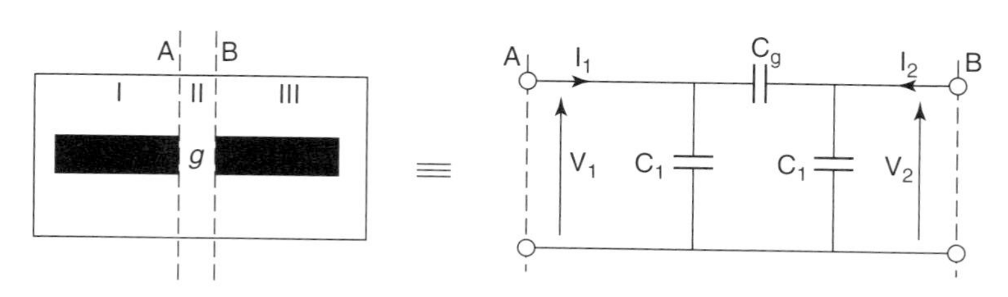

Figure 5.20 Gap and equivalent circuit.

5.5 CONCLUSIONS

The design of capacitive-gap filters using suspended substrate stripline structures has been presented. Several microwave bandpass filters have been implemented and show good agreement with the predictions. The results demonstrate that narrow- and broadband filters can readily be designed. The filter implementation is also suitable for use in hybrid circuits.

An experimental millimeter-wave filter using the suspended stripline structure has also been realized and was successful. The results suggest that it might be possible to design filters at 100 GHz or more. This is still in the realm of research, and the electromagnetic characterization using the spectral domain approach can provide a means for achieving this goal. This method provides the accuracy needed for characterizing the characteristic impedance and discontinuities in the lines at millimeter wavelengths.

Acknowledgments

The research leading to the design technique and experimental results was carried out under contract with Thomson and the French Ministry of Research and Technology.

REFERENCES

[5.1] P. Jarry, D. Lo Hine Tong, and G. Menexiadis, Design of capacitive-gap suspended substrate stripline filters, *Proc. IEEE Melecon'87*, Rome, pp. 353–355, Mar. 24–26, 1987.

[5.2] P. Jarry, D. Lo Hine Tong, and G. Menexiadis, *Computer-aided design of capacitive-gap suspended substrate strip-line filters*, presented at MIOP'87, Session 4A, Wiesbaden, Germany, May 19–21, 1987.

[5.3] D. Lo Hine Tong and P. Jarry, Synthesis and compact realization of asymmetric microwave suspended substrate filters, *Proc. 7th B. Franklin Symposium on Advances in Antenna and Microwave Technology*, IEEE AP/MTT-S, University of Pennsylvania, Philadelphia, PA, pp. 41–44, Mar. 1989.

[5.4] D. Lo Hine Tong and P. Jarry, Microwave filters on suspended substrate, *Ann. Telecommun.*, vol. 46, no. 3–4, pp. 216–222, 1991.

[5.5] Oliner, and Altschuler, Discontinuities in the center conductor of symmetric strip transmission line, *IRE Trans. Microwave Theory Tech.*, pp. 328–339, May 1960.

[5.6] P. Jarry and E. Kerhervé, Millimeter wave filter design with suspended strip line structure (SSS), *Proc. ESA Workshop on Millimeter Wave Technology and Applications*, Espoo, Finland, pp. 133–137, May 27–29, 1998.

[5.7] M. Lyakoubi and P. Jarry, Theoretical and experimental investigation of broadside-coupled suspended microstrip lines, presented at the Progress in Electromagnetic Research Symposium, PIERS'93, Pasadena, CA, July 12–16, 1993.

[5.8] M. Lyakoubi and P. Jarry, Broadside-coupled suspended line networks for passive filters, *Proc. Asia Pacific Microwave Conference*, APMC'93, Hsinchu, Taiwan, vol. 2, pp. 16-5 to 16-8, Oct. 18–22, 1993.

[5.9] M. Lyakoubi and P. Jarry, A rigorous analysis of suspended micro strip lines for passive filter applications, *Proc. Asia Pacific Microwave Conference*, APMC'93, Hsinchu, Taiwan, vol. 2, pp. 16-1 to 16-4, Oct. 18–22, 1993.

Selected Bibliography

Chang, K., and P. J. Khan, Analysis of a narrow capacitive strip in a waveguide, *IEEE Trans. Microwave Theory Tech.*, vol. 22, May 1974.

Cohn, S. B., Direct-coupled resonator filters, *Proc. IRE*, vol. 45, pp. 187–197, Feb. 1957.

Cohn, S. B., Microwave filters, an advancing art, *IEEE Trans. Microwave Theory Tech.*, vol. 13, no. 5, pp. 487–488, 1965.

Dougherty, R. M., Mm-wave filter design with suspended stripline, *Microwave J.*, pp 75–87, July 1986.

Ferrand, P., D. Baillargeat, and S. Verdeyme, Innovative technologies for quasiplanar millimeter wave filtering applications, *Int. J. RF Microwave Comput Aided Eng.*, vol. 17, no. 1, pp. 96–1103, 2007.

Glance, B., and R. Tramarulo, A waveguide to suspended stripline transition, *IEEE Trans. Microwave Theory Tech.*, vol. 21, pp. 117–119, Feb. 1973.

Hislop, A. R., and D. Rubin, Suspended substrate Ka-band multiplexer, *Microwave J.*, vol. 24, pp. 73–77, June 1981.

Konishi, Y., and K. Uenakada, The design of a bandpass filter with inductive strip: planar circuit mounted in waveguide, *IEEE Trans. Microwave Theory Tech.*, vol. 22, no. 10, pp. 869–873, 1974.

Levy, R., Theory of direct-coupled-cavity filters, *IEEE Trans. Microwave Theory Tech.*, vol. 15, no. 6, pp. 340–348, 1967.

Rubin, D., and A. R. Hislop, Millimeter wave coupled line filter design techniques for suspended substrate and microstrip, *Microwave J.*, vol. 23, pp. 67–68, Oct. 1980.

Tajima, Y., and Y. Sawayama, Design and analysis of a waveguide-sandwich microwave filter, *IEEE Trans. Microwave Theory Tech.*, vol. 22, no. 9, pp. 839–841, 1974.

Yamashita, E., B. Y. Wang, K. Atsuki, and K. R. Li, Effects of side wall grooves on transition characteristics of suspended striplines, *IEEE Trans. Microwave Theory Tech.*, vol. 33, pp. 1323–1327, Dec. 1985.

Young, L., Direct-coupled cavity filters for wide and narrow bandwidths, *IEEE Trans. Microwave Theory Tech.*, vol. 11, no. 3, pp. 162–178, 1963.

6

EVANESCENT-MODE WAVEGUIDE FILTERS WITH DIELECTRIC INSERTS

6.1 INTRODUCTION

Waveguide filters are often bulky and heavy and it is highly desirable to reduce the size and weight of these filters. Traditional waveguide filters are designed such that the fundamental mode of propagation operates above the cutoff frequency of the waveguide. Since the cutoff frequency is inverse proportional to the cross-sectional dimensions of the waveguide, there will be a limit to the reduction in size for a given frequency band of interest. In contrast, if the fundamental mode of propagation is no longer constrained to operate above the cutoff of the waveguide, the cross-sectional dimensions can be further reduced. Filters that operate using this technique are referred to as *evanescent-mode waveguide filters*.

Advanced Design Techniques and Realizations of Microwave and RF Filters,
By Pierre Jarry and Jacques Beneat

In this chapter we present the design procedure for a category of evanescent-mode waveguide filters. These filters are made of a central evanescent-mode waveguide filled with dielectric inserts. The design technique can provide minimum-phase filters with Chebyshev or Butterworth bandpass responses. It is applicable to waveguides with rectangular or circular cross sections. Furthermore, the design technique is complete in the sense that the response of the filter can be simulated prior to construction since the structure does not rely on manual tuning. Several realizations showing good agreement between simulations and measured filter responses are provided.

To achieve further reduction in size, the feasibility of folded evanescent-mode waveguide filters is examined. The structure includes mainly an additional bended waveguide for which the electromagnetic field expressions must be defined. Due to the complexity of the equations, numerical techniques are needed. Several examples are given showing that it is possible to design a folded evanescent-mode waveguide filter with Chebyshev-type responses.

6.2 EVANESCENT-MODE WAVEGUIDE FILTERS

The structure of an evanescent-mode filter is shown in Figure 6.1. It is composed of a central air-filled waveguide operating below the cutoff of the fundamental mode in which a number of dielectric inserts have been placed. The structure is terminated by larger air-filled waveguides operating above cutoff to allow for a suitable connection to the external source and load [6.1,6.2]. The permittivity of the dielectric inserts is chosen such that the fundamental mode will propagate above cutoff inside the dielectric inserts.

To analyze and subsequently to design filters, the structure of Figure 6.1 is decomposed into three types of sections:

1. A *T section* is defined as the input-terminating air-filled waveguide or the output-terminating air-filled waveguide. The input-terminating waveguide will be connected to the external source, and the output-terminating waveguide will be connected to the external load. The cross-sectional dimensions of air-filled terminating waveguides are chosen so that only the fundamental mode of propagation is kept above the cutoff frequency of the waveguide.
2. An *E section* is defined as the section of the evanescent-mode waveguide void of dielectric inserts (air-filled). The cross-sectional dimensions of air-filled evanescent waveguides are chosen so that the fundamental mode must propagate below cutoff inside an E section. The lengths of the E sections must be defined to provide the desired filtering response.
3. A *P section* is defined as the section of evanescent-mode waveguide filled with a dielectric insert of permittivity ε_r. The permittivity is chosen after selection of the cross-section dimensions of the evanescent waveguide only to allow the fundamental mode to propagate above cutoff inside a P section. The lengths of the P sections must be defined to provide the desired filtering response.

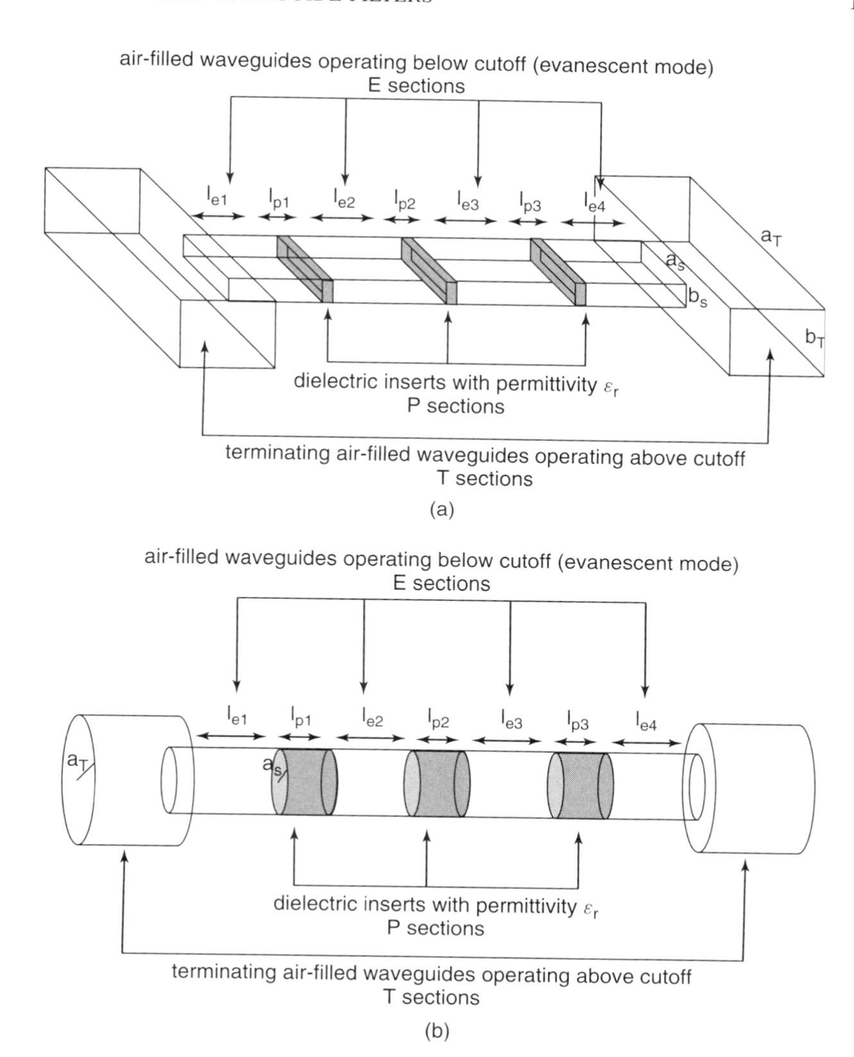

Figure 6.1 Evanescent-mode waveguide filters with (a) rectangular and (b) circular cross sections.

To characterize the structure of Figure 6.1 completely, two main types of junctions must be defined:

1. A $[T,E]$ *junction* is a junction between the input-terminating air-filled waveguide (the input T section) and the first evanescent-mode waveguide section (the first E section). [E,T] represents a similar junction between the last evanescent-mode waveguide section (the last E section) and the output-terminating air-filled waveguide (the output T section).

2. A $[E,P]$ *junction* is a junction between an evanescent-mode section (an E section) and a dielectric-filled waveguide section (a P section). [P,E] represents a similar junction between a dielectric-filled waveguide section (a P section) and an evanescent-mode section (an E section).

6.2.1 Scattering and *ABCD* Descriptions of the Structure

[T,E] and [E,T] Junctions Defining the scattering parameters of a [T,E] or a [E,T] junction is similar to that of defining the scattering parameters of what is called a *step discontinuity*. Unfortunately, such a discontinuity creates an infinite number of incident and reflected waves in the vicinity of the discontinuity, and no closed-form solution to this problem can be provided. However, the cross sections of the waveguides forming a T or an E section are chosen at the start of the design process and will remain unchanged. This means that the step discontinuity is fixed throughout the design process and a numerical characterization is sufficient.

One possible method to compute the S parameters of a step discontinuity numerically is to perform a modal analysis. The equations are rather complex and involve taking into account tens or hundreds of modes of propagation in the T and E sections. A summary of the technique is provided in Appendix 3. This method results in solving matrix equations of the type

$$[S] = [M_2]^{-1}[M_1]$$

where $[M_1]$ and $[M_2]$ are given by

$$(M_1) = \begin{pmatrix} [I] & [0] & [V_{hhuv}] & -[V_{heuv}] \\ [0] & [I] & [V_{ehuv}] & -[V_{eeuv}] \\ [V_{hhvu}] & [V_{hevu}] & -[I] & [0] \\ -[V_{ehvu}] & -[V_{eevu}] & [0] & -[I] \end{pmatrix}$$

$$(M_2) = \begin{pmatrix} [I] & [0] & [V_{hhuv}] & [V_{heuv}] \\ [0] & -[I] & [V_{ehuv}] & [V_{eeuv}] \\ -[V_{hhvu}] & [V_{hevu}] & [I] & [0] \\ [V_{ehvu}] & -[V_{eevu}] & [0] & -[I] \end{pmatrix}$$

One should note that the S parameters of the step discontinuity could also be obtained using a modern RF/microwave simulator such as HFSS.

In conclusion, the S and $ABCD$ matrices for a [T,E] section for the fundamental mode reduces to 2×2 matrices, described generically below.

$$(S) = \begin{pmatrix} S_{11} & S_{12} \\ S_{21} & S_{22} \end{pmatrix} \qquad (ABCD) = \begin{pmatrix} A & B \\ C & D \end{pmatrix}$$

The four S parameters are obtained numerically for one given set of cross-sectional dimensions and one frequency of operation. Repeating the calculations at additional frequencies can be used to simulate the frequency response of the structure. We will restrict the structure to be symmetric so that computation of the S parameters for the [E,T] section can be deduced from the S parameters of the [T,E] section.

[E,P] and [P,E] Junctions For the discontinuity created by the junction of two waveguides of different dielectric permittivities but the same cross section, an analytical solution can be found for the scattering parameters. The scattering matrix referred to the characteristic impedances of the waveguides is given in this case by

$$(S) = \begin{pmatrix} \dfrac{\gamma_{1h} - \gamma_{2h}}{\gamma_{1h} + \gamma_{2h}} & \dfrac{2\gamma_{2h}}{\gamma_{1h} + \gamma_{2h}} \sqrt{\dfrac{|\gamma_{1h}|}{|\gamma_{2h}|}} \\ \dfrac{2\gamma_{1h}}{\gamma_{1h} + \gamma_{2h}} \sqrt{\dfrac{|\gamma_{2h}|}{|\gamma_{1h}|}} & \dfrac{\gamma_{2h} - \gamma_{1h}}{\gamma_{1h} + \gamma_{2h}} \end{pmatrix}$$

where γ_{1h} and γ_{2h} represents the propagation functions of the fundamental mode in waveguides 1 and 2, respectively. The corresponding $ABCD$ matrices in the case of an [E,P] or a [P,E] junctions reduce to

$$(ABCD) = \begin{pmatrix} \sqrt{j} & 0 \\ 0 & \sqrt{j} \end{pmatrix} \qquad (ABCD) = \begin{pmatrix} \dfrac{1}{\sqrt{j}} & 0 \\ 0 & \dfrac{1}{\sqrt{j}} \end{pmatrix}$$

[E,P] junction [P,E] junction

The complex nature of these matrices is rather remarkable. It can be explained by the complex and real nature of the characteristic impedances of an E and a P section, respectively. The multiplication of these matrices results in the identity matrix, and since they always appear in pairs throughout the structure, their effect will cancel out. A similar result is found in the [T,E] and [E,T] junctions. It will be important to extract this type of matrices from the general $ABCD$ matrices of the [T,E] and [E,T] junctions before using the results in an equivalent-circuit model of the structure, as will be seen in Section 6.2.2.

P Sections The scattering and $ABCD$ matrices for a P section of length l_p referred to its real characteristic impedance $Z_p = R_c$ is given by

$$(S) = \begin{pmatrix} 0 & e^{-j\beta_h l_p} \\ e^{-j\beta_h l_p} & 0 \end{pmatrix} \qquad (ABCD) = \begin{pmatrix} \cos \beta_h l_p & jR_c \sin \beta_h l_p \\ \dfrac{j}{R_c} \sin \beta_h l_p & \cos \beta_h l_p \end{pmatrix}$$

The propagation function and characteristic impedance of a P section are given by

$$\beta_h = \sqrt{\left(\frac{\omega}{c}\right)^2 \varepsilon_r - \left(\frac{k_t}{a_S}\right)^2} \qquad Z_p = R_c = \frac{k_s \omega}{\beta_h}$$

where for the fundamental mode, $k_t = \pi$, $k_s = 2\mu_0(b_s/a_s)$ for rectangular waveguides, and $k_t = 1.841184$, $k_s = \mu_0$ for circular waveguides.

E Sections The scattering and *ABCD* matrices for an E section of length l_e referred to its purely imaginary characteristic impedance $Z_e = jX_c$ is given by

$$(S) = \begin{pmatrix} 0 & e^{-\alpha_h l_p} \\ e^{-\alpha_h l_p} & 0 \end{pmatrix} \qquad (ABCD) = \begin{pmatrix} \cosh \alpha_h l_e & jX_c \sinh \alpha_h l_e \\ \dfrac{j}{X_c} \sinh \alpha_h l_e & \cosh \alpha_h l_e \end{pmatrix}$$

The propagation function and characteristic impedance of an E section are given by

$$\alpha_h = \sqrt{\left(\frac{k_t}{a_S}\right)^2 - \left(\frac{\omega}{c}\right)^2} \qquad Z_e = jX_c = j\frac{k_s \omega}{\alpha_h}$$

where for the fundamental mode, $k_t = \pi$, $k_s = 2\mu_0(b_s/a_s)$ for rectangular waveguides, and $k_t = 1.841184$, $k_s = \mu_0$ for circular waveguides.

6.2.2 Equivalent Circuit of the Structure

The main goal after characterizing the structure is to develop a technique that can define the unknown lengths of the E and P sections so that the structure can provide a specified Chebyshev or maximally flat bandpass response. A direct approach consisting of extracting these parameters directly from the overall expression of S_{21} is not practical since the expression of S_{21} even for a few dielectrics becomes extremely cumbersome and the equations involved are strongly nonlinear. Therefore, a different approach is adopted which consists of finding an approximate lumped equivalent circuit to represent the structure. It is important to shape the equivalent circuit to resemble that of a known ladder prototype, such as those provided in Chapter 2. The ladder prototype is designed to meet the desired Chebyshev or maximally flat bandpass response. The elements of the equivalent circuit can then be computed from the elements of the ladder prototype. Since the equivalent-circuit elements do not behave as perfect lumped elements, impedances and admittances computed at the center frequency of the desired filter are used for defining the unknown parameters of the structure. This means that the structure and the ladder prototype will have exactly the same response at the center frequency, but the responses might differ at other frequencies. It is therefore important to define an equivalent circuit of the structure that is suitably complex for matching the response at these other frequencies while being simple enough so that the parameters can be extracted.

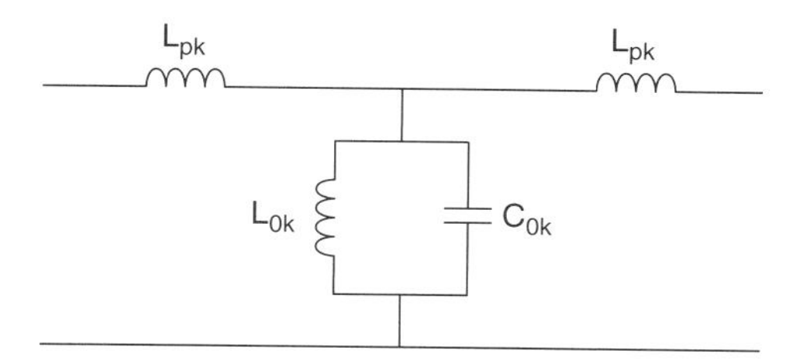

Figure 6.2 Equivalent circuit of a P section of length l_{pk}.

P Sections A P section will be modeled as a T network, as shown in Figure 6.2. It consists of two identical series inductors and one parallel *LC* cell. The series inductors are given by

$$L_{pk} = k_s \frac{l_{pk}}{2} \frac{\tan[\beta_h(l_{pk}/2)]}{\beta_h(l_{pk}/2)}$$

The parallel *LC* cell inductor and capacitor are given by

$$L_{0k} = \frac{k_s}{\varepsilon_r} \frac{c^2}{\omega_o^2} \frac{1}{l_{pk}} \frac{\beta_h l_{pk}}{\sin \beta_h l_{pk}} \qquad C_{0k} = \frac{1}{L_{0k}\omega_o^2} \qquad \omega_0 = \frac{k_t c}{a_s\sqrt{\varepsilon_r}}$$

It will be important to use the center frequency of the desired filter as the frequency in these expressions.

E Sections For E sections it was preferred to combine the *ABCD* matrix directly with other elements in the structure rather than providing a circuit equivalent of the E section and performing multiple circuit transformations. For example, the case of an E section of length l_{ek+1} placed between a P section of length l_{pk} and a P section of length l_{pk+1} is shown in Figure 6.3. In Figure 6.3 the *ABCD* matrices of the [P,E] and [E,P] junctions are also shown to be complete. However, multiplication of these matrices when taken in pairs results in the identity matrix. They can therefore be removed from the equivalent circuit. In Figure 6.3, the E section *ABCD* matrix is combined with the series inductors from the P section of length l_{pk} and the P section of length l_{pk+1} to form the admittance inverter between two parallel inductors, as shown in Figure 6.4.

The admittance inverter can be computed using

$$K_{k,k+1} = \frac{1}{J_{k,k+1}}$$
$$= \cosh(\alpha_h l_{ek+1})(L_{pk} + L_{pk+1})\omega + \sinh(\alpha_h l_{ek+1})\left(\frac{k_s}{\alpha_h} + L_{pk}L_{pk+1}\frac{\alpha_h}{k_s}\right)\omega$$

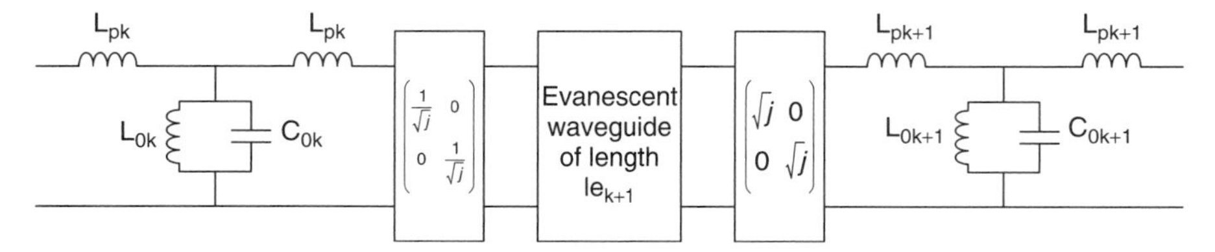

Figure 6.3 E section between two P sections.

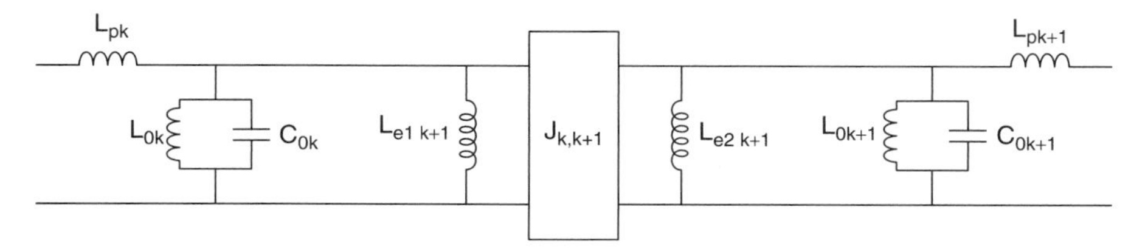

Figure 6.4 Equivalent circuit of an E section between two P sections.

The parallel inductor in front of the admittance inverter is given by

$$L_{e1k+1} = L_{pk} + \frac{L_{pk+1}\cosh(\alpha_h l_{ek+1}) + (k_s/\alpha_h)\sinh(\alpha_h l_{ek+1})}{\cosh(\alpha_h l_{ek+1}) + (\alpha_h/k_s)L_{pk+1}\sinh(\alpha_h l_{ek+1})}$$

The parallel inductor after the admittance inverter is given by

$$L_{e2k+1} = L_{pk+1} + \frac{L_{pk}\cosh(\alpha_h l_{ek+1}) + (k_s/\alpha_h)\sinh(\alpha_h l_{ek+1})}{\cosh(\alpha_h l_{ek+1}) + (\alpha_h/k_s)L_{pk+1}\sinh(\alpha_h l_{ek+1})}$$

Terminations For the case of the terminations of the structure, several transformations are needed to yield an equivalent circuit which resembles that of a ladder structure made of admittance inverters and resonating *LC* cells. First, the numerical data obtained using the modal analysis method described in Appendix 3 to compute the scattering parameters of a step discontinuity are modeled as a transformer followed by a series and parallel inductors, as shown in Figure 6.5.

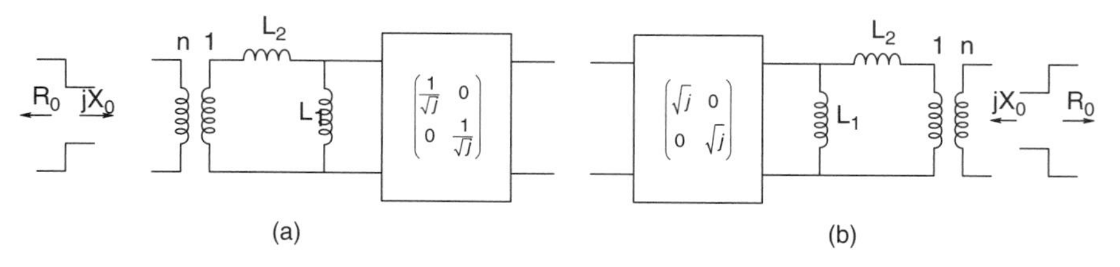

Figure 6.5 Equivalent circuit of (a) a [T,E] junction and (b) an [E,T] junction.

The complex matrices must be separated, as also shown in Figure 6.5. Since they appear in pairs, they will be removed from the equivalent circuits.

The S parameters of the step discontinuity corresponding to the [T,E] junction are transformed in $ABCD$ parameters and correspond to a complex matrix times a regular $A_N B_N C_N D_N$ matrix:

$$\begin{pmatrix} S_{11} & S_{12} \\ S_{21} & S_{22} \end{pmatrix} \Rightarrow \begin{pmatrix} \sqrt{j} & 0 \\ 0 & \sqrt{j} \end{pmatrix} \begin{pmatrix} A_N & B_N \\ C_N & D_N \end{pmatrix}$$

The transformer turn ratio n is then obtained using the S-parameter data and is given by

$$n = \frac{1}{A_N} = \frac{|S_{21}|^2}{\mathrm{Re}(S_{21}) + \mathrm{Re}(S_{11}S_{21}^*)} \sqrt{\frac{R_0}{X_c}}$$

The series inductor is also computed using the numerical S-parameter data and is given by

$$jL_1\omega = \frac{A_N}{C_N} = j\frac{\mathrm{Re}(S_{21}) + \mathrm{Re}(S_{11}S_{21}^*)}{\mathrm{Re}(S_{21}) - \mathrm{Re}(S_{11}S_{21}^*)} X_c$$

The parallel inductor is also computed using the numerical S-parameter data and is given by

$$jL_2\omega = A_N B_N = j\frac{[\mathrm{Re}(S_{21}) + \mathrm{Re}(S_{11}S_{21}^*)][-\mathrm{Im}(S_{21}) + \mathrm{Im}(S_{11}S_{21}^*)]}{|S_{21}|^4} X_c$$

where R_0 is the real characteristic impedance of the T section and jX_c is the purely imaginary characteristic impedance of the E section.

To complete the equivalent circuit of the entire structure, the situation at the input of the structure shown in Figure 6.6 must be modified. The equivalent circuit of the termination is followed by the E section of length l_{e1}. The E section is followed by the matrix of the [E,P] junction and the series inductor of the P section of length l_{p1} that has not yet recombined with an E section.

The E section of length l_{e1} is combined with the series and parallel inductors from the termination on one side and the series inductor from the P section of length l_{p1} on the other side to form an admittance inverter between two parallel

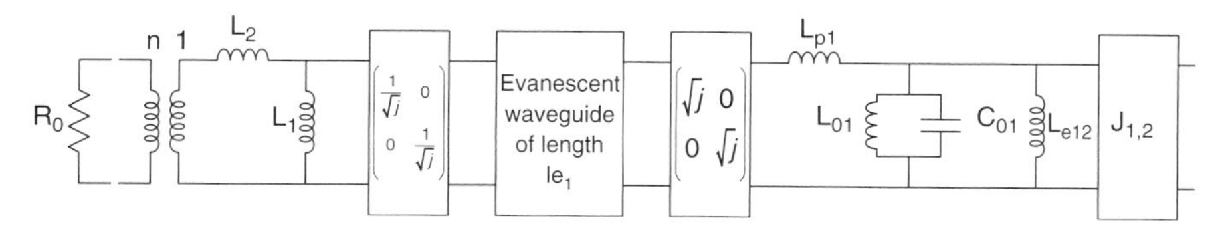

Figure 6.6 Situation at the input of the structure.

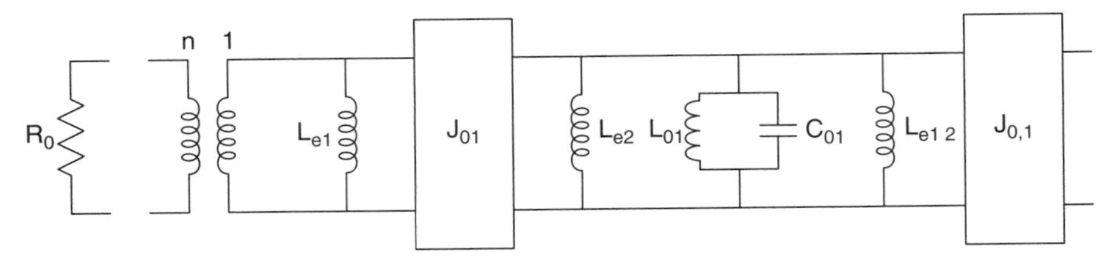

Figure 6.7 Equivalent circuit at the input of the structure.

inductors, as shown in Figure 6.7. The first admittance inverter is therefore composed of elements of the termination, the first E section and the first P section. The first admittance inverter is given by

$$K_{0,1} = \frac{1}{J_{0,1}} = \cosh(\alpha_h l_{e1}) \left[\left(1 + \frac{L_2}{L_1} \right) L_{p1} + L_2 \right] \omega + \sinh(\alpha_h l_{e1}) \left[\left(1 + \frac{L_2}{L_1} \right) \frac{k_s}{\alpha_h} + L_2 L_{p1} \frac{\alpha_h}{k_s} \right] \omega$$

The inductors surrounding the first admittance inverter are also composed of elements of the termination, the first E section and first P section. They are given by

$$L_{e1} = L_2 + \frac{L_{p1}\cosh(\alpha_h l_{e1}) + (k_s/\alpha_h)\sinh(\alpha_h l_{e1})}{(1 + L_{p1}/L_1)\cosh(\alpha_h l_{e1}) + [(k_s/\alpha_h)(1/L_1) + (\alpha_h/k_s)L_{p1}] \sinh(\alpha_h l_{e1})}$$

$$L_{e2} = L_{p1} + \frac{L_2\cosh(\alpha_h l_{e1}) + (1 + L_2/L_1)\dfrac{k_s}{\alpha_h}\sinh(\alpha_h l_{e1})}{(1 + L_2/L_1)\cosh(\alpha_h l_{e1}) + (\alpha_h/k_s)L_2\sinh(\alpha_h l_{e1})}$$

The equivalent circuit must still be modified to be compatible with the bandpass ladder prototype of Chapter 2. After some circuit manipulations, the input of the structure can be represented as the same admittance inverter followed by a parallel capacitor, as shown in Figure 6.8. Unfortunately, the source resistance changes and becomes dependent on the first E and P sections through L_{e1}.

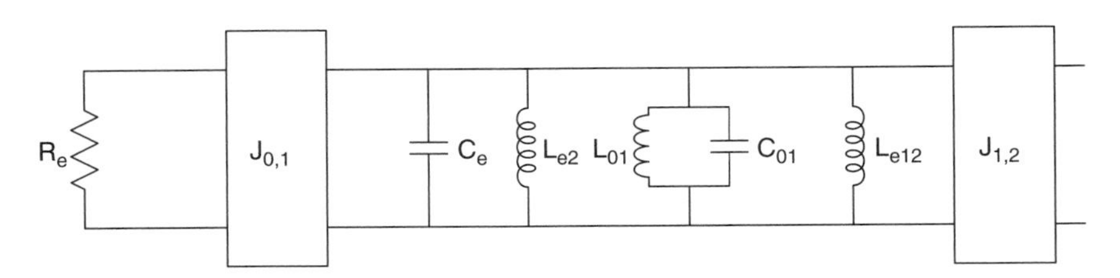

Figure 6.8 Final transformations of the input of the equivalent circuit.

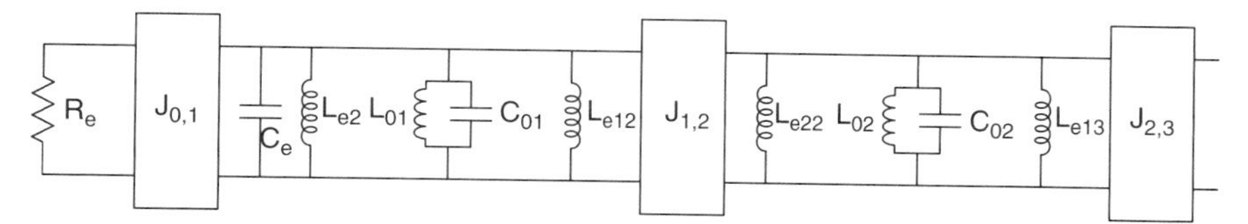

Figure 6.9 Equivalent circuit of the structure.

The termination resistance is given by

$$R_e = \frac{R_t(L_{e1}\omega)^2}{R_t^2 + (L_{e1}\omega)^2} \qquad \text{where} \quad R_t = \frac{R_0}{n^2}$$

and the parallel capacitor is given by

$$C_e = \frac{L_e}{K_{0,1}^{\ 2}} \qquad \text{where} \quad L_e = \frac{R_t^2 L_{e1}}{R_t^2 + (L_{e1}\omega)^2}$$

Summary The general form of the equivalent circuit of the structure is provided in Figure 6.9. As one can see, the admittance inverters can be compared directly to those of the ladder prototype. However, the first parallel LC cells of the ladder prototype will correspond to three parallel inductors and two parallel capacitors, while the other LC cells will correspond to three parallel inductors and one parallel capacitor. The main problem is that the terminating resistance depends on the length of the first E section and the first P section. This terminating resistance must be known in order to compute the elements of the ladder prototype. Furthermore, a given element of the equivalent circuit often depends on several unknown lengths l_e and l_p. For these reasons the design method relies on an iterative procedure where an initial guess for all the E and P section lengths is provided. The complete filter design steps as well as the iterative process algorithm are described in the next section.

6.2.3 Filter Design Procedure

The filter design technique consists of three main steps:

1. Choose the cross-sectional dimensions for the waveguides and a suitable value for the dielectric permittivity to guarantee the proper mode of operation for each section of the structure in the frequency band of interest. In the process, define the center frequency f_0 that will be used in the design.
2. Compute the S parameters of the step discontinuity for the fundamental mode at the center frequency f_0 using the cross-sectional dimensions of the terminating and evanescent waveguides.

3. Use the iterative procedure described below to extract the unknown lengths l_{ek} and the l_{pk} of the structure.

The interactive process begins by providing initial guesses for the lengths l_{ek} and l_{pk} of the E and P sections of the structure. The following iterative process is then repeated until the lengths between iterations differ by a negligible value.

1. Compute the termination resistance R_e using the lengths l_{ek} and l_{pk}. Compute the elements of the ladder prototype for this termination resistance and the desired Chebyshev response.
2. Compute the capacitors C_{Sk} of the equivalent circuit using the lengths l_{ek} and l_{pk}. Scale the admittance inverters $J_{Pk,k+1}$ of the ladder prototype so that the capacitors C_{Pk} of the ladder prototype become equal to the capacitors C_{Sk} of the equivalent circuit. This is done using the transformation

$$J_{P0,1}(\text{new}) = m_1 J_{P0,1}(\text{old})$$
$$J_{Pn,n+1}(\text{new}) = m_n J_{Pn,n+1}(\text{old})$$
$$J_{Pk,k+1}(\text{new}) = m_k m_{k+1} J_{Pk,k+1}(\text{old}) \qquad k = 1, 2, \ldots, n-1$$
$$m_k = \sqrt{\frac{C_{Sk}}{C_{Pk}}}$$

3. Extract new lengths l_{ek} by equating the admittance inverters $J_{Sk,k+1}$ of the equivalent circuit with those of the ladder prototype $J_{Pk,k+1}$. In this case,

$$J_{S0,1} = J_{Sn,n+1} = f(l_{e1}, l_{p1}, [\text{T,E}])$$
$$J_{Sk,k+1} = f(l_{pk}, l_{ek+1}, l_{pk+1}) \qquad k = 1, 2, \ldots, n-1$$

   This is done by using the old lengths l_{pk}.
4. Compute the new inductor values L_{Sk} of the equivalent circuit using the new lengths l_{ek} and the old lengths l_{pk}. Compute the new capacitor values C_{Sk} for the equivalent circuit by making the admittance of the resonating cells equal to zero at f_0.
5. Extract new lengths l_{pk} from the new values of the capacitors C_{Sk}. In this case,

$$C_{S1} = C_{Sn} = C_{01} + C_e = f(l_{e1}, l_{p1}, [\text{T,E}])$$
$$C_{Sk} = C_{0k} = f(l_{pk}) \qquad k = 2, \ldots, n-1$$

For the first capacitor, use the new value of l_{e1}.

As the process converges, all the elements of the equivalent circuit approach those of the prototype at the center frequency f_0, so the structure will have exactly the same response as that of the ladder prototype at f_0.

6.2.4 Design Examples and Realizations

Step-by-Step Design Example For a filter approximating a Chebyshev bandpass characteristic of order 6 with passband edges of $f_{\text{low}} = 13.7$ GHz and $f_{\text{high}} = 14.3$ GHz and a return loss of 15 dB, define the necessary elements of the evanescent-mode circular waveguide filter.

Step 1: Choosing the radius a_T of the terminating waveguides, the radius a_s of the evanescent-mode waveguide, and the dielectric permittivity ε_r. The value for a_T is flexible and is usually chosen to be that of a typical waveguide for operation in the frequency range of the intended filter. For example, taking $a_T = 7$ mm will produce a waveguide that has a cutoff frequency of 12.56 GHz for the fundamental mode and a cutoff frequency of 16.40 GHz for the second mode. In this case the fundamental mode will be above cutoff and other modes will be below cutoff around the frequencies of the intended filter.

Choosing a_s and ε_r is somewhat more difficult, because in addition to checking the various cutoff frequencies, it has been observed that the design technique proposed provides better results when using the frequency f_{th}, defined below as the center frequency. This frequency depends on both a_s and ε_r and should be made as close as possible to the filter passband center frequency f_0.

$$f_{\text{th}} = \frac{1}{2}\frac{x_1}{a_s}\frac{c}{2\pi}\left(1 + \frac{1}{\sqrt{\varepsilon_r}}\right) \qquad f_0 = \sqrt{f_{\text{low}} f_{\text{high}}}$$

Since dielectric values are in general less flexible than waveguide crosssection sizes, it is preferable to select the permittivity ε_r of a known dielectric material. For example, Teflon is found to be suitable for space applications and has a dielectric permittivity of $\varepsilon_r = 2.1$. It is preferable, however, to measure the exact dielectric permittivity of the Teflon to be used before designing the filter. Then using the expression of f_{th} when $f_{\text{th}} = f_0 = 13.9968$ GHz, the cross section a_s can be defined:

$$a_s = \frac{1}{2}\frac{x_1}{f_{\text{th}}}\frac{c}{2\pi}\left(1 + \frac{1}{\sqrt{\varepsilon r}}\right) = \frac{1}{2}\frac{1.841184}{13.9968\times 10^9}\frac{3\times 10^8}{2\pi}\left(1 + \frac{1}{\sqrt{2.1}}\right) = 5.307 \text{ mm}$$

It is important to check at this point if the values for a_s and ε_r will produce the necessary waveguide cutoff frequencies. Using these values, the air-filled evanescent waveguide has a cutoff frequency of 16.5649 GHz, so that all modes of propagation are evanescent in the frequency band of interest. The dielectric-filled sections have a cutoff frequency of 11.4309 GHz for the fundamental mode and a cutoff frequency of 14.9302 GHz for the second mode. The cutoff frequency of the second mode is somewhat low compared to the frequency band of interest, and reducing a_s slightly, to 5.2 mm, for example, provides a better overall choice. This leads to cutoff frequencies of 11.6661 and 15.2374 GHz for the dielectric-filled sections and 16.9058 GHz for the air-filled sections. The resulting frequency f_{th} in this case is equal to 14.2859 GHz, which is slightly higher than f_0. The purpose of the foregoing adjustments is to show that the constraint $f_{\text{th}} = f_0$ can be relaxed.

Step 2: Computing the S parameters of the step discontinuity. The S parameters of the step discontinuity are computed, for example, with a total of 10 modes of propagation on each side of the discontinuity for a terminating waveguide of radius $a_T = 7$ mm and an evanescent waveguide of radius $a_s = 5.2$ mm at the center frequency of the filter, $f_0 = 13.9968$ GHz. Note that only the S parameters for the fundamental mode are needed, so complete inversion of the matrix $[S] = [M_2]^{-1}[M_1]$ is not necessary.

Step 3: Extracting the lengths l_{ek} and l_{pk} of the E and P sections of the structure. The inputs to this process were the S parameters of the step discontinuity at a center frequency $f_0 = 13.9968$ GHz, terminating and central evanescent waveguide radii of 7.000 and 5.200 mm, dielectric permittivity of 2.1, prescribed filter passband edges of 13.7000 and 14.3000 GHz, Chebyshev bandpass response of order 6, and 15 dB of return loss.

After 100 iterations, for example, the maximum difference between iteration in lengths l_{ek} is on the order of 10^{-13} mm, while the maximum difference in lengths l_{pk} is on the order of 10^{-14} mm. In other examples using this technique, similar results are obtained except when the specified passband width is very large. In this case, the first E-section length tends to become equal to zero. The lengths of the seven E sections are found to be (in mm)

2.601 8.663 9.867 10.059 9.867 8.663 2.601

and the lengths of the six P sections are found to be (in mm)

5.371 6.028 6.009 6.009 6.028 5.371

See Figure 6.10.

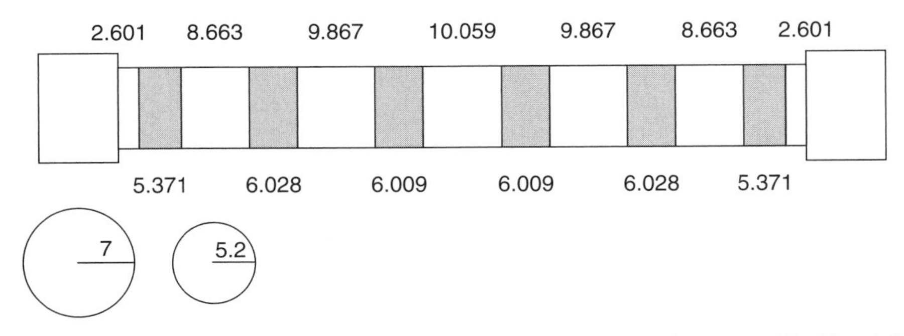

Figure 6.10 Layout of a sixth order Chebyshev evanescent-mode waveguide filter (all values in mm).

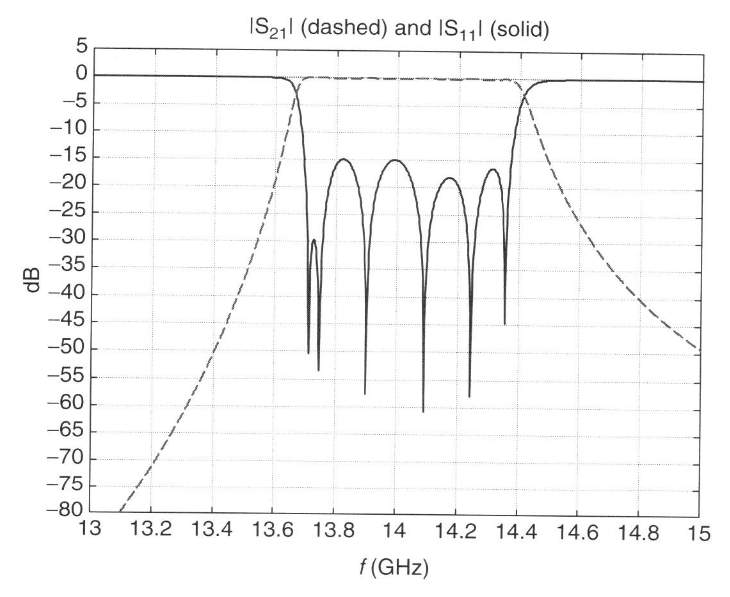

Figure 6.11 Simulated response of a sixth-order Chebyshev evanescent-mode filter.

Step 4: Simulating the structure and checking the frequency response. It is important to simulate the frequency response of the structure to verify how close the actual filtering characteristics are to the response prescribed. This can be accomplished using commercial simulation software such as HFSS, or by using the scattering matrices defined in Section 6.2.1, in which case the S parameters of the step discontinuity must be computed at additional frequencies. Figure 6.11 shows the simulated response of the structure between the frequencies of 13 and 15 GHz. It can be seen that the response of the structure is exactly that of a sixth-order Chebyshev filter at the center frequency $f_0 = 13.9968$ GHz. The response shows clearly the six zeros of reflection, and the minimum return loss of 15 dB in the entire passband is achieved. However, the reponse is not exactly equiripple, as slight differences occur at frequencies away from f_0, and the passband edges are slightly different than the prescribed passband edges of 13.7000 and 14.3000 GHz. In this case the filter had a relative bandwidth of 4.3%. The technique works well for narrowband filters since it is based on a narrowband technique. For larger bandwidths, the technique provides a very good starting point for an optimization procedure.

Realizations of Evanescent-Mode Filters The design technique and simulations have shown that it is possible to achieve evanescent-mode waveguide filters with prescribed Chebyshev responses. The next issue is how to construct such filters. An early technique described in Figure 6.12 was to think of the structure as a cascade of separate waveguide pieces. An A-type section corresponds

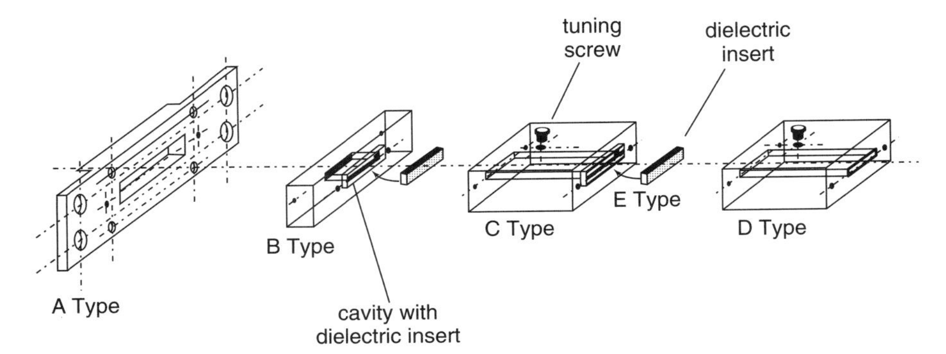

Figure 6.12 Construction of an evanescent-mode waveguide filter.

to a terminating waveguide with a surrounding metallic surface to allow for filter-length screws to maintain all the central sections of the filter in place. The B-type section corresponds to the first E section and has a grooved area on the other side to insert the first dielectric insert and maintain it in place. This technique allows the dielectric inserts to be kept at the proper locations even under mechanical stress, such as during the launching of a satellite. A C-type section is similar to a B section but will correspond to central E and P sections. A D-type section does not include any grooved area and corresponds to the last E section. In these early realizations, provisions for tuning screws were also made.

A first realization made at Alcatel Space in France consisted of a fifth-order Chebyshev filter using rectangular waveguides. The specified center frequency was 13.95 GHz, with a 900-MHz bandwidth (7%) [6.3,6.4]. The minimum passband return loss specified was 20 dB, and the dielectric inserts were made of Teflon with a relative permittivity of 2.1. Simulated and measured filter responses are shown in Figure 6.13. It can be seen that the simulations and the constructed filter responses are in close agreement.

A second realization made at CNES, a French national center for space studies, consisted of a fifth-order Chebyshev filter made of rectangular waveguides [6.5]. The prescribed center frequency was 14.25 GHz with a 500-MHz bandwidth (3.5%). The prescribed passband return loss was also 20 dB, but the dielectric inserts were made of Rexolite with a relative permittivity of 2.54. The filter, however, was constructed using a new technique. Rather than using separate waveguide pieces which were then assembled together, the filter was made of a single central evanescent-mode waveguide where the dielectric inserts were forced into place by having their dimensions slightly larger than the waveguide, as can be seen in Figure 6.14. This technique can maintain the dielectric inserts in place without the use of grooves.

Simulated and measured responses of this filter are shown in Figure 6.15. Again, it can be seen that the simulated and measured responses are in fair agreement. There is, however, some degradation in the return loss measured. At times it is seen to reach 15 dB, compared to the 20 dB specified.

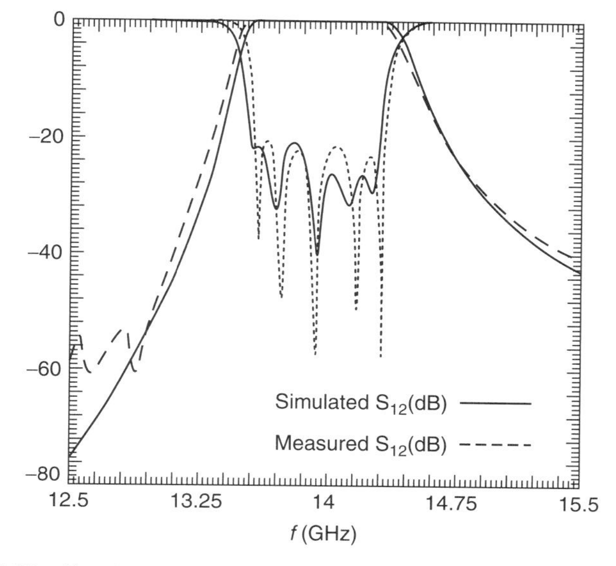

Figure 6.13 Simulated and measured responses of an evanescent-mode filter.

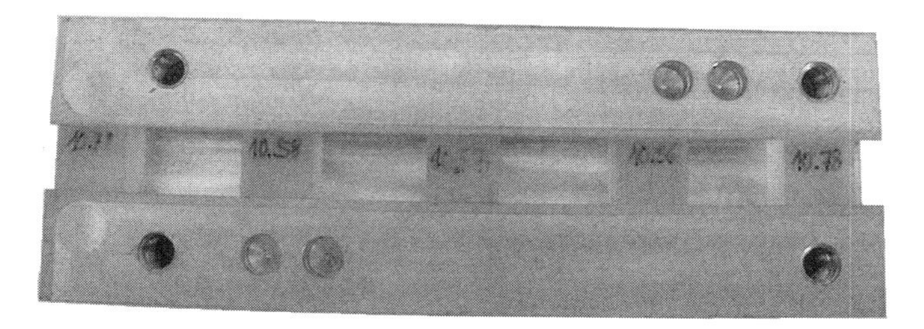

Figure 6.14 New realization technique of an evanescent-mode waveguide filter.

6.3 FOLDED EVANESCENT-MODE WAVEGUIDE FILTERS

In an attempt to reduce the form factor of the evanescent-mode waveguide filters, the folded structure shown in Figure 6.16 has been investigated [6.6,6.7]. The folded structure of Figure 6.16 is described by the same T, E, and P sections and the same [T,E], [E,T], [E,P], and [P,E], junctions as those described in Section 6.2. The main difference is that an additional section representing the bent section of an air-filled evanescent-mode waveguide and associated junctions with straight E sections must be characterized.

The new section and junctions are defined as follows:

1. The B section represents the bent section of an air-filled evanescent-mode waveguide. The cross-sectional dimensions of a B section are the same as

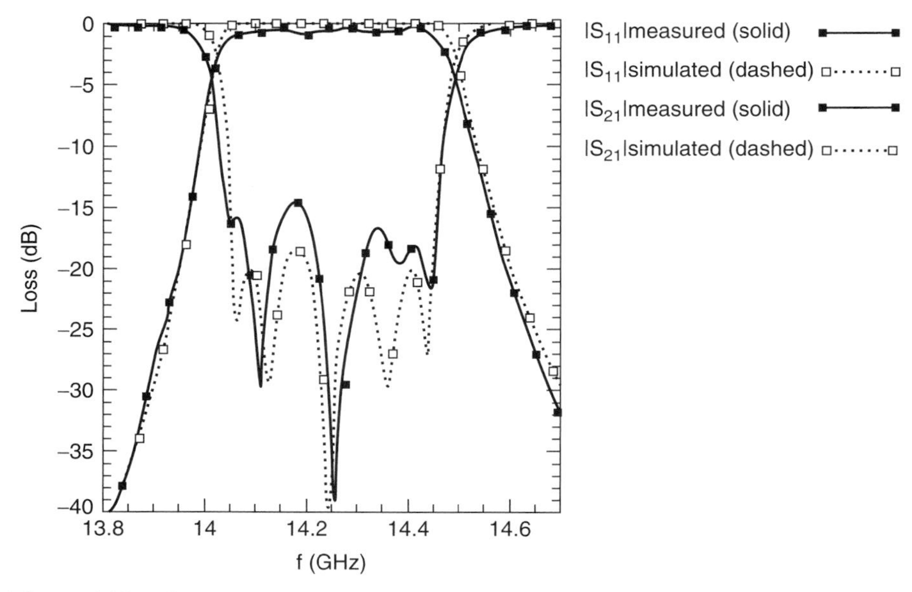

Figure 6.15 Simulated and measured responses of an evanescent-mode filter constructed using a new technique.

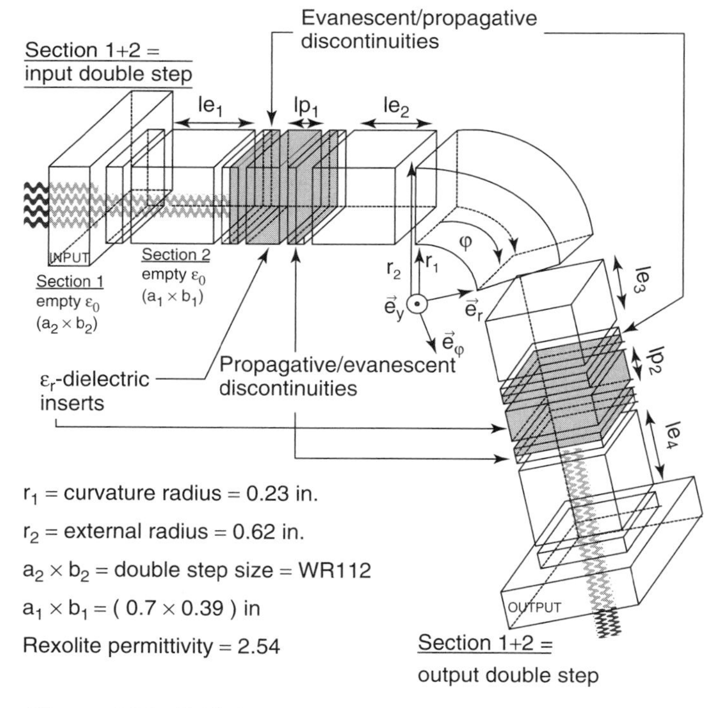

Figure 6.16 Folded evanescent-mode waveguide filter structure.

those of an E section. Furthermore, the B section is defined by an angle of curvature. The fundamental mode only should be kept above cutoff in a B section.

2. The [E,B] and the [B,E] junctions are defined as the junction between a straight E section and the start of a B section, and the junction between the end of a B section and a straight E section.

The design technique will, however, have as its goal defining the lengths l_{ek} and l_{pk} of the E and P sections of the structure once the cross-sectional dimensions, dielectric permittivity, and angle of curvature have been chosen.

6.3.1 Scattering and *ABCD* Descriptions of the Additional Elements

B Section Due to the varying nature of the bend, the electric and magnetic fields must be defined at each point using boundary conditions [6.8,6.9]. The electromagnetic waves satisfying the boundary conditions can be classified into two main categories: *LE modes*, which do not have an electric field component normal to the plane of the bend, and *LM modes*, which have a magnetic field that lies entirely in the plane of the bend. The electric and magnetic fields are specified in terms of two integers, m and n that represent the number of the null amplitude in the transverse directions $\vec{r}$ and $\vec{y}$, respectively. Applying the boundary conditions leads to the LM field equations:

$$\begin{aligned} H_r^{\mathrm{LM}} &= K\frac{\nu}{\mu\omega r}C'_\nu(hr)e^{-j\nu\varphi} \\ H_\varphi^{\mathrm{LM}} &= K\frac{j\omega\varepsilon}{h}C'_\nu(hr)e^{-j\nu\varphi} \\ E_y^{\mathrm{LM}} &= KC_\nu(hr)e^{-j\nu\varphi} \\ E_r^{\mathrm{LM}} &= E_\varphi^{\mathrm{LM}} = H_y^{\mathrm{LM}} = 0 \end{aligned}$$

A B section should operate with only the fundamental mode above cutoff. Therefore, in the rest of the analysis, only the fundamental mode will be considered, since all other modes are below cutoff. For example, in the case of the LM_{10} mode, there are only three EM components since the others fields are zero. The constant K in this case is given by

$$K = \sqrt{\frac{\mu\omega}{|\nu|Jb}} \qquad \text{with} \quad J = \int_{r_1}^{r_2} \frac{C_\nu^2(hr)}{r}\,dr$$

where h is the propagation constant of the LM_{10} mode. In general, the magnitudes of the LM_{mn} fields depend on a linear combination of Bessel functions of the first and second kinds and the functions $C_\nu(hr)$ and $C'_\nu(hr)$, given by

$$C_{-j\gamma}(hr) = J_{-j\gamma}(hr_1)Y_{-j\gamma}(hr) - Y_{-j\gamma}(hr_1)J_{-j\gamma}(hr)$$

$$C'_{-j\gamma} = \frac{1}{h}\frac{\partial}{\partial r}C_{-j\gamma}(hr)$$

The angular propagation constant γ is either real or purely imaginary and can be used to solve the boundary condition in r_2:

$$C_{-j\gamma}(hr_2) = 0 \qquad \text{for LM modes}$$

$$C'_{-j\gamma}(hr_2) = 0 \qquad \text{for LE modes}$$

The cutoff frequency can be found from the h-wave vector satisfying $C_0(hr_2) = 0$. When the mode is above cutoff, the angular propagation function is purely imaginary and equal to $\gamma = j\nu$. The cutoff frequency of a given mode of propagation therefore depends on the radius. This means that the angle of curvature should be chosen carefully to provide a cutoff for the fundamental mode and the second mode that is suitable for the frequency band of interest of the filter (e.g., the fundamental mode above cutoff and the second mode below cutoff). The LM field expressions provide the equations needed to characterize the fields in the B section in the vicinity of a junction and are used to define the scattering matrix of an [E,B] or a [B,E] junction.

[E,B] and [B,E] Junctions Due to the rather complex nature of the field equations inside a bent waveguide, no closed-form expression for the S parameters of these junctions can be proposed. Instead, the modal analysis method used for the step discontinuity of Section 6.2 is modified to this particular problem. This method results in solving the matrix equation

$$[S] = [M_2]^{-1}[M_1]$$

where in this case $[M_1]$ and $[M_2]$ are given by

$$(M_1) = \begin{pmatrix} [I] & [0] & -\left[V^{\vec{e}}_{\text{LM,TE}}\right] & -\left[V^{\vec{e}}_{\text{LE,TE}}\right] \\ [0] & [I] & -\left[V^{\vec{e}}_{\text{LM,TM}}\right] & -\left[V^{\vec{e}}_{\text{LE,TM}}\right] \\ \left[V^{\vec{h}}_{\text{TE,LM}}\right] & \left[V^{\vec{h}}_{\text{TM,LM}}\right] & -\left[V^{\vec{h}}_{\text{LM,LM}}\right] & -\left[V^{\vec{h}}_{\text{LE,LM}}\right] \\ \left[V^{\vec{h}}_{\text{TE,LE}}\right] & \left[V^{\vec{h}}_{\text{TM,LE}}\right] & -\left[V^{\vec{h}}_{\text{LM,LE}}\right] & -\left[V^{\vec{h}}_{\text{LE,LE}}\right] \end{pmatrix}$$

$$(M_2) = \begin{pmatrix} -[I] & [0] & \left[V^{\vec{e}}_{\text{LM,TE}}\right] & -\left[V^{\vec{e}}_{\text{LE,TE}}\right] \\ [0] & [I] & \left[V^{\vec{e}}_{\text{LM,TM}}\right] & -\left[V^{\vec{e}}_{\text{LE,TM}}\right] \\ \left[V^{\vec{h}}_{\text{TE,LM}}\right] & -\left[V^{\vec{h}}_{\text{TM,LM}}\right] & -\left[V^{\vec{h}}_{\text{LM,LM}}\right] & \left[V^{\vec{h}}_{\text{LE,LM}}\right] \\ \left[V^{\vec{h}}_{\text{TE,LE}}\right] & -\left[V^{\vec{h}}_{\text{TM,LE}}\right] & -\left[V^{\vec{h}}_{\text{LM,LE}}\right] & \left[V^{\vec{h}}_{\text{LE,LE}}\right] \end{pmatrix}$$

6.3.2 Filter Design Procedure

The design technique follows the general lines of the original evanescent-mode waveguide filter design. However, additional optimization is often needed to provide the desired filter response. The method used is based on the genetic algorithm [6.14–6.13] that is presented in Chapter 14.

6.3.3 Design Examples and Realizations

A first example consists of a rectangular waveguide filter with an angle of curvature of 57° designed to provide a third-order Chebyshev response centered at 8.4 GHz. The dielectric inserts were taken as Rexolite with a relative permittivity of 2.54. The specified bandpass edges were 8.4 and 8.5 GHz (1% relative bandwidth). It was found that only two dielectric inserts were needed to obtain a response with three zeros of reflection. Figure 6.17 provides a picture of the folded evanescent-mode waveguide filter. The bend had curvature radii of $r_1 = 6$ mm, $r_2 = 22$ mm and a φ-curvature angle of 57°.

Figure 6.18 shows the simulated and measured frequency responses [6.14,6.15]. There are several differences. The measured return loss is worse than the simulations, but one can recognize the three zeros of reflection. There is also about a 40-MHz frequency shift between the two responses. It is believed that the differences are due to approximations related to the numerical Bessel function zeros used in the programs and unaccounted imperfections introduced during construction. Nevertheless, one should note that no manual tuning for final adjustments were performed, and a remarkable feature of this structure has been verified. The structure provides three zeros of reflection even though only two dielectric inserts were used. This means that the bend acts as an additional resonator to provide the third zero of reflection.

A second example consists of a rectangular waveguide filter with an angle of curvature of 161° designed to provide a third-order Chebyshev response centered at 14.6 GHz. Two Rexolite dielectric inserts were used with a relative permittivity

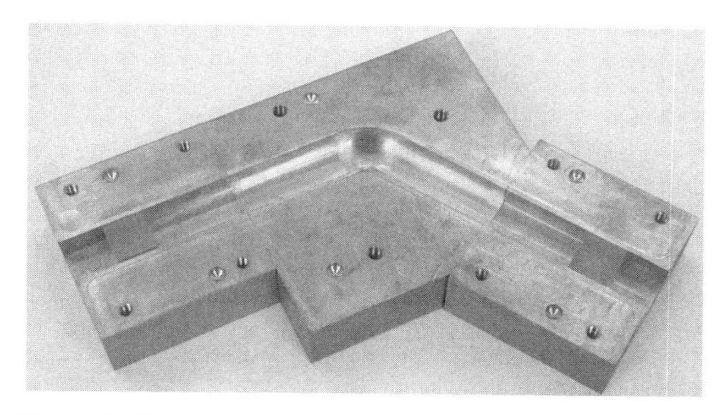

Figure 6.17 Folded evanescent-mode waveguide filter (57°).

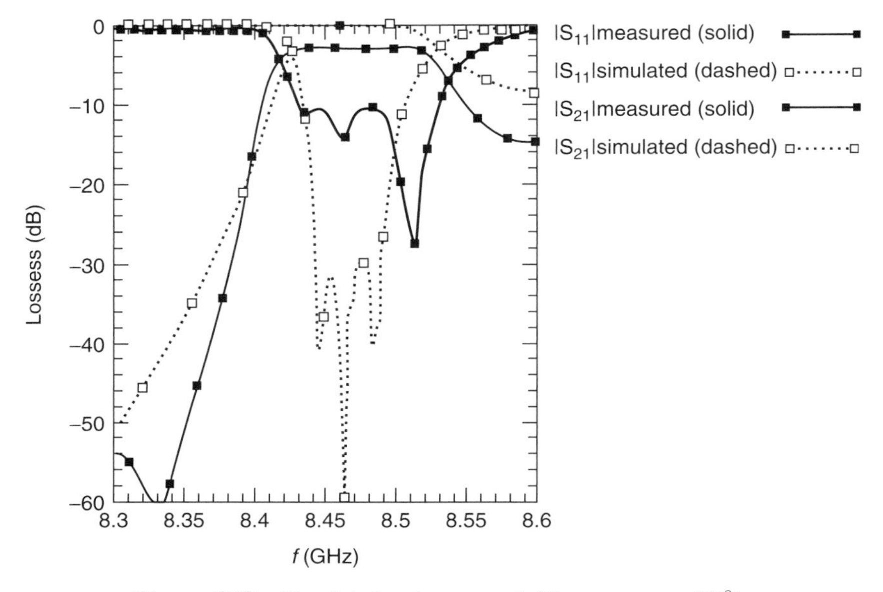

Figure 6.18 Simulated and measured filter responses (57°).

of 2.54. The specified bandpass edges were 14.7 and 14.9 GHz. A picture of the folded evanescent-mode waveguide filter is provided in Figure 6.19.

Figure 6.20 provides the simulated and measured filter responses. In this figure there are two simulated responses [6.16,6.17]. The first simulation is performed using the scattering descriptions and modal analysis methods presented in this chapter, and the second simulation is performed using the commercial microwave simulator HFSS (high-frequency structure simulator). One can see that the "in-house" simulations are closer than HFSS to the actual measured

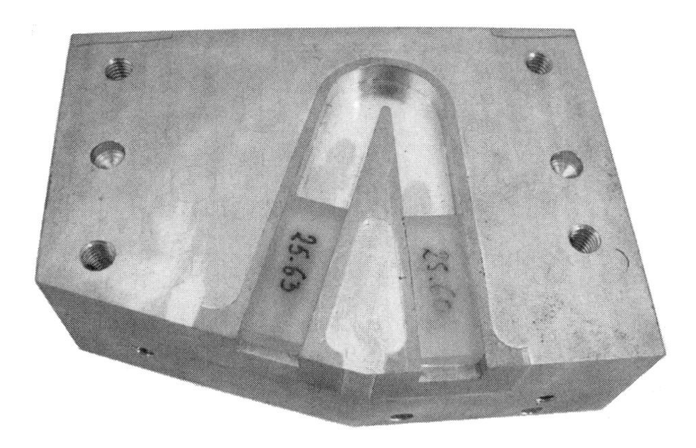

Figure 6.19 Folded evanescent-mode waveguide filter (161°).

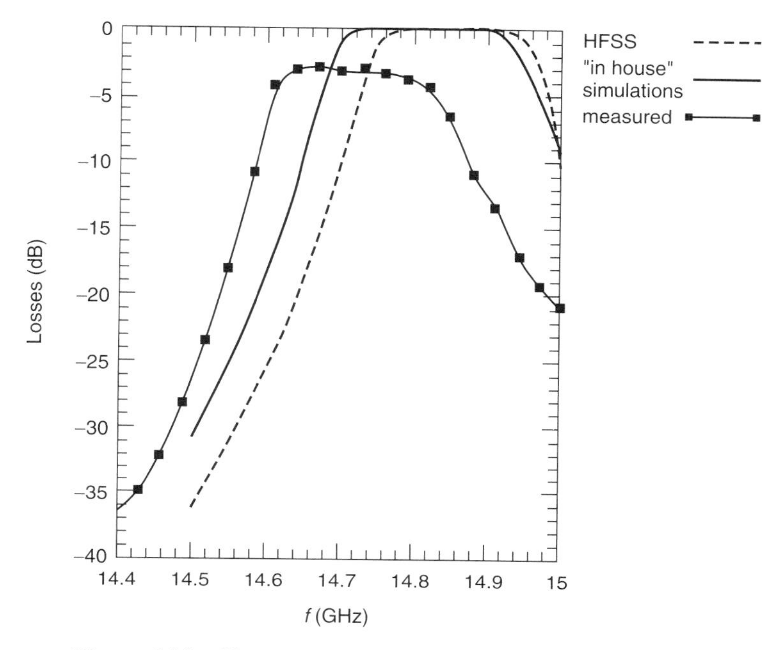

Figure 6.20 Simulated and measured filter responses (161°).

response. The difference between simulations appears as a frequency shift of the response. A similar phenomenon was observed in the case of the original evanescent-mode waveguide filter, due to the level of resolution to be specified in HFSS. In addition, one can see that the insertion loss of the actual filter is rather high and that a frequency shift between responses is also present, but the structure provides the type of filtering characteristics desired.

It is clear from the two examples that additional work is needed to either improve the construction technique or to include the imperfections in the simulations. The design technique, which already relies heavily on numerical and iterative processes, could be augmented without major difficulties.

6.4 CONCLUSIONS

Evanescent-mode waveguide filters, by their nature, result in a reduction in size and weight compared to traditional waveguide filters. The evanescent-mode waveguide filters presented in this chapter consist mainly of a central waveguide operating below cutoff of the fundamental mode, in which dielectric inserts are carefully positioned. A core design technique was proposed suitable for designing narrowband evanescent-mode waveguide filters with prescribed Chebyshev responses. The characterization and modeling steps presented can serve as a framework for other microwave structures.

A key issue when using an equivalent circuit of a microwave structure is to make sure that it is accurate enough so that the structure designed will achieve the desired filtering characteristics while limiting its complexity so that the unknown parameters can be defined. Several examples were provided showing simulations of the designed structure that meet the filtering specifications. Realization examples performed at CNES and Alcatel Space showed good agreement between simulated and measured responses. The framework allowed extending the design method to folded evanescent-mode waveguide filters [6.10–6.13]. Due to the complex nature of the field expressions in a bent waveguide, the technique had to rely even more heavily on a numerical process. A remarkable feature of these structures was that a bent waveguide could act as an additional resonator, reducing the number of dielectric inserts needed to achieve a given number of zeros of reflection. Several realizations proved that three zeros of transmissions were possible using only two dielectric inserts. There were, however, some noticeable differences between simulated and measured responses.

In modern microwave filter design, it should be noted that differences can occur between the simulated response of the designed structure and the prescribed response. There can also be difference between the realization and the simulated structure. The second type of difference is the most critical and should be resolved by improving the construction technique or the simulation model. Once simulated and measured responses are in agreement, the problem reduces to that of "tweaking" design parameters. This is done using numerical optimization techniques. Due to the increased availability and performance of computers, this last step is increasingly becoming a part of microwave filter design.

REFERENCES

[6.1] H. Baher, J. Beneat, and P. Jarry, A new class of evanescent mode waveguide filters, *Proc. IEEE International Conference on Electronics, Circuits and Systems*, ICECS'96, Rhodes Dreece, pp. 335–338, Oct. 13–16, 1996.

[6.2] P. Jarry, J. Beneat, E. Kerhervé, and H. Baher, New class of rectangular and circular evanescent-mode waveguide filter using dielectric inserts, *Int. J. Microwave Millimeter Wave Comput Aided Engi.*, vol. 8, no. 2, pp. 161–192, 1998.

[6.3] F. Le Pennec, P. Jarry, B. Theron, and S. Vigneron, Analysis and original synthesis of evanescent mode band-pass filters using dielectric resonators, *Proc. 8th B. Franklin Symposium on Advances in Antenna and Microwave Technology*, IEEE AP/MTT–S, University of Pennsylvania, Philadelphia, PA, pp. 77–80, Mar. 1990.

[6.4] F. Le Pennec, P. Jarry, B. Theron, and S. Vigneron, Original synthesis of evanescent mode band pass filters using dielectric resonators in a rectangular waveguide, *Proc. ESA/ESTEC Workshop on Microwave Filters for Space Applications*, Noordwijk, The Netherlands, pp. 303–312, June 7–8, 1990.

[6.5] M. Lecouve, P. Jarry, E. Kerhervé, N. Boutheiller, and F. Marc, Genetic algorithm optimisation for evanescent mode waveguide filter design, *Proc. IEEE International Symposium on Circuits and Systems*, Geneva, Switzerland, pp. 441–414, May 28–31, 2000.

[6.6] H. Tertuliano, P. Jarry, and L. A. Bermudez, Use of a bend waveguide to construct Ka band-pass filters, presented at the IEEE AP-S International Symposium and URSI Radio Science Meeting, University of Michigan, Ann Arbor, MI, June 27–July 2, 1993.

[6.7] H. Tertuliano, P. Jarry, and L. A. Bermudez, An original filtering process of a bandpass filter using a curved waveguide, presented at the 23rd European Microwave Conference, EMC'93, Madrid, Spain, Sept. 6–9, 1993.

[6.8] H. T. S. Filho and P. Jarry, and Microwave filtering using dielectric resonators in a curved waveguide, *Proc. Asia Pacific Microwave Conference*, APMC'93, Hsinchu, Taiwan, vol. 2, pp. 16–21 to 16–25 Oct. 18–22, 1993.

[6.9] H. Tertuliano, P. Jarry, and L. A. Bermudez, Synthesis of a curved waveguide evanescent mode filter, *Proc. IEEE AP-S International Symposium and URSI Radio Science Meeting*, Washington, DC, pp. 467–468, June 19-24, 1994.

[6.10] M. Lecouve, N. Boutheiller, P. Jarry, and E. Kerhervé, Genetic algorithm based CAD of microwave waveguide filters with resonating bent, presented at the European Microwave Conference, EMC'00, Paris Oct. 2–6, 2000.

[6.11] M. Lecouve, N. Boutheiller, P. Jarry, E. Kerhervé, J. M. Pham, C. Zanchi, and J. Puech, CAD of high power microwave filters with resonating bend using the genetic algorithm, presented at the Asia Pacific Microwave Conference, APMC'02, Kyoto, Japan, pp. 731–734, Nov. 19–22, 2002.

[6.12] P. Jarry, H. T. Dos Santos, E. Kerhervé, J. M. Pham, M. Lecouve, and N. Boutheiller, Realizations of microwave filters with resonating bend: CAD using a genetic algorithm, *Proc. IEEE International Microwave and Optoelectronics Conference*, IMOC'03, Iguaçu, Brazil, pp. 5–8, Sept. 19–23, 2003.

[6.13] N. Boutheiller, M. Lecouve, P. Jarry, and J. M. Pham, Synthesis and optimization using the genetic algorithm (GA) of straight and curved evanescent microwave filters, invited paper, *Transworld Research Network*, Special Issue on Microwave Theory and Techniques, pp. 1–16, June 2004.

[6.14] H. Tertuliano, P. Jarry, and L. A. Bermudez, Calculation of a microwave filter: modelling and optimization of a curved evanescent Ku band-pass filter, *Proc. Progress in Electromagnetic Research Symposium and ESA/ESTEC*, PIERS'94, European Space Research Symposium Technology Center, Noordwijk, The Netherlands, pp. 1139–1143, July 11–15, 1994.

[6.15] H. Tertuliano, P. Jarry, and L. A. Bermudez, The radius limitations: influence on the waves propagation in a curved rectangular waveguide, *Proc. Progress in Electromagnetic Research, European Microwave Conference*, EMC'94, Cannes, France, pp. 1343–1347, Sept. 5–6, 1994.

[6.16] H. Tertuliano, P. Jarry, and E. Kerhervé, The equivalent circuit for the junction between curved and straight waveguides, presented at the IEEE AP-S International Symposium and USNC/URSI Radio Science Meeting, Newport Beach, CA, June 18–23, 1995.

[6.17] H. Tertuliano, H. T. S. Filho, E. Kerhervé, and P. Jarry, Modelling of an original curved waveguide for microwave filtering applications, *Proc. SBMO IEEE MTT-S International Microwave and Optoelectronics Conference*, IMOC'97, Natal, Brazil, pp. 509–516, Aug. 11–14, 1997.

Selected Bibliography

Accatino, L., and G. Bertin, Modal analysis of curved waveguides, *Proc. 20th European Microwave Conference*, Budapest, Hungary, pp. 1246–1250, Sept. 1990.

Akers, N. P., and P. D. Allan, An evanescent mode waveguide bandpass filter at Q band, *IEEE Trans. Microwave Theory Tech.*, vol. 32, no. 11, pp. 1487–1489, 1984.

Cochran, J. A., and R. G. Pecina, Mode propagation in continuously curved waveguides, *Radio Sci.*, vol. 1, no. 6, pp. 679–696, 1966.

Cornet, P., and R. Dusseaux, Wave propagation in curved waveguides of rectangular cross section, *IEEE Trans. Microwave Theory Tech.*, vol. 47, no. 7, 1999.

Craven, G. F., and C. K. Mok, The design of evanescent mode waveguide bandpass filters for a prescribed insertion loss characteristic, *IEEE Trans. Microwave Theory Tech.*, vol. 19, no. 3, pp. 295–308, 1971.

Gimeno, B., and M. Guglielmi, Multimode equivalent network representation for H- and E-plane uniform bends in rectangular waveguide, *IEEE Trans. Microwave Theory Tech.*, vol. 44, no. 10, 1996.

Howard, J., and W. C. Lin, Evanescent mode filter: design and implementation, *Microwave J.*, pp. 121–135, Oct. 1989.

Igushi, K., M. Tsuji, and H. Shigesawa, Negative coupling between TE_{10} and TE_{20} modes for use in evanescent-mode bandpass filters and their field-theoretic CAD, *IEEE-MTT-S Dig.*, pp. 727–730, 1994.

Rice, S. O., Reflections from circular bends in rectangular waveguides-matrix theory, *Bell Syst. Tech. J.*, vol. 27, no. 2, pp. 305–349, 1948.

Shih, Y. C., and K. G. Gray, Analysis and design of evanescent-mode waveguide dielectric resonator filters, *IEEE MTT-S Int. Microwave Symp. Dig.*, vol. 84, no. 1, pp. 238–239, 1984.

Snyder, R. V., New application of evanescent mode wave-guide to filter design, *IEEE Trans. Microwave Theory Tech.*, vol. 25, no. 12, pp. 1013–1021, 1977.

Snyder, R. V., Broadband waveguide filters with wide stopbands using a stepped-wall evanescent mode approach, *IEEE MTT-S Int. Microwave Symp. Dig.*, vol. 83, no. 1, pp. 151–153, 1983.

Weisshaar, A., and V. K. Tripathi, Perturbation analysis and modeling of curved micro strip bends, *IEEE Trans. Microwave Theory Tech.*, vol. 38, no. 10, 1990.

7

INTERDIGITAL FILTERS

7.1 INTRODUCTION

In this chapter we analyze interdigital filters based on the method of graphs and a design method based on a low-pass prototype circuit and a special frequency transformation. The technique works well for wideband filters, where only n lines are needed for an nth-degree filter. However, in the case of narrowband filters, the technique produces unrealizable element values, and two additional lines are needed to solve this problem. The chapter concludes with a description of two interdigital filters implemented in suspended substrate stripline at 1 and 2 GHz.

Advanced Design Techniques and Realizations of Microwave and RF Filters,
By Pierre Jarry and Jacques Beneat

7.2 INTERDIGITAL FILTERS

An interdigital filter consists of N resonant lines where the coupling is considered only between adjacent lines [7.1,7.2]. Each line is shorted to the ground at one end while the other end is left open-circuit. For interdigital filters, all lines have the same length of $\lambda_0/4$, as shown in Figure 7.1. Due to a periodicity in frequency, these filters have a stopband at $2f_0$. Since only coupling between two adjacent lines is considered, the structure can be modeled from the case of two-coupled lines, as shown in Figure 7.2. The admittance matrix equation of the two-coupled line is given by

$$\begin{bmatrix} I_1 \\ I_2 \\ I_3 \\ I_4 \end{bmatrix} = \frac{v}{t} \begin{bmatrix} [C] & -\sqrt{1-t^2}[C] \\ -\sqrt{1-t^2}[C] & [C] \end{bmatrix} \begin{bmatrix} V_1 \\ V_2 \\ V_3 \\ V_4 \end{bmatrix}$$

where $t = j \tan j\beta l$ with $\beta = \omega/v$, where v is the velocity of light in the medium, t is Richard's variable, β is the propagation constant, and l is the length of the line segment. In addition, the capacitance matrix $[C]$ is given by

$$[C] = \begin{bmatrix} C_{11} & -C_{12} \\ -C_{21} & C_{22} \end{bmatrix}$$

where $Y_{ij} = vC_{ij}$ are the unnormalized admittances of each line. Also note that $C_{ij} = C_{ji}$.

Analysis of the complete interdigital structure becomes quickly cumbersome using these matrix formulations. Sato and Cristal [7.3,7.4] provided a technique based on graphs to ease the modeling of these structures. For example, the

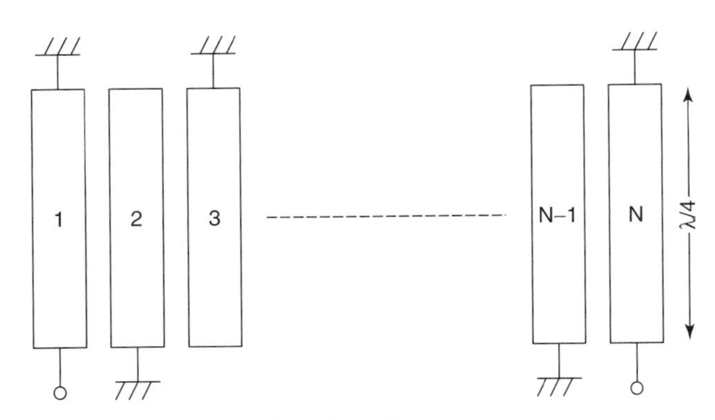

Figure 7.1 Interdigital structure.

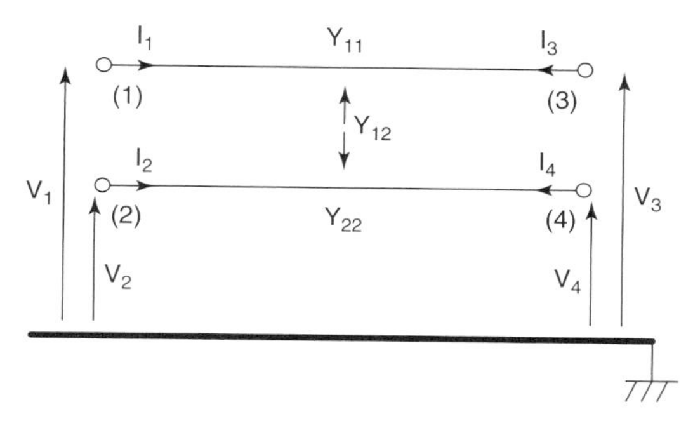

Figure 7.2 Two-coupled lines with a reference ground plane.

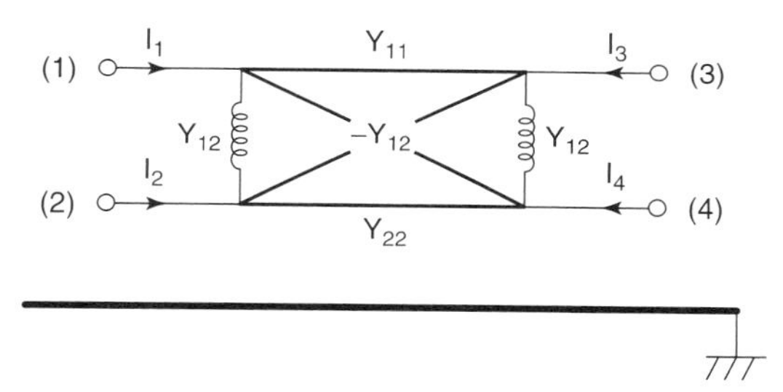

Figure 7.3 Graph representation of a two-coupled lines.

two-line structure of Figure 7.2 can be represented as the graph shown in Figure 7.3, defined as in Table 7.1.

For the case of interdigital structures, the lines alternatively are terminated with a ground or an open circuit. Figure 7.4 shows the case for the first two lines. In this case $V_2 = V_3 = 0$ and the admittance matrix equation reduces to

$$\begin{bmatrix} I_1 \\ I_4 \end{bmatrix} = \frac{v}{t} \begin{bmatrix} C_{11} & -\sqrt{1-t^2}C_{12} \\ -\sqrt{1-t^2}\,C_{12} & C_{22} \end{bmatrix} \begin{bmatrix} V_1 \\ V_4 \end{bmatrix}$$

The graph representation is shown in Figure 7.5. At this stage, a unit element that is shorted at one end will act as a distributed admittance. For example, the case of Y_{11} is shown in Figure 7.6. Using this property and redrawing Figure 7.5 differently, we have the equivalent circuit of two-coupled lines, as shown in Figure 7.7. The procedure can then be generalized to n-coupled lines. The graph representation of this case is shown in Figure 7.8, and the equivalent circuit of the entire interdigital structure is given in Figure 7.9. This representation is valid as long as the coupling between nonadjacent lines is negligible.

TABLE 7.1

Shunt inductor Y_{12}	$ABCD$ matrix[a]: $\begin{pmatrix} 1 & 0 \\ -jY_k \cot\theta & 1 \end{pmatrix}$ This represents a distributed admittance, $Y = -jY_{12}\cot\theta$.
Line $-Y_{12}$ ≡ $-Y_{12}$ u.e.	$ABCD$ matrix of a unit element: $\begin{pmatrix} \cos\theta & j\frac{1}{-Y_{12}}\sin\theta \\ j(-Y_{12})\sin\theta & \cos\theta \end{pmatrix}$
Line Y_{11} ≡ Y_{11} u.e.	$ABCD$ matrix of a unit element: $\begin{pmatrix} \cos\theta & j\frac{1}{Y_{11}}\sin\theta \\ jY_{11}\sin\theta & \cos\theta \end{pmatrix}$ The result is similar for Y_{22}.

[a] $\theta = \beta l$, $\beta = \omega/c$, $l = \lambda_0/4$, and $\lambda_0 = c/f_0$.

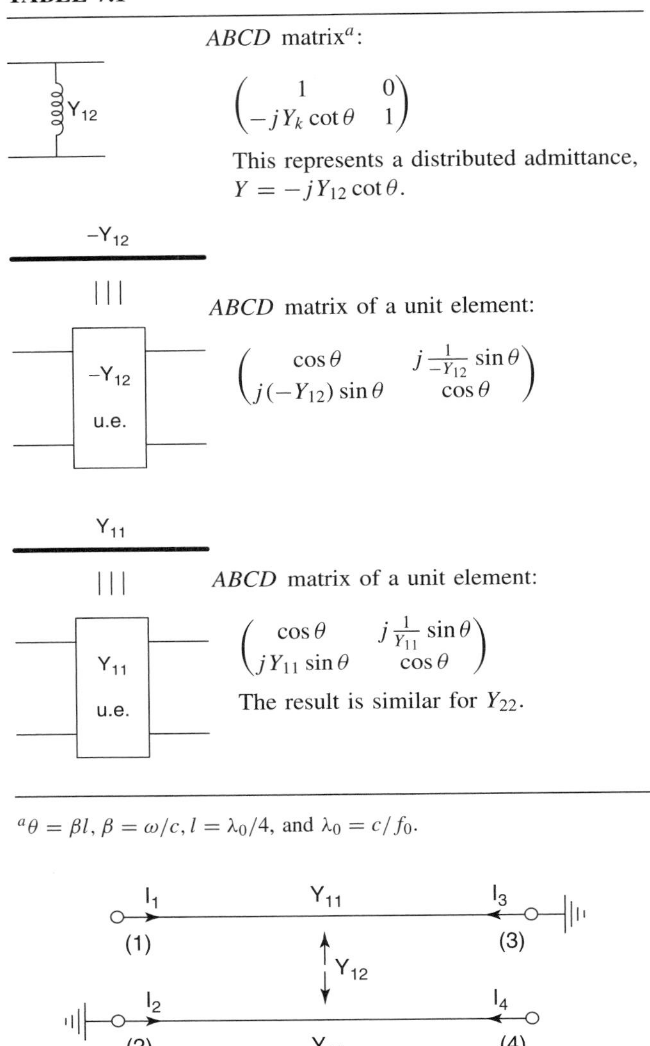

Figure 7.4 Two-coupled lines used in interdigital structures.

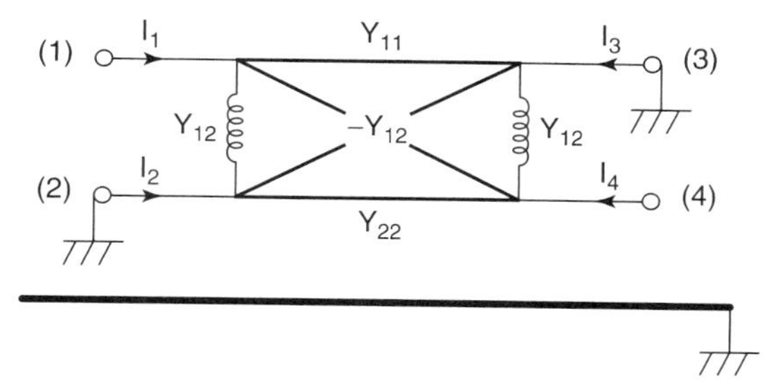

Figure 7.5 Graph representation of two-coupled lines used in interdigital structures.

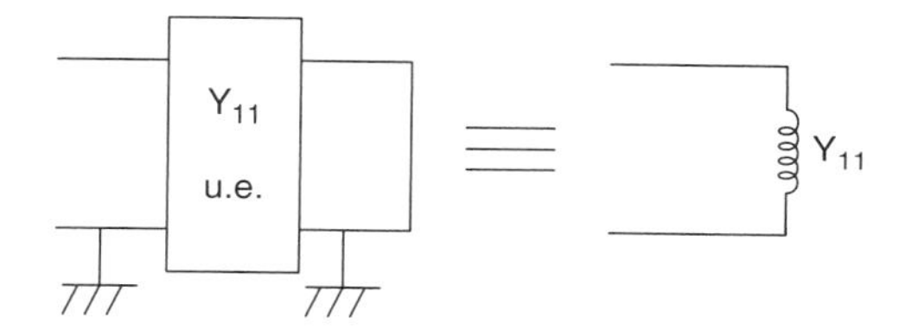

Figure 7.6 Unit element with a shorted output.

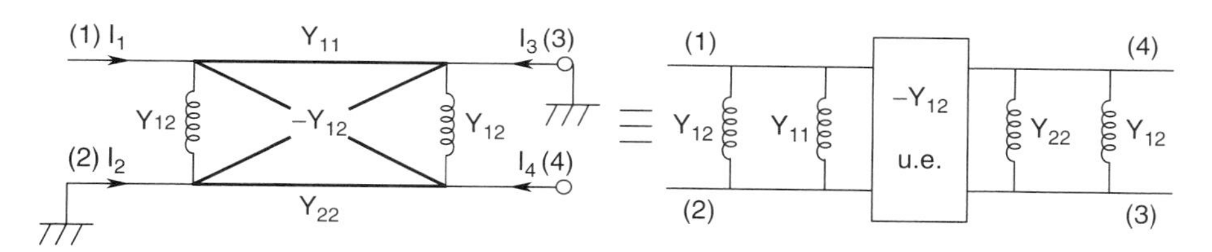

Figure 7.7 Equivalent circuit of two-coupled lines used in interdigital filters.

7.3 DESIGN METHOD

7.3.1 Prototype Circuit

Due to the periodic nature of the response of an interdigital filter through the angle θ, a low-pass ladder prototype circuit is used. A low-pass prototype circuit based on admittance inverters is shown in Figure 7.10. From Chapter 2 the capacitors can be arbitrary. In the case of interdigital filters, the capacitors of the prototype circuit will all have the same value (e.g., $C_k = 1/\delta$). The admittance inverters are given by

$$J_{k,k+1} = \frac{1}{\delta}\frac{1}{\sqrt{g_k g_{k+1}}} \quad \text{for } k = 1, 2, \ldots, n-1 \quad \text{and} \quad J_{01} = \frac{1}{\sqrt{\delta g_1}} \quad \text{and}$$

$$J_{n,n+1} = \frac{1}{\sqrt{\delta g_n}}$$

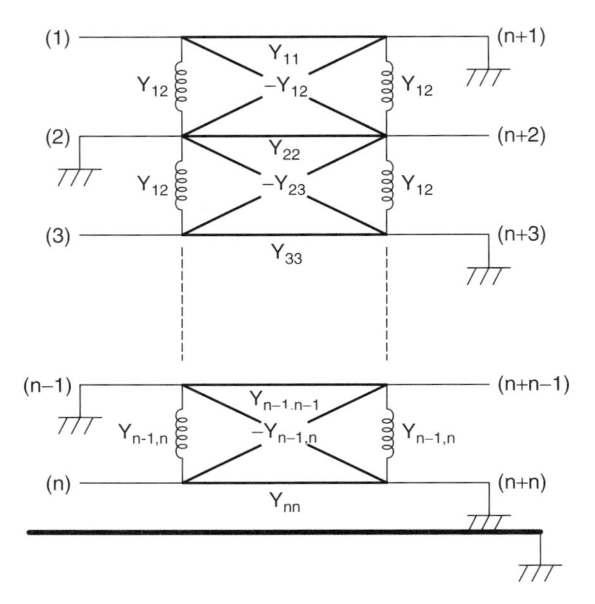

Figure 7.8 Graph representation of an interdigital structure.

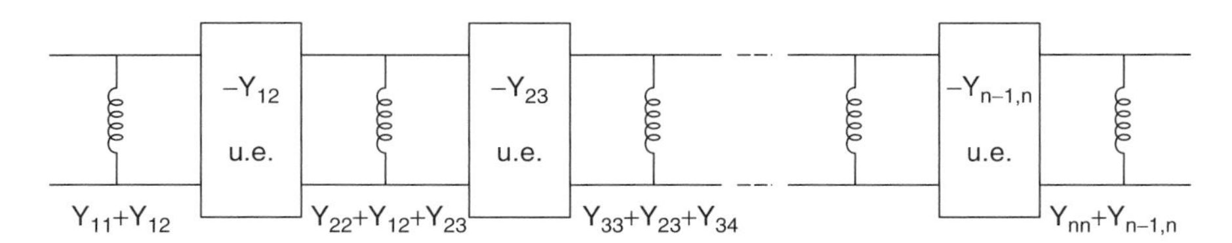

Figure 7.9 Equivalent circuit of an interdigital structure.

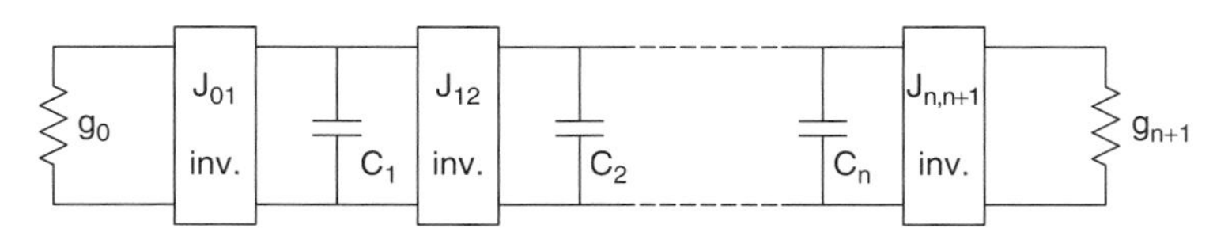

Figure 7.10 Low-pass prototype circuit with admittance inverters.

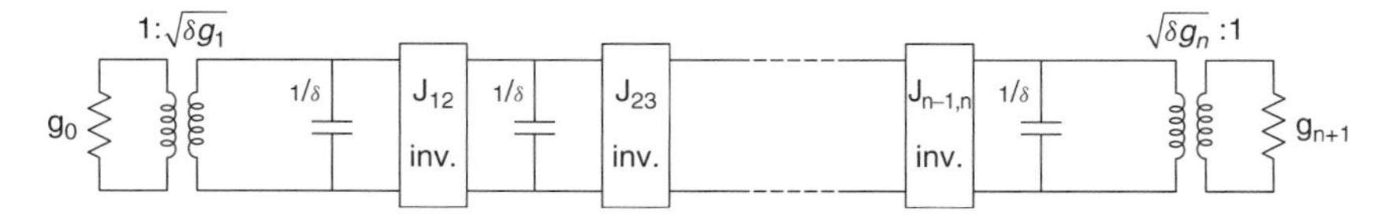

Figure 7.11 Final low-pass prototype circuit with admittance inverters.

The g_k are provided in Chapter 2 for Butterworth or Chebyshev responses with terminations g_0 and g_{n+1}. To facilitate identification with the equivalent circuit of the interdigital filter, admittance inverters J_{01} and $J_{n,n+1}$ are replaced by ideal transformers, as shown in Figure 7.11.

7.3.2 Equivalent Circuit

The first step in modifying the equivalent circuit of Figure 7.9 consists of making the unit elements have positive values. This is achieved through the correspondence shown in Figure 7.12. With the suppression of the ideal transformers, the equivalent circuit of the interdigital circuit with positive unit elements is shown in Figure 7.13.

The equivalent circuit is modified further to remove the frequency dependence on θ in the unit elements and to transform these unit elements into admittance inverters. These modifications are explained below. The unit elements are distributed elements and they depend on the frequency through the angle θ [7.5,7.6]. From the *ABCD* matrix equation

$$\begin{pmatrix} \cos\theta & j\dfrac{\sin\theta}{Y_{k,k+1}} \\ jY_{k,k+1}\sin\theta & \cos\theta \end{pmatrix}$$
$$= \begin{pmatrix} 1 & 0 \\ -jY_{k,k+1}\cot\theta & 1 \end{pmatrix} \begin{pmatrix} 0 & j\dfrac{\sin\theta}{Y_{k,k+1}} \\ j\dfrac{Y_{k,k+1}}{\sin\theta} & 0 \end{pmatrix} \begin{pmatrix} 1 & 0 \\ -jY_{k,k+1}\cot\theta & 1 \end{pmatrix}$$

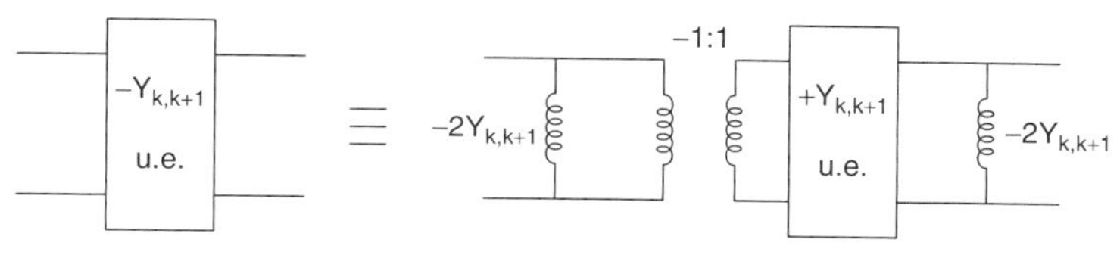

Figure 7.12 Equivalent circuit of a negative unit element.

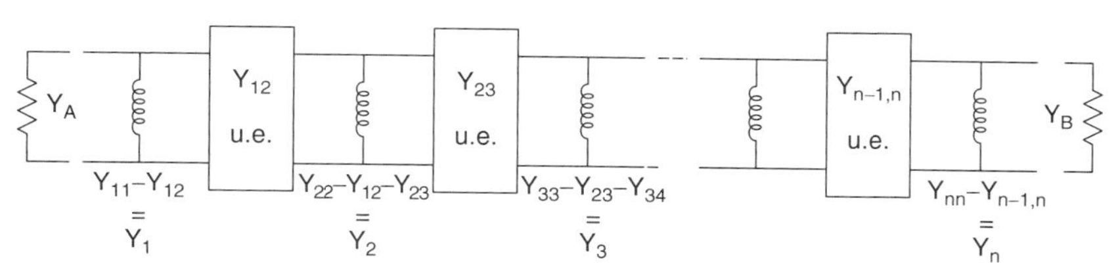

Figure 7.13 Equivalent circuit of an interdigital structure with positive u.e.'s.

we can represent a unit element as an admittance inverter surrounded by two distributed admittances, as shown in Figure 7.14. This simplifies the expressions of the admittances between admittance inverters. For example, in general the admittance between admittance inverters will be $Y_k + Y_{k-1,k} + Y_{k,k+1}$, due to the transformation of Figure 7.14, but $Y_k = Y_{kk} - Y_{k-1,k} - Y_{k,k+1}$ from Figure 7.13. Therefore, the admittance between admittance inverters reduces to Y_{kk}. One can check that a similar result is found at the input and output of the structure as well. The new equivalent circuit of the interdigital filter is shown in Figure 7.15.

The frequency dependence $\sin\theta$ of the admittance inverter can be removed based on the following matrix equation:

$$\begin{pmatrix} 0 & j\dfrac{\sin\theta}{Y_{k,k+1}} \\ j\dfrac{Y_{k,k+1}}{\sin\theta} & 0 \end{pmatrix} = \begin{pmatrix} \sqrt{\sin\theta} & 0 \\ 0 & \dfrac{1}{\sqrt{\sin\theta}} \end{pmatrix} \begin{pmatrix} 0 & \dfrac{j}{Y_{k,k+1}} \\ jY_{k,k+1} & 0 \end{pmatrix} \begin{pmatrix} \dfrac{1}{\sqrt{\sin\theta}} & 0 \\ 0 & \sqrt{\sin\theta} \end{pmatrix}$$

This means that the admittance inverter of Figure 7.14 is equivalent to a frequency-independent admittance inverter surrounded by two ideal transformers, as shown in Figure 7.16.

The equivalent circuit between two consecutive inverters is shown in Figure 7.17. The *ABCD* matrix equation below shows that a distributed admittance Y_{kk} surrounded by ideal transformers $1:\sqrt{\sin\theta}$ and $\sqrt{\sin\theta}:1$ is equivalent

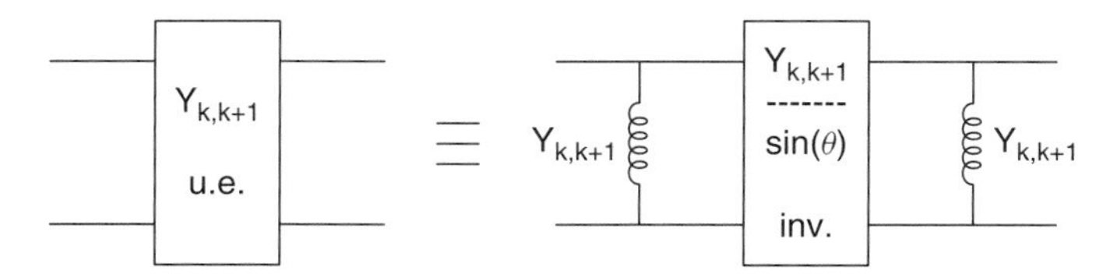

Figure 7.14 Transforming a unit element into an admittance inverter.

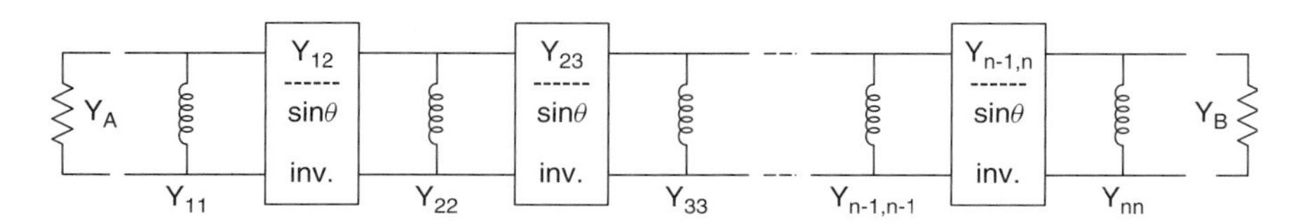

Figure 7.15 Equivalent circuit of the interdigital structure with admittance inverters.

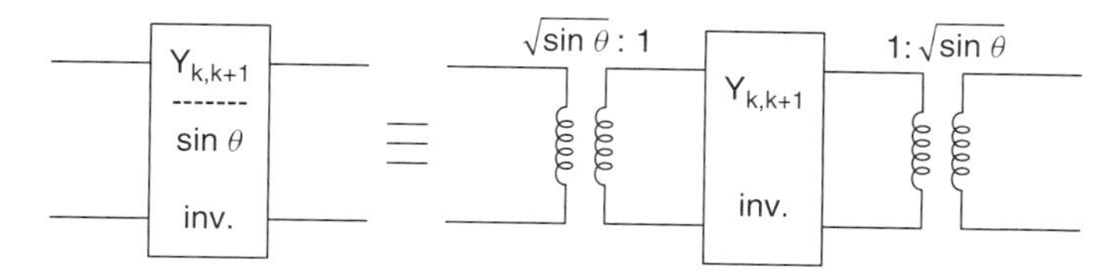

Figure 7.16 Admittance inverter independent of frequency.

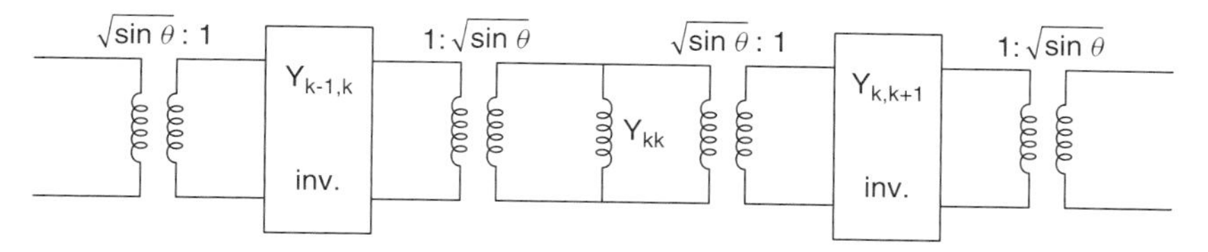

Figure 7.17 Equivalent circuit between two admittance inverters.

to a new distributed admittance $Y_{kk}\sin\theta$:

$$\begin{pmatrix} \dfrac{1}{\sqrt{\sin\theta}} & 0 \\ 0 & \sqrt{\sin\theta} \end{pmatrix} \begin{pmatrix} 1 & 0 \\ -jY_{kk}\cot\theta & 1 \end{pmatrix} \begin{pmatrix} \sqrt{\sin\theta} & 0 \\ 0 & \dfrac{1}{\sqrt{\sin\theta}} \end{pmatrix}$$
$$= \begin{pmatrix} 1 & 0 \\ -j(Y_{kk}\sin\theta)\cot\theta & 1 \end{pmatrix}$$

At the input of the circuit, we have the situation shown in Figure 7.18 (left). By moving the admittance to the other side of the transformer, we get a new distributed admittance $Y_{11}\sin\theta$, as shown in Figure 7.18 (right).

A similar result is found at the output of the circuit, and the final equivalent circuit of the interdigital structure is shown in Figure 7.19. The next step is to match the prototype circuit of Figure 7.11 and the equivalent circuit of Figure 7.19. By comparing the admittance inverters, we find that

$$Y_{k,k+1} = J_{k,k+1} \qquad k = 1, 2, \ldots, n-1$$

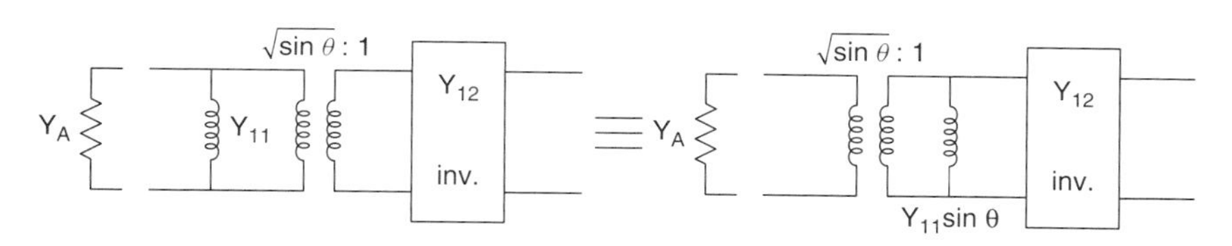

Figure 7.18 Modifications at the input of the structure.

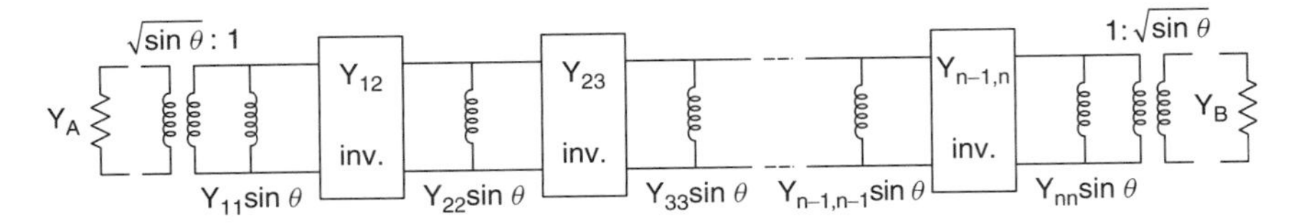

Figure 7.19 Final equivalent circuit of the interdigital structure.

By comparing the admittances, we find that

$$-j(Y_{kk}\ \sin\theta)\cot\theta = -jY_{kk}\cos\theta = jC\omega = j\frac{\omega}{\delta}$$

If the frequency transformation $\omega \to -\delta\cos\theta$ is used, the admittances are simply given by

$$Y_{kk} = 1 \qquad k = 1, 2, \ldots, n$$

7.3.3 Input and Output

For the input and output, we match the admittances seen after the transformer. For example, the case of the input is shown in Figure 7.20. By matching the admittances seen after the transformer, the interdigital filter will require an input admittance Y_A such that

$$Y_A = \frac{g_0}{\delta\ g_1}\frac{1}{\sin\theta}$$

Using a similar approach, the interdigital filter will require an output admittance Y_B such that

$$Y_B = \frac{g_{n+1}}{\delta\ g_n}\frac{1}{\sin\theta}$$

Note that for $\lambda/4$ lines, $\theta \approx \pi/2$ and $\sin\theta \approx 1$. For example, when the prototype is defined using $g_0 = g_{n,n+1} = 1$, the terminations will be proportional to δ and g_1 or g_n. For the interdigital filter to be terminated on a traditional 50 Ω, the values of $Y_{k,k+1}$ and Y_{kk} will need to be scaled.

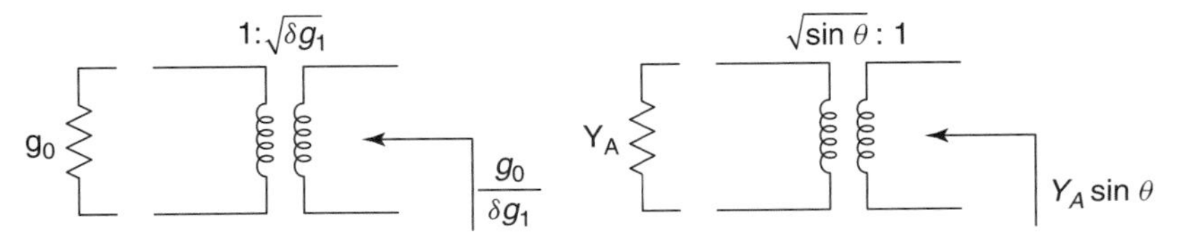

Figure 7.20 Comparing the input of the circuits.

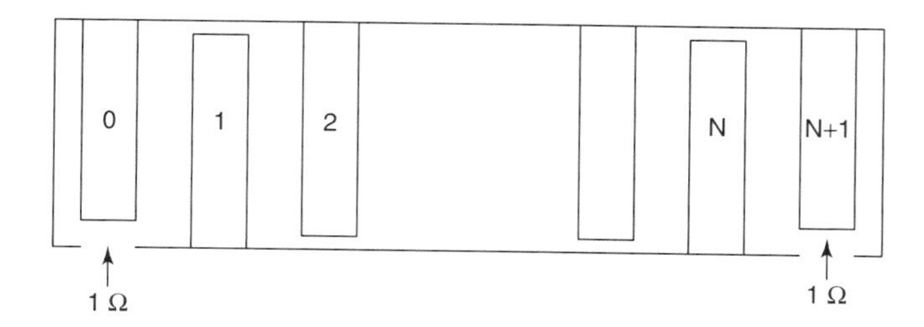

Figure 7.21 Using additional lines to adapt the terminations.

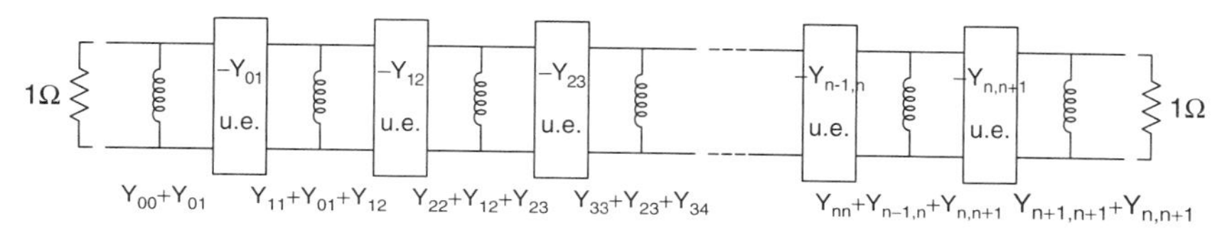

Figure 7.22 Equivalent circuit of a 0-to-$n+1$ interdigital filter.

7.3.4 Case of Narrowband Filters

For wideband applications, the value of δ is small and the scaled values of $Y_{k,k+1}$ and Y_{kk} are realizable. In the case of narrowband applications, however, δ becomes large and the scaled values of $Y_{k,k+1}$ and Y_{kk} become large and can no longer be realizable. One solution consists of adding a line 0 and a line $N+1$, as shown in Figure 7.21. Figure 7.9 is then increased to account for the two additional lines, and the equivalent circuit of the 0-to-$n+1$ interdigital filter is shown in Figure 7.22. By using the same transformations as described for the 1-to-n interdigital filter, we now have additional admittance inverters to adapt the terminations. By comparing the equivalent circuit and the prototype circuits at the input/outputs, it is found that these additional lines should be taken as

$$Y_{00} = 1 - \frac{2}{\sqrt{\delta g_1}} \qquad Y_{01} = \frac{1}{\sqrt{\delta g_1}}$$

$$Y_{n+1,n+1} = 1 - \frac{2}{\sqrt{\delta g_n}} \qquad Y_{n,n+1} = \frac{1}{\sqrt{\delta g_n}}$$

The 1-to-n lines are given using the same equations as before.

7.3.5 Frequency Transformation

The parameter δ will be used for the low pass-to-bandpass transformation. We assume that the prototype elements are given for a low-pass cutoff frequency of Ω. This low-pass cutoff frequency is to be associated with the upper and lower cutoff frequencies f_1 and f_2 of the target bandpass filter centered at f_0. First, let's express the frequency dependence of θ:

$$\theta = \beta\frac{\lambda_0}{4} = \frac{\omega}{c}\frac{\lambda_0}{4} = \frac{2\pi f}{4c}\frac{c}{f_0} = \frac{\pi}{2}\frac{f}{f_0}$$

Then the association of Ω with f_1 and f_2 is as follows:

$$-\delta\cos\left(\frac{\pi}{2}\frac{f_1}{f_0}\right) = -\Omega$$

$$-\delta\cos\left(\frac{\pi}{2}\frac{f_2}{f_0}\right) = \Omega$$

Since f_0 is taken as the center of f_1 and f_2, $f_0 = (f_1 + f_2)/2$ and

$$-\delta\cos\left(\frac{\pi}{2}\frac{2f_1}{f_1+f_2}\right) = -\delta\sin\left[\frac{\pi}{2}\left(1-\frac{2f_1}{f_1+f_2}\right)\right] = -\delta\sin\left(\frac{\pi}{2}\frac{f_2-f_1}{f_1+f_2}\right) = -\Omega$$

$$-\delta\cos\left(\frac{\pi}{2}\frac{2f_2}{f_1+f_2}\right) = -\delta\sin\left[\frac{\pi}{2}\left(1-\frac{2f_2}{f_1+f_2}\right)\right] = -\delta\sin\left(\frac{\pi}{2}\frac{f_1-f_2}{f_1+f_2}\right) = \Omega$$

These two equations result in a formula that provides the value of δ:

$$\delta = \Omega/\sin\left(\frac{\pi}{2}\frac{f_2-f_1}{f_1+f_2}\right)$$

7.3.6 Physical Parameters of the Interdigital Filter

The design technique proposed in the preceding sections does not provide the final physical properties of the lines, such as the width of the lines and the spacing between the lines. However, as will be shown, knowledge of $Y_{k,k+1}$ and Y_{kk} will be sufficient for defining the physical properties. The technique relies on relating $Y_{k,k+1}$ and Y_{kk} with the capacitances defined in Figure 7.23, where

t is the thickness of a bar

b is the distance between the two main ground planes

W_k is the width of bar k

$S_{k,k+1}$ is the distance between bars k and $k+1$

C_{pk} is the parallel-plate capacitance of bar k

C'_f is the fringing capacitance for an isolated bar (0 or $n+1$)

$(C'_{fe})_{k,k+1}$ is the even-mode fringing capacitance for bars k and $k+1$

$(C'_{fo})_{k,k+1}$ is the odd-mode fringing capacitance for bars k and $k+1$

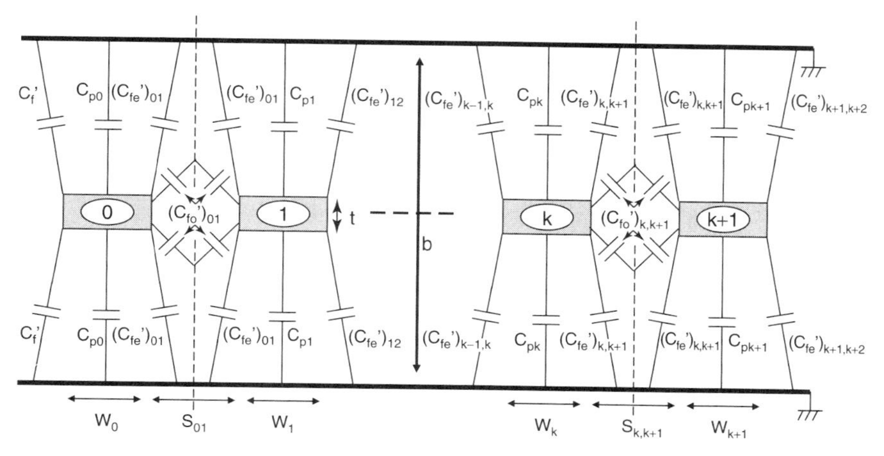

Figure 7.23 Definition of the capacitances in coupled bar structures.

Getsinger [7.7] computed the various capacitances of Figure 7.23 and provided a set of curves relating the capacitances, the distance between ground planes b, the thickness of the bars t, and the distance S between consecutive bars. The curve shown in Figure 7.24(a) provides C'_{fe}/ε and $\Delta C/\varepsilon = C'_{fo}/\varepsilon - C'_{fe}/\varepsilon$ as functions of the spacing between bars S/b, with the bar thickness t/b as a parameter. The curve shown in Figure 7.24(b) provides the fringing capacitance C'_f/ε of an isolated bar as a function of bar thickness t/b.

In these curves, the ratios C/ε of static capacitance per unit length to the permittivity of the dielectric of the medium are related to the characteristic impedance Z_c of a lossless transmission line by

$$Z_c\sqrt{\varepsilon_r} = \frac{\eta}{C/\varepsilon} \quad \text{or} \quad \frac{C}{\varepsilon} = \frac{\eta Y_c}{\sqrt{\varepsilon_r}}$$

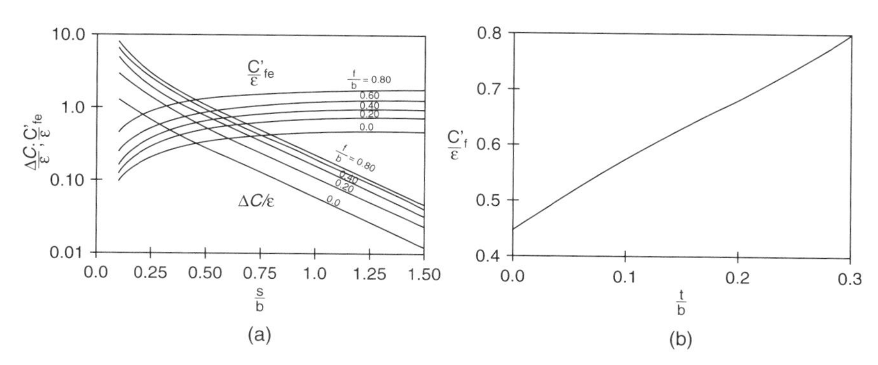

Figure 7.24 Capacitance values for Figure 7.23 as a function of (a) S/b and (b) t/b.

where η is the impedance of free space per square meter ($\eta = 376.7\ \Omega/\text{m}^2$) and ε_r is the relative permittivity of the dielectric medium. These curves allow us to find the distance between bars $S_{k,k+1}$ when the $\Delta C_{k,k+1}/\varepsilon$ are given. In addition, the parallel-plate capacitance will not be directly available, and additional curves will be used to compute its final value. Once the parallel-plate capacitance is known, the width of the bars can be computed using the following relation:

$$\frac{C_{pk}}{\varepsilon} = 2\frac{W_k/b}{1 - t/b} \quad \text{or} \quad \frac{W_k}{b} = \frac{1}{2}\left(1 - \frac{t}{b}\right)\frac{C_{pk}}{\varepsilon}$$

In order to relate the capacitances in Figure 7.23 to the interdigital structure, the capacitance model of Figure 7.25 is used. Comparing Figures 7.23 and 7.25, provides the following relations between the two sets of capacitances:

$$\begin{aligned}
C_{k,k+1} &= \left[C'_{fo} - C'_{fe}\right]_{k,k+1} = \Delta C_{k,k+1} \\
C_k &= 2\left[C_{pk} + (C'_{fe})_{k,k+1} + (C'_{fe})_{k-1,k}\right] \\
C_0 &= 2\left[C_{p0} + (C'_{fe})_{01} + C'_f\right] \\
C_n &= 2\left[C_{pn} + (C'_{fe})_{n-1,n} + C'_f\right]
\end{aligned}$$

and

$$\begin{aligned}
\frac{\Delta C_{k,k+1}}{\varepsilon} &= \frac{C_{k,k+1}}{\varepsilon} \\
\frac{C_{pk}}{\varepsilon} &= \frac{1}{2}\frac{C_k}{\varepsilon} - \frac{(C'_{fe})_{k,k+1}}{\varepsilon} - \frac{(C'_{fe})_{k-1,k}}{\varepsilon} \\
\frac{C_{p0}}{\varepsilon} &= \frac{1}{2}\frac{C_0}{\varepsilon} - \frac{(C'_{fe})_{01}}{\varepsilon} - \frac{C'_f}{\varepsilon} \\
\frac{C_{pn}}{\varepsilon} &= \frac{1}{2}\frac{C_n}{\varepsilon} - \frac{(C'_{fe})_{n-1,n}}{\varepsilon} - \frac{C'_f}{\varepsilon}
\end{aligned}$$

The capacitances of Figure 7.25 are computed from the $Y_{k,k+1}$ and Y_{kk} of the equivalent circuit of the interdigital structure:

$$\begin{aligned}
\frac{C_{k,k+1}}{\varepsilon} &= \frac{\eta Y_c}{\sqrt{\varepsilon_r}} Y_{k,k+1} \qquad k = 1, 2, \ldots, n-1 \\
\frac{C_k}{\varepsilon} &= \frac{\eta Y_c}{\sqrt{\varepsilon_r}} Y_{kk} \qquad k = 1, 2, 3, \ldots, n
\end{aligned}$$

where Y_c is the desired characteristic admittance at the input/output of the structure (e.g., $Y_C = 1/50\ \Omega$) and assuming that $Y_{k,k+1}$ and Y_{kk} are provided for 1 Ω terminations. At this stage, the various capacitance ratios are computed from $Y_{k,k+1}$ and Y_{kk} of the equivalent circuit of the interdigital structure. Then the

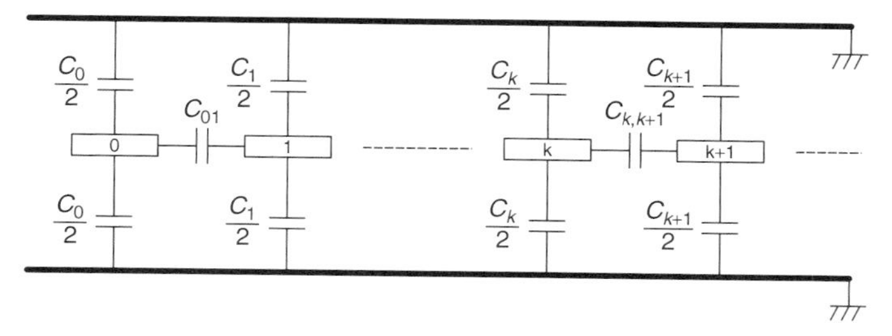

Figure 7.25 Capacitance model of the interdigital structure.

$\Delta C_{k,k+1}/\varepsilon$ are used with the Getsinger curves to find the $S_{k,k+1}/b$ distances for a fixed t/b ratio. Note that the t/b ratio is decided on at the very start. Knowledge of $S_{k,k+1}/b$ then provides the capacitances $(C'_{fe})_{k,k+1}/\varepsilon$. Using only t/b, the capacitances C'_f/ε can be found using the second curve. At this stage the parallel-plate capacitances C_{pk}/ε can finally be computed, from which the widths W_k of the bars can be defined.

7.4 DESIGN EXAMPLES

7.4.1 Wideband Example

The filter should have a third-order Chebyshev response with a −20-dB return loss. The filter has a center frequency of $f_0 = 2000$ MHz with a bandwidth of 1000 MHz. The first step is to compute the value of δ. This is done using

$$\delta = \Omega/\sin\left(\frac{\pi}{2}\frac{f_2 - f_1}{f_1 + f_2}\right)$$

where $\Omega = 1$ rad/s, the frequency at which the Chebyshev low-pass prototype crosses the specified return loss for the last time. This gives $\delta = 2.6131$. The low-pass prototype filter is computed for $g_0 = g_{n+1} = 1$ (e.g., prototype terminated on 1 Ω). The simulated response of the prototype filter with the terminations of $R_G = R_L = 1\ \Omega$, with the first and last admittance inverters replaced by the proposed ideal transformers, is given in Figure 7.26. The low-pass prototype values are given by

$$g_1 = 0.8534 \qquad g_2 = 1.1039 \qquad g_3 = 0.8534$$

$$\sqrt{\delta g_1} = 1.4934 \qquad J_{12} = 0.3943 \qquad J_{23} = 0.3943 \qquad \sqrt{\delta g_3} = 1.4934$$

The interdigital filter parameters are given by

$$Y_{11} = 1 \qquad Y_{12} = 0.3943 \qquad Y_{22} = 1 \qquad Y_{23} = 0.3943 \qquad Y_{33} = 1$$

The interdigital filter requires terminations $R_G = \delta g_1 = 2.23\ \Omega$ and $R_L = \delta g_n = 2.23\ \Omega$.

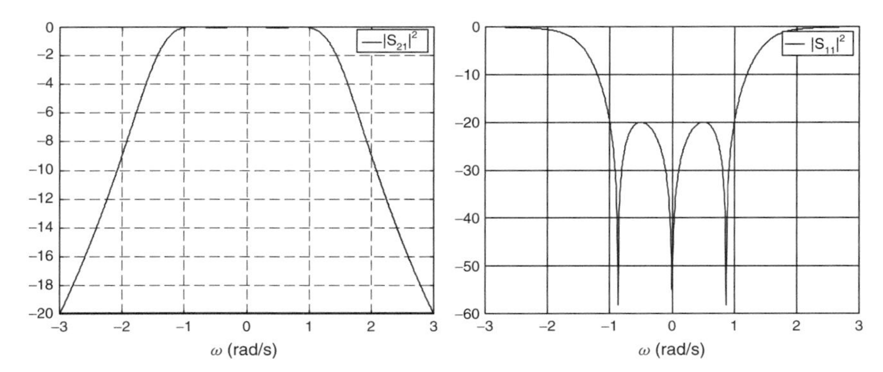

Figure 7.26 Simulated response of the prototype circuit with input/output transformers.

Impedance scaling is used to bring the terminations of the equivalent circuit to 1 Ω. This is done by multiplying the admittances Y_{kk} and $Y_{k,k+1}$ by $\delta g_1 = \delta g_n$. The admittance values for terminations normalized to 1 Ω are

$$Y_{11} = 2.2302 \quad Y_{12} = 0.8793 \quad Y_{22} = 2.2302 \quad Y_{23} = 0.8793 \quad Y_{33} = 2.2302$$

The simulated response of the equivalent circuit of the interdigital filter with 1-Ω terminations is given in Figure 7.27. One can see that some frequency dependences are still not accounted for in the design method since the return loss of the equivalent circuit is slightly different from the prototype.

The static capacitances are then computed for actual filter terminations $Y_c = 1/50\ \Omega$ using

$$\begin{aligned} \frac{C_{k,k+1}}{\varepsilon} &= \frac{\eta Y_c}{\sqrt{\varepsilon_r}} Y_{k,k+1} \qquad k = 0, 1, 2, \ldots, n \\ \frac{C_k}{\varepsilon} &= \frac{\eta Y_c}{\sqrt{\varepsilon_r}} Y_{kk} \qquad k = 0, 1, 2, 3, \ldots, n+1 \end{aligned} \qquad \text{with} \quad \varepsilon_r = 1$$

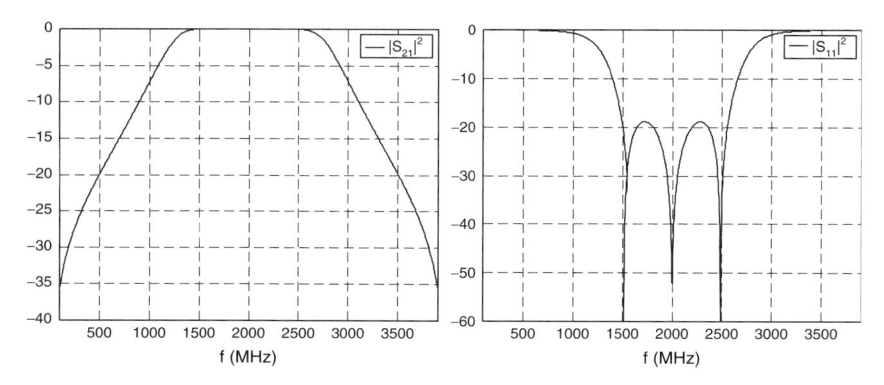

Figure 7.27 Simulated response using the equivalent circuit of Figure 7.9.

This provides the capacitances values

$$\frac{C_{1,2}}{\varepsilon} = 6.6245 \qquad \frac{C_{2,3}}{\varepsilon} = 6.6245$$

$$\frac{C_1}{\varepsilon} = 16.8021 \qquad \frac{C_2}{\varepsilon} = 16.8021 \qquad \frac{C_3}{\varepsilon} = 16.8021$$

Using the Getsinger plots with, for example, $t/b = 0.2$, we find the distances between lines:

$$\frac{S_{1,2}}{\varepsilon} = 0.04 \qquad \frac{S_{2,3}}{\varepsilon} = 0.04$$

$$\frac{C_{fe1,2}}{\varepsilon} = 0.05 \qquad \frac{C_{fe2,3}}{\varepsilon} = 0.05$$

Also, when using $t/b = 0.2$, the fringing capacitance is found to be $C'_f = 0.68$. The parallel-plate capacitances are then computed and the width of the lines are then found.

$$\frac{C_{p1}}{\varepsilon} = 7.6710 \qquad \frac{C_{p2}}{\varepsilon} = 8.3010 \qquad \frac{C_{p3}}{\varepsilon} = 7.6710$$

$$\frac{W_1}{b} = 3.0684 \qquad \frac{W_2}{b} = 3.3204 \qquad \frac{W_3}{b} = 3.0684$$

7.4.2 Narrowband Example

The filter should have a third-order Chebyshev response with a −20-dB return loss. The filter has a center frequency $f_0 = 2000$ MHz with a bandwidth of 100 MHz (Figure 7.28). In this case, $\delta = 25.4713$. The low-pass prototype values are given by

$$g_1 = 0.8534 \qquad g_2 = 1.1039 \qquad g_3 = 0.8534$$

$$\sqrt{\delta g_1} = 4.6625 \qquad J_{12} = 0.0404 \qquad J_{23} = 0.0404 \qquad \sqrt{\delta g_3} = 4.6625$$

The interdigital filter parameters for 1-Ω terminations are given by

$$Y_{00} = 0.5710 \qquad Y_{11} = 1 \qquad Y_{22} = 1 \qquad Y_{33} = 1 \qquad Y_{44} = 0.5710$$

$$Y_{01} = 0.2145 \qquad Y_{12} = 0.0404 \qquad Y_{23} = 0.0404 \qquad Y_{34} = 0.2145$$

The static capacitances are then computed for actual filter terminations $Y_c = 1/50\ \Omega$ using

$$\frac{C_{k,k+1}}{\varepsilon} = \frac{\eta Y_c}{\sqrt{\varepsilon_r}} Y_{k,k+1} \qquad k = 0, 1, 2, \ldots, n$$

$$\frac{C_k}{\varepsilon} = \frac{\eta Y_c}{\sqrt{\varepsilon_r}} Y_{kk} \qquad k = 0, 1, 2, 3, \ldots, n+1 \qquad \text{with} \quad \varepsilon_r = 1$$

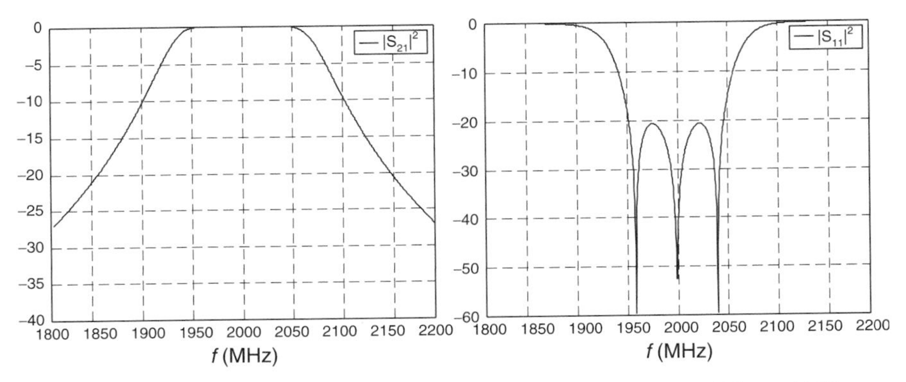

Figure 7.28 Simulated response using the equivalent circuit of Figure 7.22.

This provides the capacitance values

$$\frac{C_{0,1}}{\varepsilon} = 1.6159 \qquad \frac{C_{1,2}}{\varepsilon} = 0.3047 \qquad \frac{C_{2,3}}{\varepsilon} = 0.3047$$
$$\frac{C_{3,4}}{\varepsilon} = 1.6159 \qquad \frac{C_0}{\varepsilon} = 4.3022 \qquad \frac{C_1}{\varepsilon} = 7.5340$$
$$\frac{C_2}{\varepsilon} = 7.5340 \qquad \frac{C_3}{\varepsilon} = 7.5340 \qquad \frac{C_4}{\varepsilon} = 4.3022$$

One then uses the Getsinger plots to define the physical parameters of the filter. However, the values are realizable.

7.5 REALIZATIONS AND MEASURED PERFORMANCE

A first interdigital filter implemented in suspended substrate stripline [7.8] had the following specifications:

- Center frequency: 1 GHz
- Passband bandwidth: 40 MHz
- Degree of the filter: 3
- Permittivity of the substrate: $\varepsilon_r = 2.54$
- Dimensions: $b = 3.18$ mm, $t = 0.2$ mm

This corresponds to an intermediate-band filter with a 4% relative bandwidth. The element values were determined using a dielectric constant $\varepsilon_r = 2.54$. The constructed filter is shown in Figure 7.29. The filter was based on a prototype circuit with improved group delay performance. The theoretical and measured responses are shown in Figure 7.30.

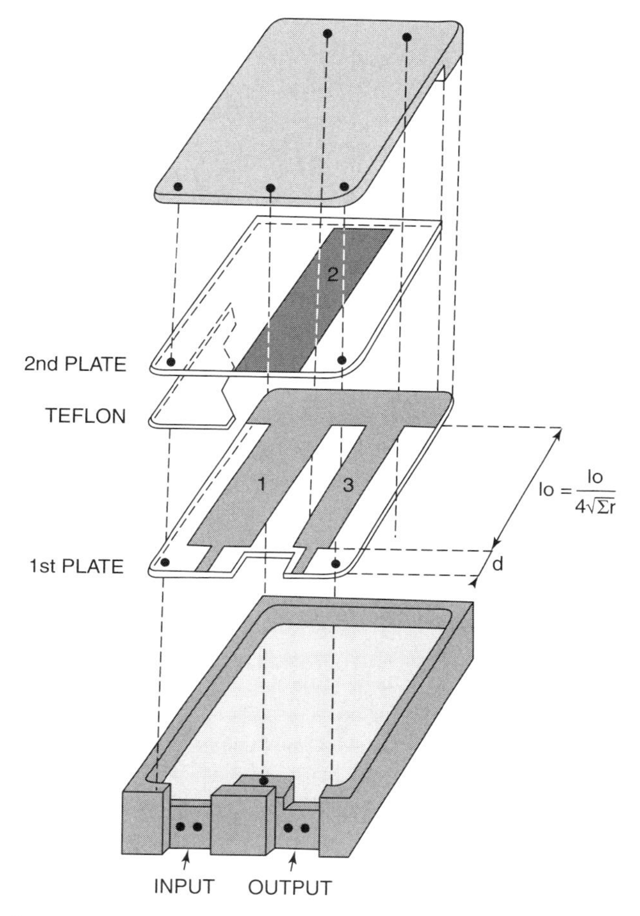

Figure 7.29 Construction of the third-order interdigital filter.

A second filter implemented in suspended substrate stripline at 2 GHz had the following specifications:

- Center frequency: 1.970 GHz
- Passband bandwidth: 20 MHz
- Maximum passband delay distortion: 0.2 ns
- Degree of the filter: 3
- Permittivity of the substrate: $\varepsilon_r = 2.54$
- Dimensions: $b = 3.18$ mm, $t = 0.2$ mm

This corresponds to a filter with a 1% relative bandwidth. The theoretical and measured responses of the filter are shown in Figure 7.31. Realizable physical

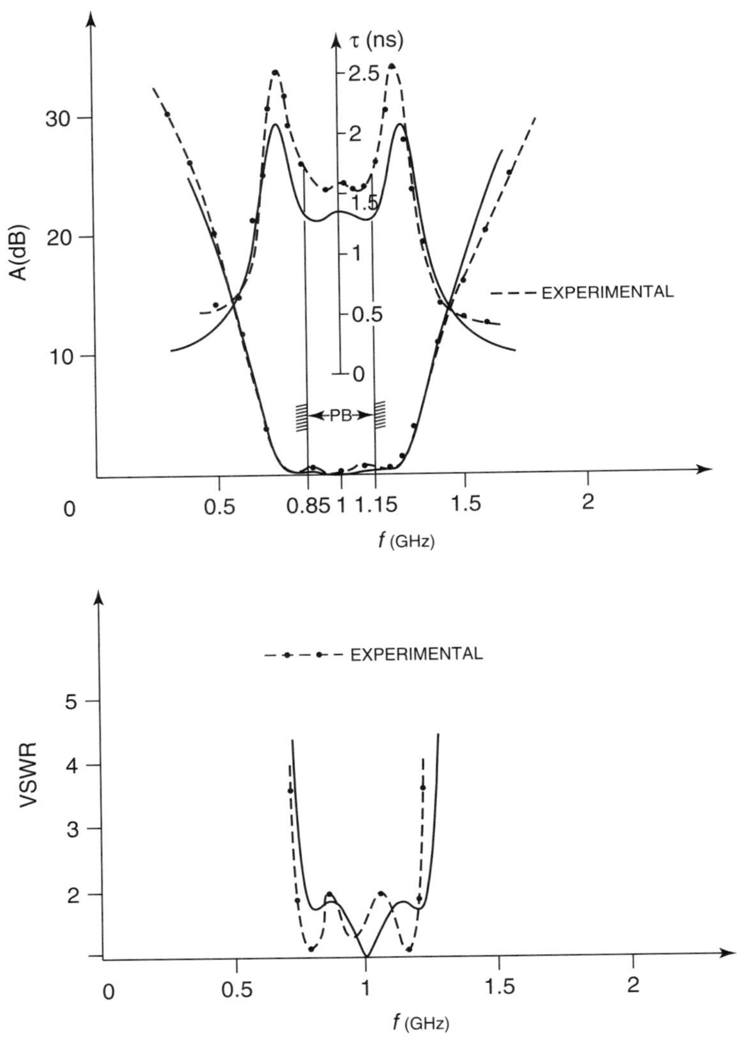

Figure 7.30 Attenuation and delay of the filter and VSWR around 1 GHz.

dimensions were obtained by introduction of additional lines at the input/output. We were particularly interested in the phase distortion.

7.6 CONCLUSIONS

Interdigital filters are very compact and economical. They are still the object of research efforts, particularly for determining the length of the lines [7.9] (determination of resonant frequency) in integrated systems using MMIC technology [7.10] and defining interdigital filters with transmission zeros [7.11].

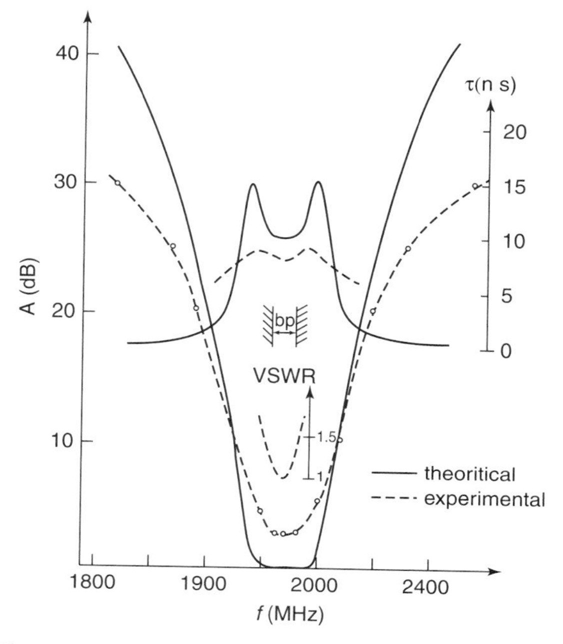

Figure 7.31 Responses of the 2-GHz interdigital filter.

REFERENCES

[7.1] G. L. Matthaei, Interdigital band-pass filters, *IEEE Trans. Microwave Theory Tech.*, vol. 10, no. 11, pp. 479–491, 1962.

[7.2] R. J. Wenzel, Exact theory of interdigital band-pass filters and related coupled structures, *IEEE Trans. Microwave Theory Tech.*, vol. 13, no. 9, pp. 559–595, 1965.

[7.3] R. Sato and E. G. Cristal, Simplified analysis of coupled transmission-line networks, *IEEE Trans. Microwave Theory and Tech.*, vol. 18, pp. 122–131, Mar. 1970.

[7.4] E. G. Cristal, Microwave filters, in *Modern Filter Theory Design*, ed. G. C. Temes and S. K. Mitra, pp. 273–231, Wiley, New York, 1973.

[7.5] J. D. Rhodes, The generalized interdigital linear phase filter, *IEEE Trans. Microwave Theory Tech.*, vol. 18, no. 6, pp. 301–307, 1970.

[7.6] S. O. Scanlan, Theory of microwave coupled-lines networks, *Proc. IEEE*, vol. 68, pp. 209–231, Feb. 1980.

[7.7] W. J. Getsinger, Coupled rectangular bars between parallel plates, *IRE Trans. Microwave Theory Tech.*, vol. 10, no. 1, pp. 65–72, 1962.

[7.8] Y. Garault and P. Jarry, Interdigital filters with a linear phase and a minimum attenuation in the pass-band, *Int. J. Circuit Theory Appl.*, vol. 3, pp. 183–191, Oct. 1976.

[7.9] J. H. Cloete, The resonant frequency of rectangular interdigital filter element, *IEEE Trans. Microwave Theory Tech.*, pp. 772–774, Sept. 1983.

[7.10] M. I. Herman, S. Valas, D. M. McNay, R. Knust-Graichen, and J. C. Chen, Investigation of passive bandpass filters using MMIC technology, *Microwave Guided Wave Lett.*, pp. 228–230, June 1992.

[7.11] C. Ernst and V. Postoyalko, Tapped-line interdigital filter equivalent circuits, *Proc. MTT-S International Microwave Symposium*, pp. 801–804, 1997.

Selected Bibliography

Cristal, E. G., Coupled circular cylindrical rods between parallel ground planes, *IEEE Trans. Microwave Theory Tech.*, vol. 12, no. 4, pp. 428–439, 1964.

Robinson, L. A., Wideband interdigital filters with capacitively loaded resonators, *IEEE G-MTT Symp. Program Dig.*, vol. 65, no. 1, pp. 33–38, 1965.

8

COMBLINE FILTERS IMPLEMENTED IN SSS

8.1 INTRODUCTION

In this chapter we provide an analysis of combline filters based on the method of graphs and a design method based on a bandpass prototype circuit. In the case of narrowband filters, two nonresonant lines are used to provide the input/outputs to the filter. The chapter concludes with a description of a combline filter implemented in suspended substrate stripline at 1.2 GHz.

8.2 COMBLINE FILTERS

A combline filter consists of n resonant $\lambda/8$ lines where the coupling is considered only between adjacent lines. Each line is shorted to ground at the same end while the opposite ends are terminated in lumped capacitors C_k. These capacitors

Advanced Design Techniques and Realizations of Microwave and RF Filters,
By Pierre Jarry and Jacques Beneat

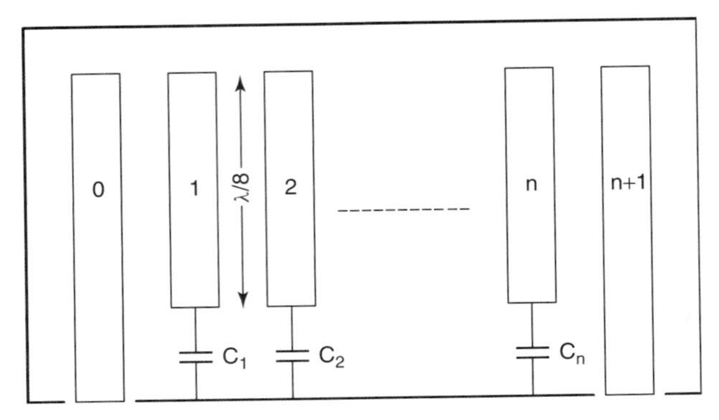

Figure 8.1 Combline filter with nonresonant input and output.

allow reducing the length of the resonators to $\lambda/8$ compared to $\lambda/4$ needed for interdigital filters. The filter is more compact, has better stopband characteristics, and has good selectivity [8.1,8.2]. One could connect the filter to the rest of the system directly on lines 1 and n. This works well for wideband applications. For narrowband applications, the input and output are provided by using nonresonant lines 0 and $n+1$, as shown in Figure 8.1.

A network of n coupled transmission lines can be described by its admittance matrix [8.3]:

$$\begin{bmatrix} I_1 \\ I_2 \\ \vdots \\ I_n \\ \vdots \\ I_{2n} \end{bmatrix} = \frac{v}{t} \begin{bmatrix} [C] & -\sqrt{1-t^2}[C] \\ -\sqrt{1-t^2}[C] & [C] \end{bmatrix} \begin{bmatrix} V_1 \\ V_2 \\ \vdots \\ V_n \\ \vdots \\ V_{2n} \end{bmatrix}$$

where v is the propagation speed and $t = j\ \tan\theta$ with θ the electric length of the transmission lines. The capacitance matrix $[C]$ for n coupled lines is given by

$$[C] = \begin{bmatrix} C_{11} & -C_{12} & \cdots & -C_{1n} \\ -C_{21} & C_{22} & \cdots & -C_{2n} \\ \vdots & \vdots & \cdots & \vdots \\ -C_{n1} & -C_{n2} & \cdots & C_{nn} \end{bmatrix}$$

where $C_{ij} = C_{ji}$ F/m, $C_{ii} \geq 0$, and $C_{ii} \geq \sum_{j \neq i} C_{ij}$. The analysis of the combline structure becomes quickly cumbersome using these matrix formulations. Sato and Cristal [8.4] derived a means of modeling these structures through the use of graphs. The graph representation of n coupled lines is shown in Figure 8.2, defined as in Table 8.1.

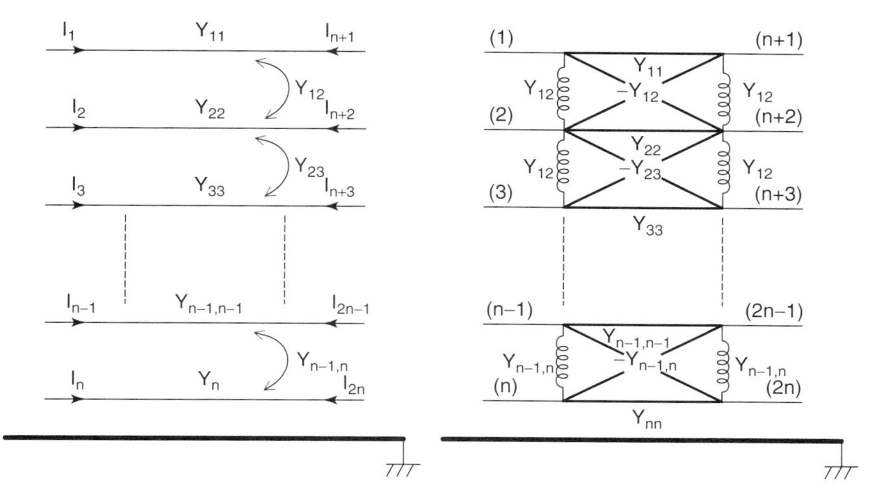

Figure 8.2 *n* coupled lines and associated graph representation.

TABLE 8.1

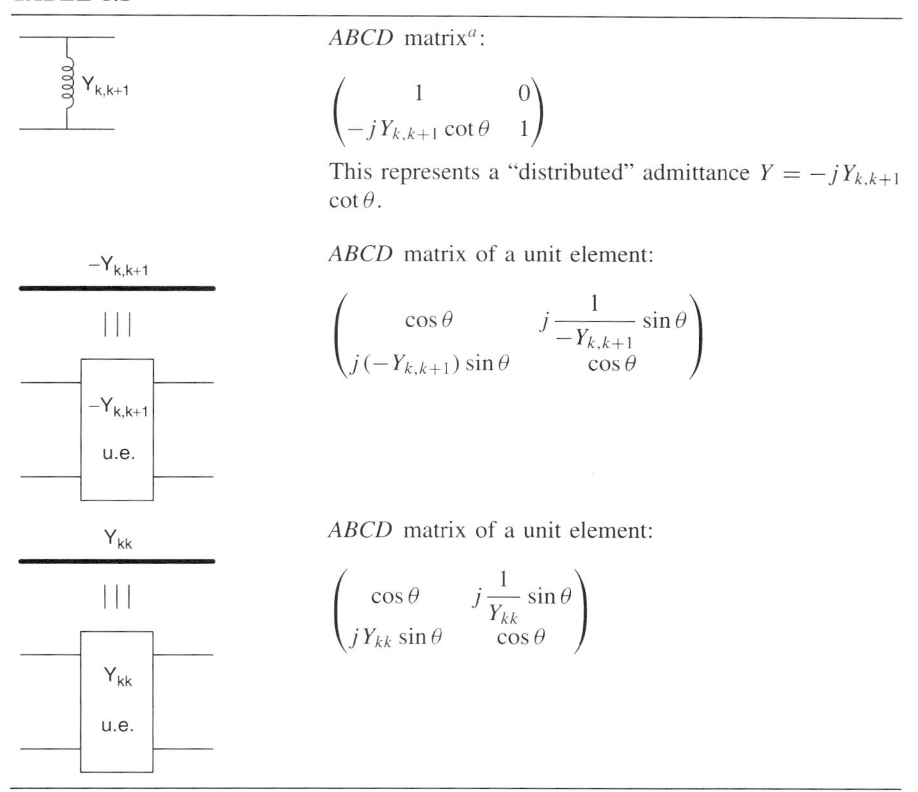

ABCD matrix[a]:

$$\begin{pmatrix} 1 & 0 \\ -jY_{k,k+1}\cot\theta & 1 \end{pmatrix}$$

This represents a "distributed" admittance $Y = -jY_{k,k+1} \cot\theta$.

ABCD matrix of a unit element:

$$\begin{pmatrix} \cos\theta & j\dfrac{1}{-Y_{k,k+1}}\sin\theta \\ j(-Y_{k,k+1})\sin\theta & \cos\theta \end{pmatrix}$$

ABCD matrix of a unit element:

$$\begin{pmatrix} \cos\theta & j\dfrac{1}{Y_{kk}}\sin\theta \\ jY_{kk}\sin\theta & \cos\theta \end{pmatrix}$$

[a] $\theta = \beta l$, $\beta = \omega/c$, $l = \lambda_{0/4}$, and $\lambda_0 = c/f_0$.

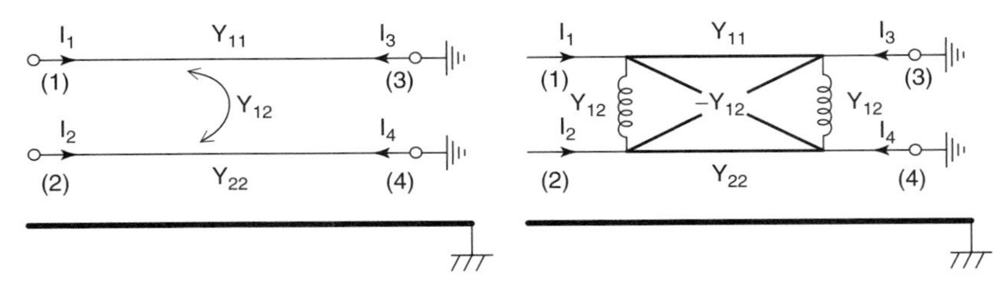

Figure 8.3 Basic element and graph for combline filters.

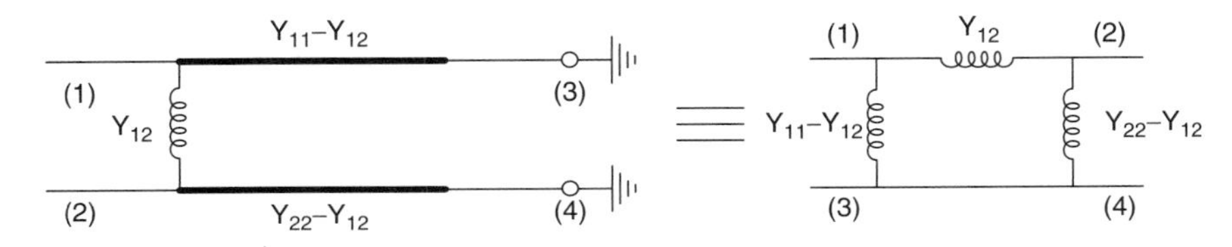

Figure 8.4 Circuit equivalent of the basic element.

In the case of combline filters, the lines are all grounded on the same side of the line. The basic element of the combline structure is shown in Figure 8.3 with associated graph representation. With the potentials 3 and 4 at zero, one unit element $-Y_{12}$ is in parallel with the unit element Y_{11}, while the other unit element $-Y_{12}$ is in parallel with the unit element Y_{22}. These combinations result in new unit elements of value $Y_{11} - Y_{12}$ and $Y_{22} - Y_{12}$, as shown in Figure 8.4 (left). Furthermore, a unit element that is shorted at one end will act as a distributed admittance. This occurs for $Y_{11} - Y_{12}$, and $Y_{22} - Y_{12}$ and explains why they are replaced by admittances in Figure 8.4 (right).

A generalization of this procedure gives the equivalent circuit for an n-degree combline filter (without input and output). The resonators are composed of lumped capacitors C_k of admittances $j\omega C_k$ in parallel with shunt distributed inductors [8.5] separated by series distributed inductors as shown in Figure 8.5.

8.3 DESIGN METHOD

8.3.1 Prototype Circuit

The prototype circuit that will be used to design combline filters is the admittance inverter bandpass circuit of Figure 8.6, where

$$J_{k,k+1} = \frac{\omega_0}{\alpha}\sqrt{\frac{C_{ak}C_{ak+1}}{g_k g_{k+1}}} \qquad \text{for } k = 1, 2, \ldots, n-1$$

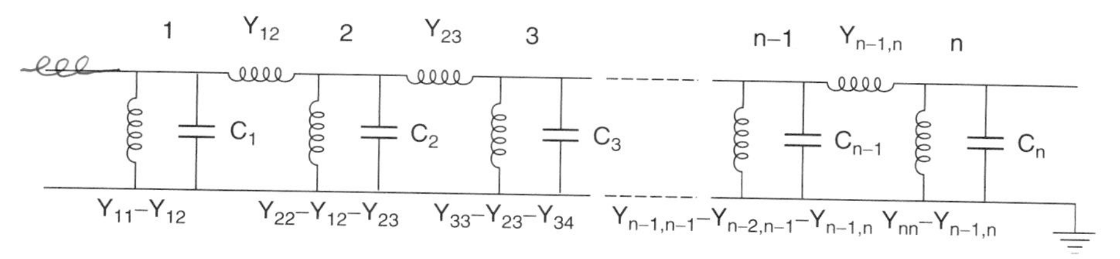

Figure 8.5 Equivalent circuit of a combline structure (without input and output).

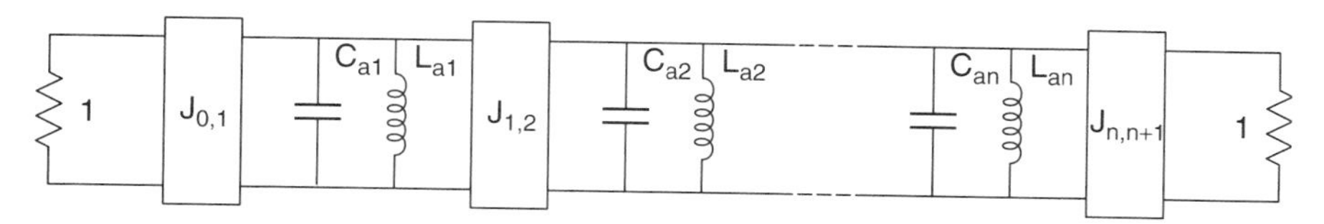

Figure 8.6 Admittance inverter bandpass prototype.

$$J_{0,1} = \sqrt{\frac{\omega_0}{\alpha}\frac{G_G C_{a1}}{g_0 g_1}} \quad \text{and} \quad J_{n,n+1} = \sqrt{\frac{\omega_0}{\alpha}\frac{C_{an} G_L}{g_n g_{n+1}}}$$

with $\omega_0 = \sqrt{\omega_1\omega_2}$ and where $\alpha = \omega_0/(\omega_2 - \omega_1)$, and where the g_k's are given from the low-pass *LC* ladder prototype elements of Chapter 2. Note that $g_0 = g_{n+1} = 1$.

8.3.2 Equivalent Circuit

The first step in modifying the equivalent circuit of the combline structure of Figure 8.5 is to introduce the admittance inverters through the equivalence of Figure 8.7. This is shown by using the *ABCD* matrix of a Π network where the admittances are given by $Y_1 = Y_3 = +jY_{k,k+1}\cot\theta$ and $Y_2 = -jY_{k,k+1}\cot\theta$. The matrix is then compared with the *ABCD* matrix of an ideal admittance inverter of value J:

$$\begin{pmatrix} 1 + \dfrac{Y_3}{Y_2} & \dfrac{1}{Y_2} \\ Y_1 + Y_3 + \dfrac{Y_1 Y_3}{Y_2} & 1 + \dfrac{Y_1}{Y_2} \end{pmatrix} = \begin{pmatrix} 0 & \dfrac{1}{-jY_{k,k+1}\cot\theta} \\ jY_{k,k+1}\cot\theta & 0 \end{pmatrix} = \begin{pmatrix} 0 & \dfrac{j}{J} \\ jJ & 0 \end{pmatrix}$$

This gives the admittance inverter value $J = Y_{k,k+1}\cot\theta$ and is used in Figure 8.7. The equivalent circuit of the combline structure can then be represented as shown in Figure 8.8.

The admittance inverters are, however, dependent on frequency through the angle θ. This frequency dependence of the inverters is moved to the generator and load by scaling the entire network admittances by the factor $\tan\theta/\tan\theta_0$ and also by multiplying the resistance of the generator and load by the factor $\tan\theta/\tan\theta_0$

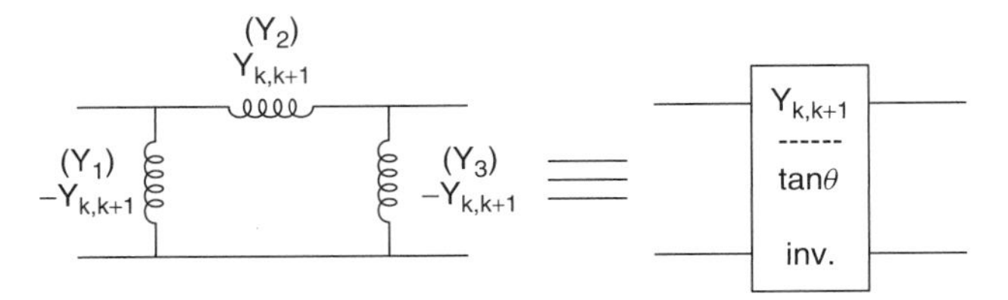

Figure 8.7 Equivalent circuit of an admittance inverter.

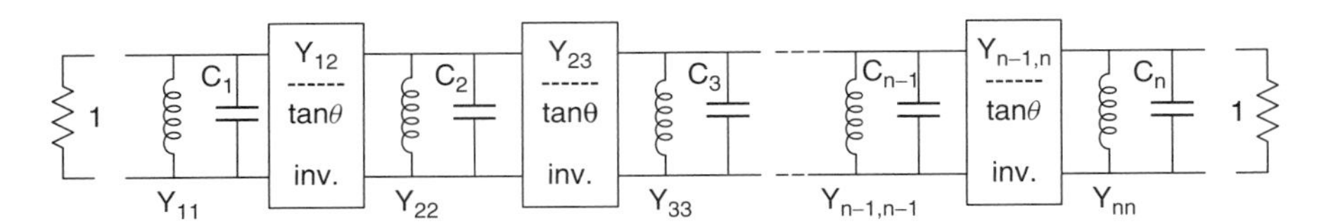

Figure 8.8 Equivalent circuit using admittance inverters (without input and output).

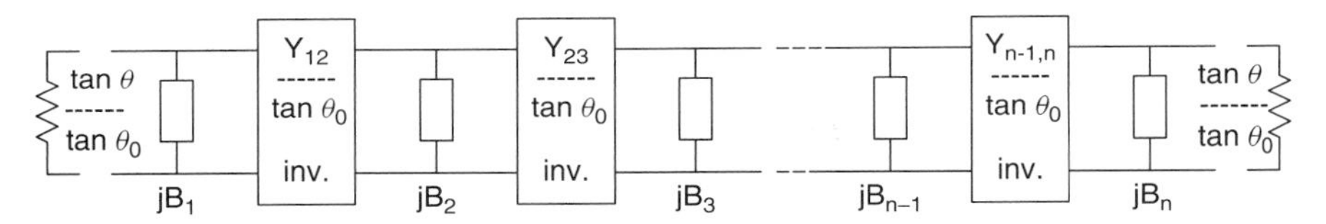

Figure 8.9 Equivalent circuit using admittance inverters independent of frequency (without input and output).

and where $\theta_0 = \pi/4$ corresponds to a $\lambda/8$ line at the center frequency of the filter. The resulting equivalent circuit is shown in Figure 8.9. The admittance inverters of Figure 8.9 are given by

$$J_{k,k+1} = \frac{Y_{k,k+1}}{\tan\theta_0} \qquad k = 1, 2, \ldots, n-1$$

and the shunt resonators of Figure 8.9 are given by

$$jB_k = j\omega C_k \frac{\tan\theta}{\tan\theta_0} - j\frac{Y_{kk}}{\tan\theta_0} \qquad k = 1, 2, \ldots, n$$

The input line is now added to the equivalent circuit as shown in Figure 8.10.

The same procedure as before is used to replace the Π network Y_{01} by the frequency-dependent admittance inverter and then removing the frequency dependence by scaling by the factor $\tan\theta/\tan\theta_0$. The same technique is applied for the output line and we have the equivalent circuit with input and output in Figure 8.11, where $jB_0 = -j(Y_{00}/\tan\theta_0)$ and $jB_{n+1} = -j(Y_{n+1,n+1}/\tan\theta_0)$. Comparing the equivalent circuit of Figure 8.11 to the bandpass prototype circuit

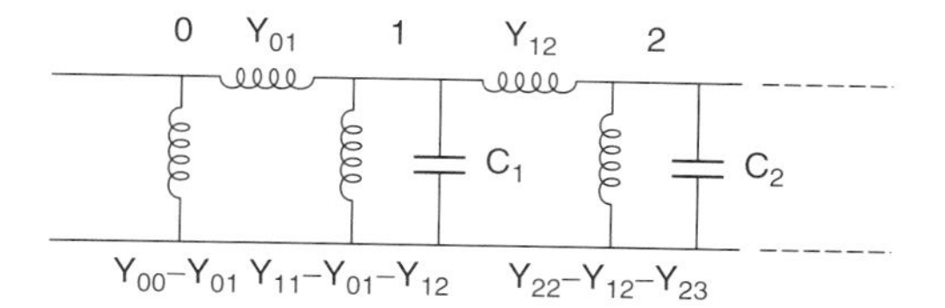

Figure 8.10 Adding the input line.

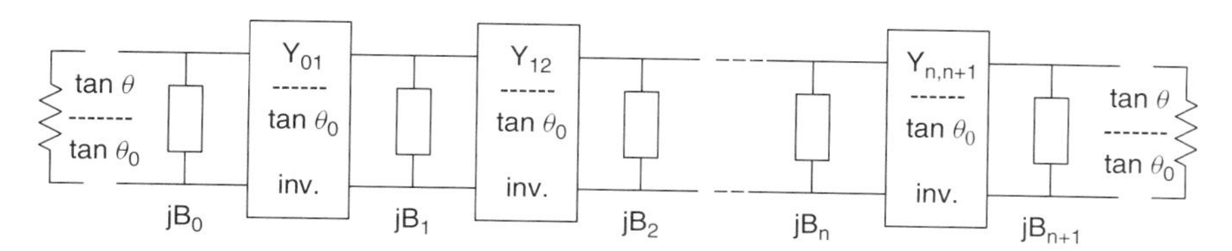

Figure 8.11 Equivalent circuit with input and output lines.

of Figure 8.6, it seems impossible to match the two structures at the input and outputs [8.6]. Several techniques have been proposed to solve this problem, but they were not very good. The technique proposed below gives the best results.

8.3.3 Input and Output

The main idea is to match the admittance seen after the first inverter in the prototype circuit and the admittance seen aftersome reactance B_1' in the equivalent circuit, as shown in Figure 8.12. Therefore, the first shunt reactance jB_1 is split into two parts such that

$$jB_1 = jB_1' + jB_1'' = j\omega C_1 \frac{\tan\theta}{\tan\theta_0} - j\frac{Y_{11}}{\tan\theta_0}$$

$$jB_1' = -j\frac{Y}{\tan\theta_0}$$

$$jB_1'' = j\omega C_1 \frac{\tan\theta}{\tan\theta_0} - j\frac{Y_{11} - Y}{\tan\theta_0}$$

and since $jB_0 = -j(Y_{00}/\tan\theta_0)$, we have for the admittance Y_G of the equivalent circuit,

$$Y_G = \frac{(Y_{01}/\tan\theta_0)^2}{\tan\theta/\tan\theta_0 - j(Y_{00}/\tan\theta_0)} - j\frac{Y}{\tan\theta_0}$$

but the admittance Y_G of the prototype circuit is given by

$$Y_G = J_{01}^2$$

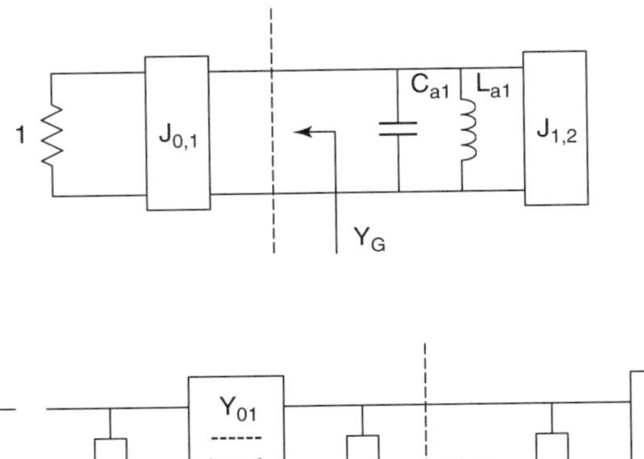

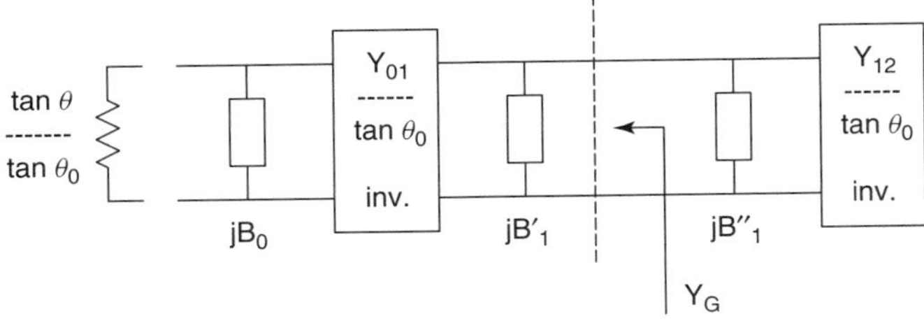

Figure 8.12 Matching the inputs of the prototype and the equivalent circuit.

The circuits are equivalent if the two values of Y_G are equal for all θ. This is not possible for all the values of θ, so we will restrict the equality to take place at $\theta = \theta_0$ or the center frequency of the filter. Since the admittance Y_G from the equivalent circuit can be complex valued, and since there are three unknowns Y_{00}, Y_{01}, Y, we need to satisfy the three equations

$$\begin{aligned} \mathrm{Re}(Y_G)|_{\theta_0} &= J_{01}^2 \\ \mathrm{Im}(Y_G)|_{\theta_0} &= 0 \\ \frac{d}{d\theta}\mathrm{Re}(Y_G)|_{\theta_0} &= 0 \end{aligned} \qquad \text{with} \qquad \begin{aligned} \mathrm{Re}(Y_G) &= \frac{Y_{01}^2}{Y_{00}^2 + \tan^2\theta}\frac{\tan\theta}{\tan\theta_0} \\ \mathrm{Im}(Y_G) &= -\frac{Y}{\tan\theta_0} + \frac{Y_{00}}{\tan\theta}\mathrm{Re}(Y_G) \end{aligned}$$

The solution of these three equations is remarkably simple [8.6]:

$$\begin{aligned} Y_{00} &= \tan\theta_0 \\ Y_{01} &= \sqrt{2}J_{01}\tan\theta_0 \\ Y &= J_{01}^2\tan\theta_0 \end{aligned}$$

and the admittance Y_G of the equivalent circuit is given by

$$Y_G = J_{01}^2\, e^{-j2\phi} \qquad \text{with} \quad \tan\phi = \frac{\sin(\theta - \theta_0)}{\sin(\theta + \theta_0)}$$

This means that the magnitude of Y_G is equal to J_{01}^2 for all values of θ:

$$|Y_G| = J_{01}^2$$

Using the same technique, we find for the output,

$$Y_{n+1,n+1} = \tan\theta_0$$
$$Y_{n,n+1} = \sqrt{2} J_{n,n+1} \tan\theta_0$$
$$Y = J_{n,n+1}^2 \tan\theta_0$$

and

$$jB_n = jB_n' + jB_n'' = j\omega C_n \frac{\tan\theta}{\tan\theta_0} - j\frac{Y_{nn}}{\tan\theta_0}$$
$$jB_n' = -j\frac{Y}{\tan\theta_0}$$
$$jB_n'' = j\omega\, C_n \frac{\tan\theta}{\tan\theta_0} - j\frac{Y_{nn} - Y}{\tan\theta_0}$$

and

$$Y_L = J_{n,n+1}^2\, e^{-j2\phi} \qquad \text{with} \quad \tan\phi = \frac{\sin(\theta - \theta_0)}{\sin(\theta + \theta_0)}$$

It is now possible to match the equivalent circuit of Figure 8.10 with the bandpass prototype circuit of Figure 8.11. The admittances and their derivative are identified at the center frequency of the filter (θ_0 and ω_0):

$$B_k = \frac{1}{\tan\theta_0}(\omega C_k \tan\theta - Y_{kk}) = C_{ak}\omega - \frac{1}{L_{ak}\omega}\bigg|_{\omega=\omega_0,\theta=\theta_0}$$
$$\frac{dB_k}{d\omega} = \frac{C_k}{\tan\theta_0}\left(\tan\theta + \omega\frac{d\tan\theta}{d\omega}\right) = C_{ak} + \frac{1}{L_{ak}\omega^2}\bigg|_{\omega=\omega_0,\theta=\theta_0} \qquad k = 1, 2, 3, \ldots, n$$

where Y_{11} and Y_{nn} are actually $Y_{11}'' = Y_{11} - Y$ and $Y_{nn}'' = Y_{nn} - Y$, respectively. From these equations the values of the lumped capacitors C_k and the admittances Y_{kk} can be computed using

$$C_k = \left(C_{ak} + \frac{1}{\omega_0^2 L_{ak}}\right)\left(1 + \frac{2\theta_0}{\sin 2\theta_0}\right)^{-1}$$
$$Y_{kk} = \left(\frac{1}{L_{ak}\omega_0} - \omega_0 C_{ak} + \omega_0 C_k\right)\tan\theta_0 \qquad k = 1, 2, 3, \ldots, n \text{ and} \qquad Y_{00} = \tan\theta_0, \quad Y_{n+1,n+1} = \tan\theta_0$$

where Y_{11} and Y_{nn} are actually $Y_{11}'' = Y_{11} - Y$ and $Y_{nn}'' = Y_{nn} - Y$, respectively, so that the final Y_{11} and Y_{nn} are $Y_{11} = Y_{11}'' + Y = Y_{11}'' + J_{01}^2 \tan\theta_0$ and $Y_{nn} = Y_{nn}'' + Y = Y_{nn}'' + J_{n.n+1}^2 \tan\theta_0$, respectively. Identification of the admittance

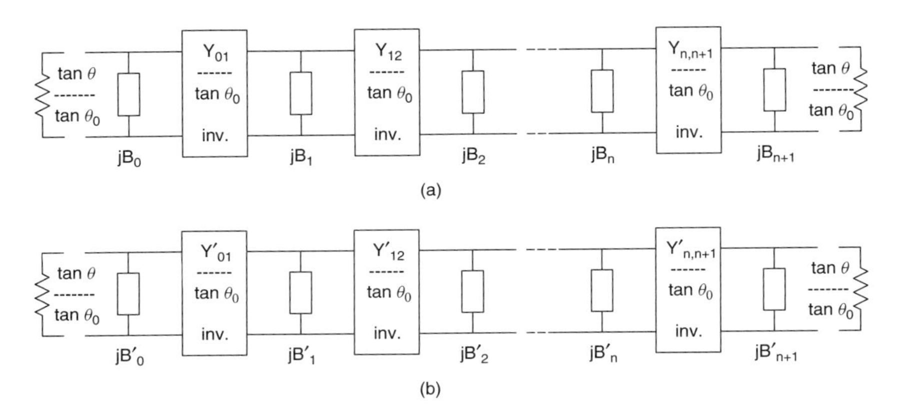

Figure 8.13 Equivalence of two equivalent circuits.

inverters provides the values of $Y_{k,k+1}$:

$$Y_{k,k+1} = J_{k,k+1} \tan\theta_0 \qquad k = 1, 2, \ldots, n-1 \quad \text{and} \quad \begin{aligned} Y_{01} &= \sqrt{2}\, J_{01} \tan\theta_0 \\ Y_{n,n+1} &= \sqrt{2}\, J_{n,n+1} \tan\theta_0 \end{aligned}$$

8.3.4 Feasibility

In some cases the values of $Y_{k,k+1}$ and Y_{kk} are not physically feasible. The technique described below can be used to scale these values without changing the response of the filter. The two circuits of Figure 8.13 have the same inputs and outputs and will have the same response if

$$B'_k = n_k^2 B_k \qquad k = 1, 2, \ldots, n$$

$$Y'_{k,k+1} = n_k n_{k+1}\, Y_{k,k+1} \qquad k = 0, 2, \ldots, n$$

The scaling factors n_k are arbitrary except for n_0 and n_{n+1}, which must be equal to 1.

8.3.5 Physical Parameters of the Combline Structure

The design technique proposed in the preceding sections does not provide the final physical properties of the lines, such as the width of the lines, and the spacing between lines. However, knowledge of $Y_{k,k+1}$ and Y_{kk} will be sufficient for defining the physical properties. The technique relies on relating $Y_{k,k+1}$ and Y_{kk} with the capacitances defined in Figure 8.14, where

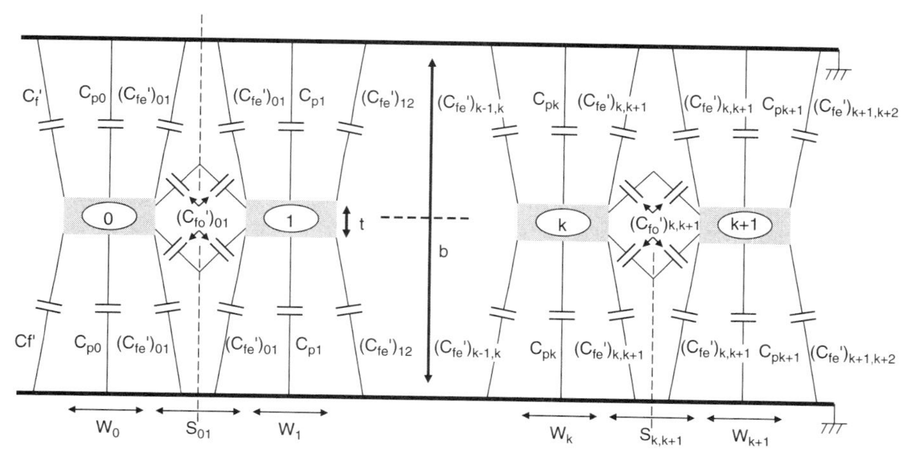

Figure 8.14 Definition of the capacitances in coupled bar structures.

t is the thickness of a bar

b is the distance between the two main ground planes

W_k is the width of bar k

$S_{k,k+1}$ is the distance between bars k and $k+1$

C_{pk} is the parallel-plate capacitance of bar k

C'_f is the fringing capacitance for an isolated bar (0 or $n+1$)

$(C'_{fe})_{k,k+1}$ is the even-mode fringing capacitance for bars k and $k+1$

$(C'_{fo})_{k,k+1}$ is the odd-mode fringing capacitance for bars k and $k+1$

Getsinger [8.7] computed the various capacitances of Figure 8.13 and provided a set of curves relating the capacitances, the distance between ground planes b, the thickness of the bars t, and the distance S between consecutive bars. The first curve provides C'_{fe}/ε and $\Delta C/\varepsilon = C'_{fo}/\varepsilon - C'_{fe}/\varepsilon$ as functions of the spacing between bars S/b with the bar thickness t/b as a parameter. The second curve provides the fringing capacitance C'_f/ε of an isolated bar as a function of bar thickness t/b. In these curves the ratios C/ε of static capacitance per unit length to the permittivity of the dielectric of the medium are related to the characteristic impedance Z_c of a lossless transmission line by

$$Z_c\sqrt{\varepsilon_r} = \frac{\eta}{C/\varepsilon} \quad \text{or} \quad \frac{C}{\varepsilon} = \frac{\eta Y_c}{\sqrt{\varepsilon_r}}$$

where η is the impedance of free space per square meter ($\eta = 376.7\ \Omega/\text{m}^2$) and ε_r is the relative permittivity of the dielectric medium. These curves allow us to find the distance between bars $S_{k,k+1}$ when the $\Delta C_{k,k+1}/\varepsilon$ are given. In addition, the parallel-plate capacitance will not be directly available and the additional curves will be used to compute its final value. Once the parallel-plate capacitance is known, the width of the bars can be computed using the relation

$$\frac{C_{pk}}{\varepsilon} = 2\frac{W_k/b}{1 - t/b} \quad \text{or} \quad \frac{W_k}{b} = \frac{1}{2}\left(1 - \frac{t}{b}\right)\frac{C_{pk}}{\varepsilon}$$

To relate the capacitances in Figure 8.14 to the combline structure, the capacitance model of Figure 8.15 is used. Comparing Figures 8.14 and 8.15, provides the following relations between the two sets of capacitances:

$$\begin{aligned} C_{k,k+1} &= [C'_{fo} - C'_{fe}]_{k,k+1} = \Delta C_{k,k+1} \\ C_k &= 2[C_{pk} + (C'_{fe})_{k,k+1} + (C'_{fe})_{k-1,k}] \\ C_0 &= 2[C_{p0} + (C'_{fe})_{01} + C'_f] \\ C_n &= 2[C_{pn} + (C'_{fe})_{n-1,n} + C'_f] \end{aligned}$$

and

$$\begin{aligned} \frac{\Delta C_{k,k+1}}{\varepsilon} &= \frac{C_{k,k+1}}{\varepsilon} \\ \frac{C_{pk}}{\varepsilon} &= \frac{1}{2}\frac{C_k}{\varepsilon} - \frac{(C'_{fe})_{k,k+1}}{\varepsilon} - \frac{(C'_{fe})_{k-1,k}}{\varepsilon} \\ \frac{C_{p0}}{\varepsilon} &= \frac{1}{2}\frac{C_0}{\varepsilon} - \frac{(C'_{fe})_{01}}{\varepsilon} - \frac{C'_f}{\varepsilon} \\ \frac{C_{pn}}{\varepsilon} &= \frac{1}{2}\frac{C_n}{\varepsilon} - \frac{(C'_{fe})_{n-1,n}}{\varepsilon} - \frac{C'_f}{\varepsilon} \end{aligned}$$

The capacitances of Figure 8.15 are computed from the $Y_{k,k+1}$ and Y_{kk} of the equivalent circuit of the combline structure:

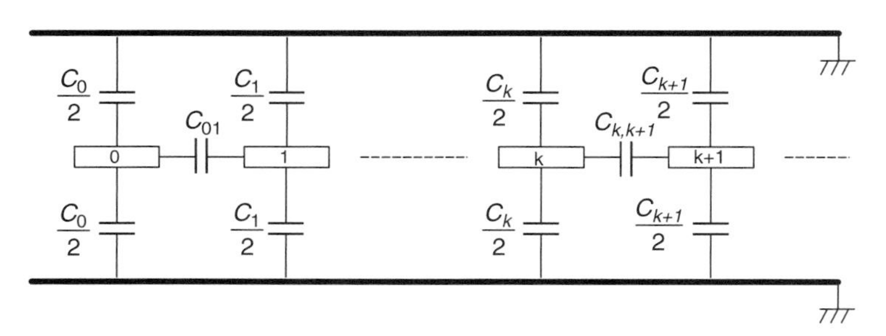

Figure 8.15 Capacitance model of the combline structure.

$$\begin{cases} \dfrac{C_{k,k+1}}{\varepsilon} = \dfrac{\eta Y_c}{\sqrt{\varepsilon_r}} Y_{k,k+1} & k = 0, 1, 2, \ldots, n \\ \dfrac{C_k}{\varepsilon} = \dfrac{\eta Y_c}{\sqrt{\varepsilon_r}} Y_{kk} & k = 0, 1, 2, 3, \ldots, n+1 \end{cases}$$

where Y_c is the desired characteristic admittance at the input/output of the structure (e.g., $Y_C = 1/50\ \Omega$) and assuming that $Y_{k,k+1}$ and Y_{kk} are provided for 1-Ω terminations. At this stage, the various capacitance ratios are computed from the $Y_{k,k+1}$ and Y_{kk} of the equivalent circuit of the combline structure. Then the $\Delta C_{k,k+1}/\varepsilon$ are used with the Getsinger curves to find the $S_{k,k+1}/b$ distances for a fixed t/b ratio. Note that the t/b ratio is decided on at the very start. Knowledge of $S_{k,k+1}/b$ then provides the capacitances $(C'_{fe})_{k,k+1}/\varepsilon$. Using only t/b, the capacitances C'_f/ε can be found using the other curve. At this stage the parallel-plate capacitances C_{pk}/ε can finally be computed, from which the widths W_k of the bars can be defined using the equation. In the case of suspended substrate stripline realizations, ε_r should be taken as the effective relative dielectric permittivity ε_{eff}.

8.4 DESIGN EXAMPLE

The filter should have a third-order Chebyshev response with a return loss of −20 dB. The passband edges are $f_1 = 960$ MHz and $f_2 = 1040$ MHz, leading to a center frequency $f_0 = \sqrt{f_1 f_2} \approx 1$ GHz. The bandpass prototype ladder elements are given by

$$J_{01} = 0.3063 \qquad J_{12} = 0.0825 \qquad J_{23} = 0.0825 \qquad J_{34} = 0.3063$$

$$C_{a1} = 0.1593\ \text{nF} \qquad C_{a2} = 0.1593\ \text{nF} \qquad C_{a3} = 0.1593\ \text{nF}$$

$$L_{a1} = 0.1593\ \text{nH} \qquad L_{a2} = 0.1593\ \text{nH} \qquad L_{a3} = 0.1593\ \text{nH}$$

The arbitrary capacitors of the original low-pass filter are taken as $C_{ak} = 1/\omega_0$ to have an admittance of about 1. The inductors are computed from $L_{ak} = 1/C_{ak}\omega_0^2$. The response of the prototype bandpass filter for 1-Ω terminations is shown in Figure 8.16. The elements of the equivalent circuit of the combline filter are given by

$$C_1 = 0.1239\ \text{nF} \qquad C_2 = 0.1239\ \text{nF} \qquad C_3 = 0.1239\ \text{nF}$$

$$Y_{00} = 1 \qquad Y_{11} = 0.8718 \qquad Y_{22} = 0.7780 \qquad Y_{33} = 0.8718 \qquad Y_{44} = 1$$

$$Y_{01} = 0.4332 \qquad Y_{12} = 0.0825 \qquad Y_{23} = 0.0825 \qquad Y_{34} = 0.4332$$

Note that $Y = 0.0938$ at the input and output.

The simulated response of the equivalent circuit of the combline filter of Figure 8.5 with the additional input/output lines for 1-Ω terminations is shown

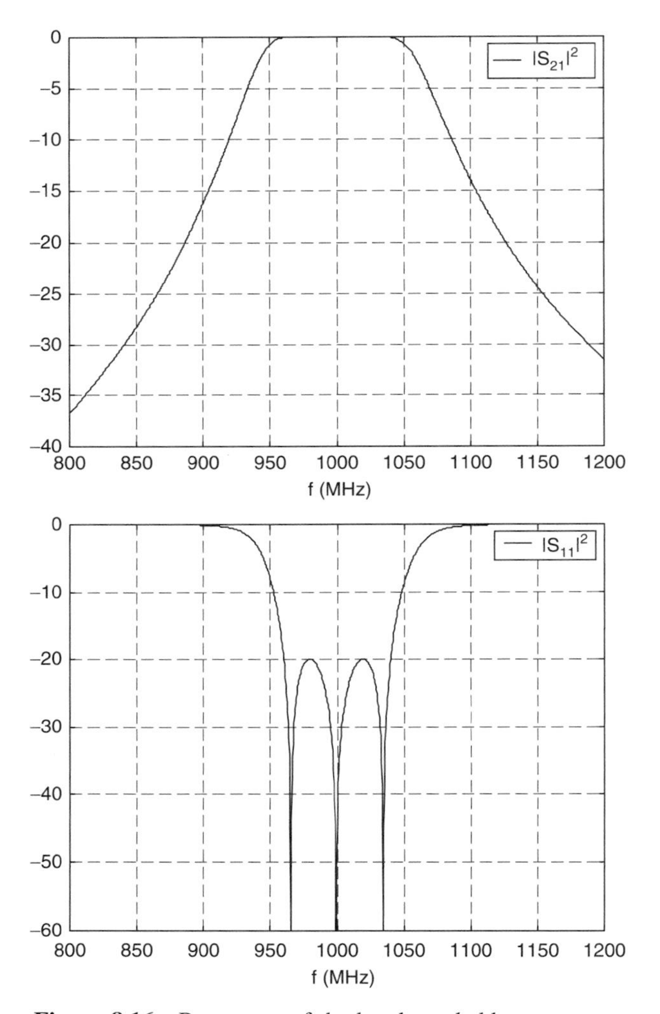

Figure 8.16 Responses of the bandpass ladder prototype.

in Figure 8.17. One can see that there are some differences between the equivalent circuit and the prototype circuit, such as a greater return loss and a slight frequency shift. The physical parameters are then defined starting with the normalized capacitances

$$\frac{C_{k,k+1}}{\varepsilon} = \frac{\eta Y_c}{\sqrt{\varepsilon_r}} Y_{k,k+1} \qquad k = 0, 1, 2, \ldots, n$$

$$\frac{C_k}{\varepsilon} = \frac{\eta Y_c}{\sqrt{\varepsilon_r}} Y_{kk} \qquad k = 0, 1, 2, 3, \ldots, n+1$$

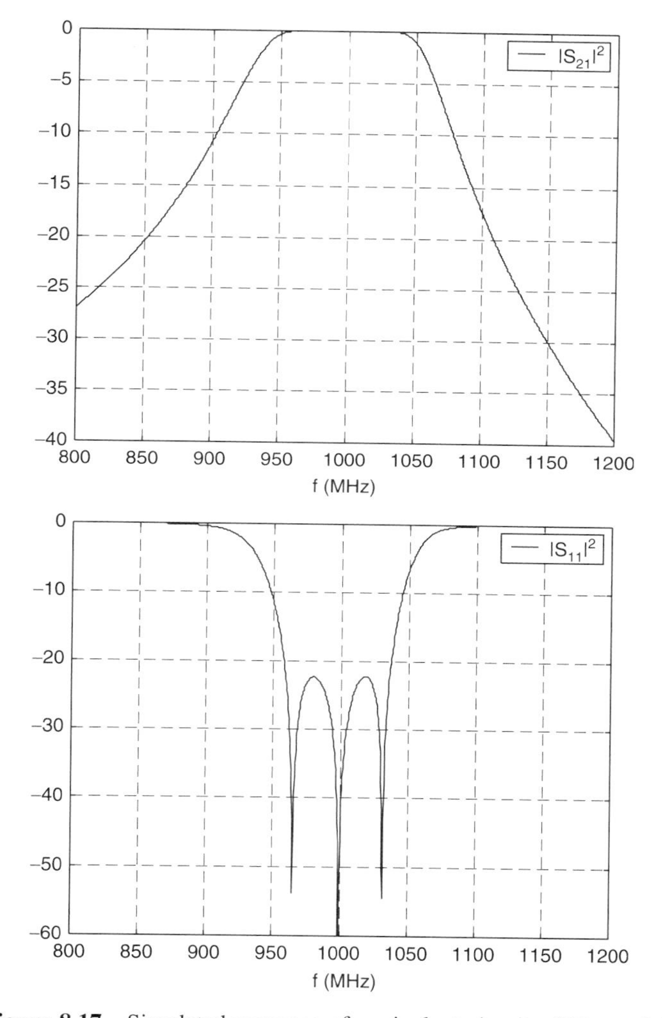

Figure 8.17 Simulated response of equivalent circuit of Figure 8.5.

With $\varepsilon_r = \varepsilon_{\text{eff}} = 1.16$,

$$\frac{C_{0,1}}{\varepsilon} = 3.0300 \qquad \frac{C_{1,2}}{\varepsilon} = 0.5770 \qquad \frac{C_{2,3}}{\varepsilon} = 0.5770$$

$$\frac{C_{3,4}}{\varepsilon} = 3.0300 \qquad \frac{C_0}{\varepsilon} = 6.9951 \qquad \frac{C_1}{\varepsilon} = 6.0982$$

$$\frac{C_2}{\varepsilon} = 5.4420 \qquad \frac{C_3}{\varepsilon} = 6.0982 \qquad \frac{C_4}{\varepsilon} = 6.9951$$

With $t/b = 0.1$ we get

$$\frac{S_{0,1}}{b} = 0.06 \qquad \frac{S_{1,2}}{\varepsilon} = 0.39 \qquad \frac{S_{2,3}}{\varepsilon} = 0.39 \qquad \frac{S_{3,4}}{b} = 0.06$$

$$\frac{C_{fe0,1}}{\varepsilon} = 0.06 \qquad \frac{C_{fe1,2}}{\varepsilon} = 0.35 \qquad \frac{C_{fe2,3}}{\varepsilon} = 0.35 \qquad \frac{C_{fe3,4}}{\varepsilon} = 0.06$$

With $t/b = 0.1$ we have $C_f' = 0.58$ and

$$\frac{C_{p0}}{\varepsilon} = 2.8576 \qquad \frac{C_{p1}}{\varepsilon} = 2.6391 \qquad \frac{C_{p2}}{\varepsilon} = 2.0210 \qquad \frac{C_{p3}}{\varepsilon} = 2.6391$$

$$\frac{C_{p4}}{\varepsilon} = 2.8576 \qquad \frac{W_0}{b} = 1.2859 \qquad \frac{W_1}{b} = 1.1876 \qquad \frac{W_2}{b} = 0.9095$$

$$\frac{W_3}{b} = 1.1876 \qquad \frac{W_4}{b} = 1.2859$$

8.5 REALIZATIONS AND MEASURED PERFORMANCE

A combline filter implemented in suspended substrate stripline had the following specifications [8.6].

- Center frequency: 1.219 GHz
- Passband bandwidth: 24 MHz
- Degree of the filter: 3
- Permittivity of the substrate: $\varepsilon_r = 2.55$
- Dimensions: $b = 8.8$ mm, $t = 0.79$ mm

The element values were determined with an effective dielectric constant $\varepsilon_{\text{eff}} = 1.16$. The suspended substrate stripline and the mounting of the lumped capacitances are shown in Figure 8.18. The overall filter is shown in Figure 8.19. The experimental performance of the filter is shown in Figure 8.20. The filter performs according to theoretical expectations.

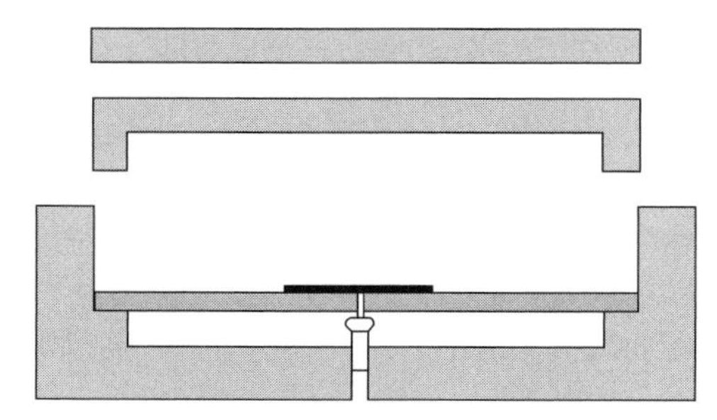

Figure 8.18 Suspended substrate stripline and mounting of the lumped capacitors.

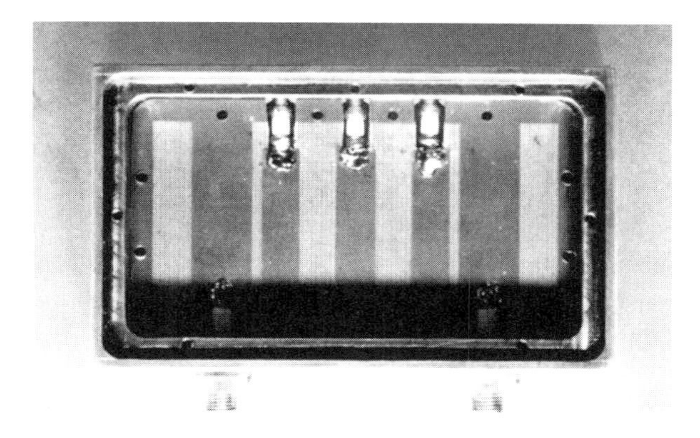

Figure 8.19 Combline filter with lumped capacitances.

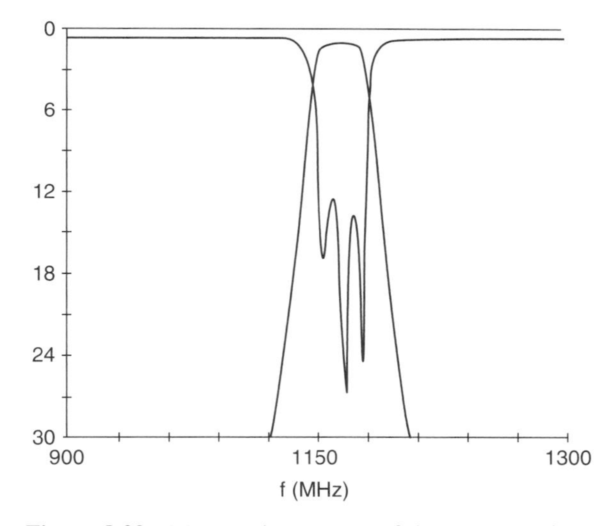

Figure 8.20 Measured responses of the combline filter.

8.6 CONCLUSIONS

Combline filters are very popular, due to their compact size and broad stopbands and can be very selective. They are still the object of research efforts, particularly in the domain of the input/output lines [8.8], in using dielectric resonators [8.9], and in the realization of transmission zeros [8.10].

Acknowledgments

This work was supported by Thomson and by the French Ministry of Research and Technology.

REFERENCES

[8.1] G. L. Matthei, Combline bandpass filters of narrow or moderate bandwidth, *Microwave.*, vol. 1, pp. 82–91 Aug. 1963.

[8.2] R. Levy and J. D. Rhodes, A combline elliptic filter, *IEEE Trans. Microwave Theory Tech.*, vol. 19, pp. 26–29 Jan. 1971.

[8.3] E. G. Cristal, Microwave filters, in *Modern Filter Theory and Design*, ed, G. C. Temes and S. K. Mitra, pp. 273–231, Wiley, New York, 1973.

[8.4] R. Sato and E. G. Cristal, Simplified analysis of coupled transmission-line networks, *IEEE Trans. Microwave Theory Tech.*, vol. 18, pp. 122–131, Mar. 1970.

[8.5] I. C. Hunter and J. D. Rhodes, Electronically tunable microwave bandpass filters, *IEEE Trans. Microwave Theory Tech.*, vol. 30, pp. 1354–1360, Sept. 1982.

[8.6] A. Pochat, P. Jarry, and G. Menexiadis, Computer-aided design of combline suspended substrate strip-line filters, *Proc. URSI'86*, Gerona, Italy, pp. 254–256, Sept. 1986.

[8.7] W. J. Getsinger, Coupled rectangular bars between parallel plated, *IRE Trans. Microwave Theory Tech.*, vol. 10, no. 1, pp. 65–72, 1962.

[8.8] S. Caspi and J. Aldeman, Design of combline and interdigital filters with tapped line input, *IEEE Trans. Microwave Theory Tech.*, vol. 36, pp. 759–763, Apr. 1988.

[8.9] Chi Wang, K. A. Zaki, A. E. Atia, and T. G. Dolan, Dielectric combline resonators and filters, *IEEE Trans. Microwave Theory Tech.*, vol. 46, pp. 2501–2506 Dec. 1998.

[8.10] Ching-Wen Tang, Yin-Ching Lin, and Chi-Yang Chang, Realization of transmission zeros in combline filters using an auxiliary inductively coupled ground plane, *IEEE Trans. Microwave Theory Tech.*, vol. 51, pp. 2112–2118, Oct. 2003.

Selected Bibliography

Rhodes, J. D., The stepped digital elliptic filter, *IEEE Trans. Microwave Theory Tech.*, vol. 17, no. 4, pp. 178–184, Apr. 1969.

Rhodes, J. D., The half-wave stepped digital elliptic filter, *IEEE Trans. Microwave Theory Tech.*, vol. 17, no. 12, pp. 1102–1107, 1969.

Wenzel, R. J., Synthesis of combline and capacitively loaded interdigital bandpass filters of arbitrary bandwidth, *IEEE Trans. Microwave Theory Tech.*, vol. 19, pp. 678–686, Aug. 1971.

Cristal, E.G., Tapped-line coupled transmission lines with applications to interdigital and combline filters, *IEEE Trans. Microwave Theory Tech.*, vol. 23, pp. 1007–1012, Dec. 1975.

PART III

NON-MINIMUM-PHASE SYMMETRICAL RESPONSE FILTERS

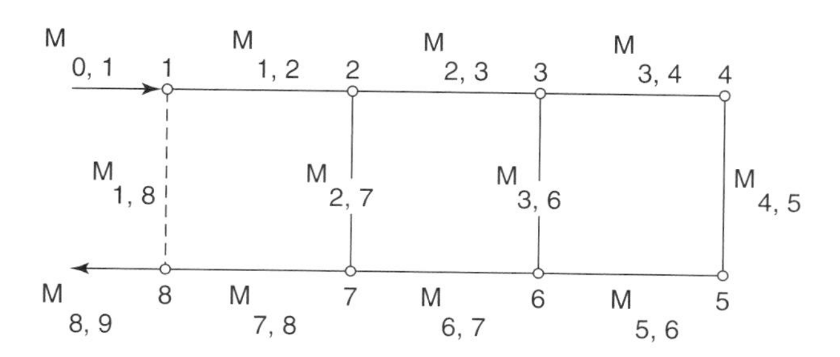

9

GENERALIZED INTERDIGITAL FILTERS WITH CONDITIONS ON AMPLITUDE AND PHASE

9.1 INTRODUCTION

The traditional technique for obtaining both amplitude and phase characteristics consists of treating the amplitude and phase requirements independently. This approach has the inconvenience of requiring an overall filtering system that has a high order. An alternative solution is to consider the case where the amplitude and phase conditions are considered simultaneously. Such filters, with optimized amplitude and phase responses, require non-minimum-phase transfer functions. In

Advanced Design Techniques and Realizations of Microwave and RF Filters,
By Pierre Jarry and Jacques Beneat

this chapter we present an exact method for designing non-minimum-phase transfer functions with maximally flat amplitude response, with an additional perfect transmission point at an arbitrary frequency ω_0 while preserving the phase linearity. This transfer function is based on a polynomial combining two successive Bessel polynomials.

A suitable structure to implement these non-minimum-phase transfer functions is the generalized interdigital structure [9.1,9.2]. This structure consists of two interdigital structures with additional slot coupling to achieve non-minimum-phase functions. The experimental results presented at the end of this chapter show that these non-minimum-phase filters can be built in practice and offer an improved attenuation response in the rejection band.

9.2 GENERALIZED INTERDIGITAL FILTER

Compared to an interdigital filter with a number r of $\lambda/4$ lines, the generalized interdigital filter consists of two sets of r $\lambda/4$ lines that are placed one on top of the other and separated by a ground plane with slots the length of the lines to allow for slot coupling. Figure 9.1 shows a general view of the filter along with a side view to better see the location of the slots and the lines. The side view shows that this structure is enclosed in a metallic box. Since it is based on interdigital filters, the ends of the lines will alternate between being left open or being shorted to the ground.

Figure 9.2 provides a first equivalent circuit of the generalized interdigital filter. The input of the system is taken on the far side of the bottom line 1, while the output is taken on the near side of the top line 1. Also shown are which ends of the lines are shorted to ground and the additional slot coupling. Each block is equivalent to a unit element. Since a shorted unit element k can be represented as an admittance Y_k, the equivalent circuit can be redrawn as in Figure 9.3. The circuit in Figure 9.3 can then be redrawn as in Figure 9.4. This equivalent circuit, according to the theory of distributed networks and the Richards variable, is defined as in Table 9.1.

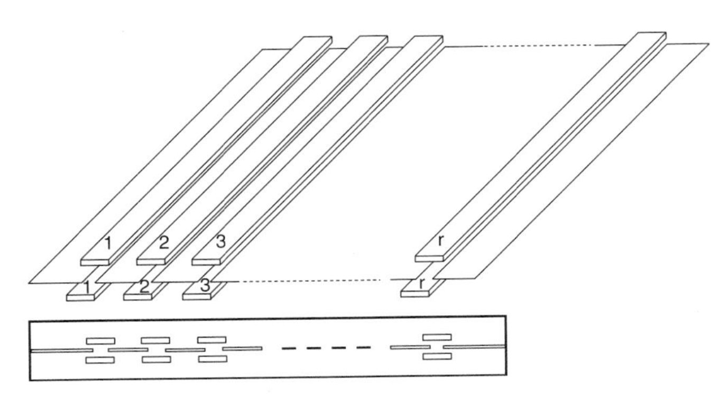

Figure 9.1 Generalized interdigital filter of degree $2r$.

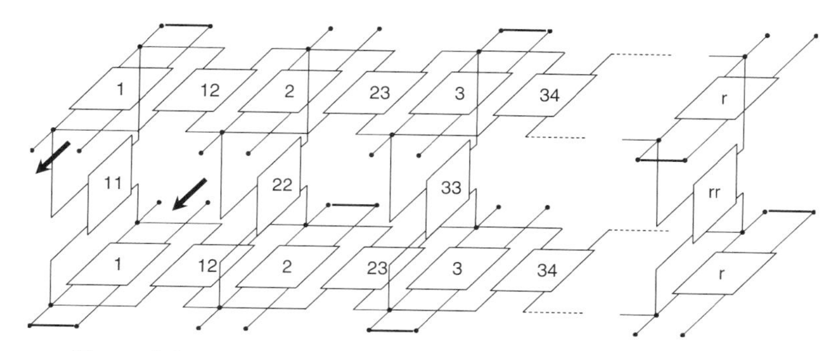

Figure 9.2 Equivalent circuit of the generalized interdigital filter.

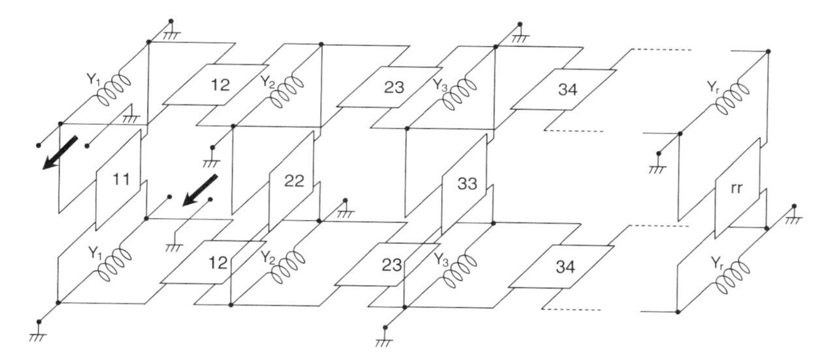

Figure 9.3 Equivalent circuit of the generalized interdigital filter.

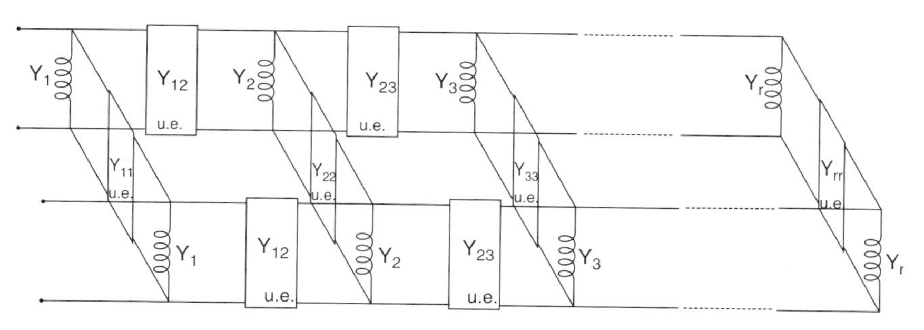

Figure 9.4 Equivalent circuit of the generalized interdigital filter.

9.3 SIMULTANEOUS AMPLITUDE AND PHASE FUNCTIONS

9.3.1 Minimum-Phase Functions with Linear Phase

Consider the minimum-phase transfer function

$$S_{12}(s) = \frac{1}{H_n(s)}$$

TABLE 9.1

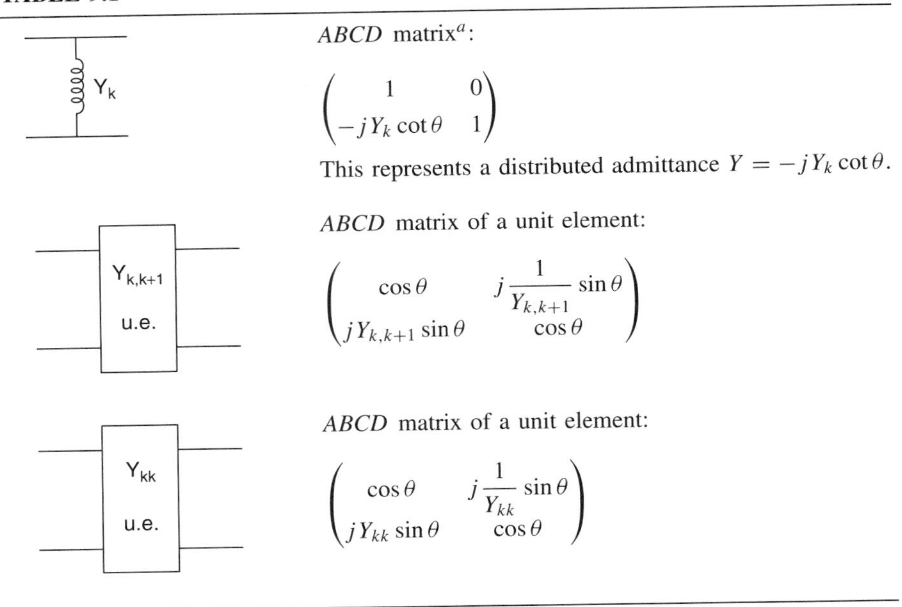

Y_k	*ABCD* matrix[a]: $\begin{pmatrix} 1 & 0 \\ -jY_k \cot\theta & 1 \end{pmatrix}$ This represents a distributed admittance $Y = -jY_k \cot\theta$.
$Y_{k,k+1}$ u.e.	*ABCD* matrix of a unit element: $\begin{pmatrix} \cos\theta & j\dfrac{1}{Y_{k,k+1}}\sin\theta \\ jY_{k,k+1}\sin\theta & \cos\theta \end{pmatrix}$
Y_{kk} u.e.	*ABCD* matrix of a unit element: $\begin{pmatrix} \cos\theta & j\dfrac{1}{Y_{kk}}\sin\theta \\ jY_{kk}\sin\theta & \cos\theta \end{pmatrix}$

[a] $\theta = \beta l$, $\beta = \omega/c$, $l = \lambda_0/4$, and $\lambda_0 = c/f_0$.

The group delay of this transfer function is given by

$$T(s) = \frac{1}{2}\left[\frac{H_n'(s)}{H_n(s)} + \frac{H_n'(-s)}{H_n(-s)}\right]$$

It is desired that the group delay have $n-1$ derivatives that are equal to zero at $\omega = 0$ and one derivative that is equal to zero at $\omega = \omega_0$. In this case the group delay will be of the form

$$T(s) = 1 + k\frac{s^{2(n-1)}(s^2 + \omega_0^2)}{H_n(s)H_n(-s)}$$

A polynomial $H_n(s)$ that satisfies this condition [9.3] is given in terms of the Bessel polynomials $Q_n(s)$ as

$$H_n(s) = Q_{n-1}(s) + \frac{s^2}{(2n-3)(2n-1-\alpha)}Q_{n-2}(s)$$

where $\omega_0^2 = \alpha(2n-1-\alpha)$. For stability, this polynomial will be strictly Hurwitz if $\alpha \leq 2n-1$. The polynomial $H_n(s)$ can also be computed using the recurrence formula

$$H_{n+1}(s) = \left(1 - \frac{\alpha}{2n-1}\right) H_n(s) + \frac{s^2 + \alpha(2n+1-\alpha)}{(2n-1)(2n+1-\alpha)} Q_{n-1}(s)$$

The first $H_n(s)$ polynomials are given by

$$H_1(s) = 1 + \frac{1}{1-\alpha} s$$

$$H_2(s) = 1 + s + \frac{1}{3-\alpha} s^2$$

$$H_3(s) = 1 + s + \frac{6-\alpha}{3(5-\alpha)} s^2 + \frac{1}{3(5-\alpha)} s^3$$

$$H_4(s) = 1 + s + \frac{15-2\alpha}{5(7-\alpha)} s^2 + \frac{10-\alpha}{15(7-\alpha)} s^3 + \frac{1}{15(7-\alpha)} s^4$$

When $\alpha = 0$, $H_n(s)$ will be the same as the Bessel polynomial. The polynomial $H_n(s)$ is used as the base for defining a non-minimum-phase transfer function with simultaneous conditions on the magnitude and phase.

9.3.2 Non-Minimum-Phase Functions with Simultaneous Conditions on the Amplitude and Phase

Simultaneous optimization of amplitude and phase in the same frequency band requires about twice as many amplitude conditions as phase conditions. Thus, let's consider the transfer function [9.6]

$$S_{12}(s) = \frac{\mathrm{Re}[D_n(s)D_n(-s)]}{D_n(s)D_n^*(s)}$$

where Re stands for the real part, and

$$D_n(s) = H_n(s) + j\frac{(2n-3-\alpha)s}{(2n-3)(2n-1-\alpha)} H_{n-1}(s)$$

$$D_n^*(s) = H_n(s) - j\frac{(2n-3-\alpha)s}{(2n-3)(2n-1-\alpha)} H_{n-1}(s)$$

The transfer function will have the following magnitude and phase properties:

$$S_{12}(j\omega)S_{12}(-j\omega) = \frac{1}{1 + [2\beta_n \omega^{2n-2}(\omega^2 - \omega_0^2)]^2/|E(j\omega)|^2} \quad \text{and}$$

$$-\angle S_{12}(j\omega) = 2\omega + b_1\omega^{2n-1} + b_2\omega^{2n+1} + \cdots$$

In other words, the magnitude has a full transmission point at $\omega = \omega_0$ with $\omega_0^2 = \alpha(2n - 1 - \alpha)$, and the phase exhibits the same behavior as a Bessel filter of order $n - 1$.

Figures 9.5 and 9.6 give examples of magnitude and group delay responses for this type of transfer function. In both cases, the full transmission point occurs at $\omega = \omega_0$, and the group delay is flat around $\omega = 0$. One notes that by varying α it is possible to vary the passband ripple and the selectivity of the filter. Figure 9.6 shows that a zero of transmission is possible when increasing α enough. Unfortunately, the passband ripple degrades when increasing α. Note that the order of these transfer functions is always even and equal to $N = 2n$.

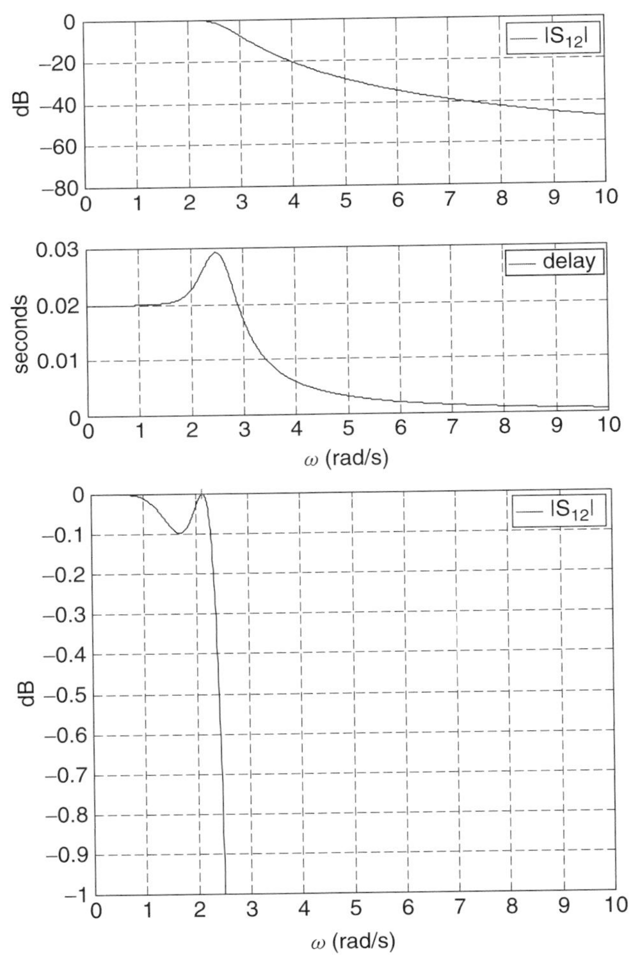

Figure 9.5 Simultaneous amplitude and phase transfer function responses for $\alpha = 1.2$, $n = 3$, giving $\omega_0 = 2.14$.

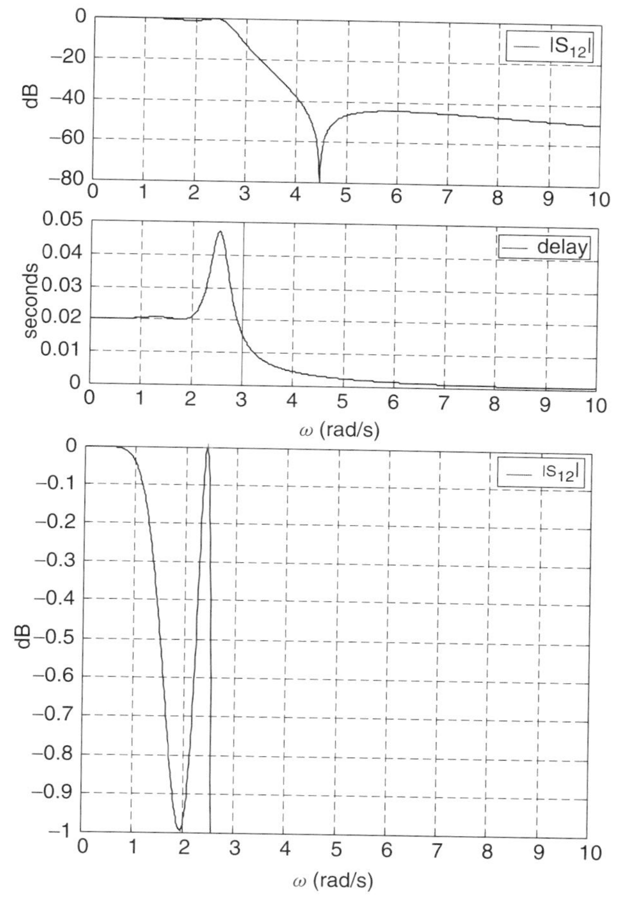

Figure 9.6 Simultaneous amplitude and phase transfer function responses for $\alpha = 1.85$, $n = 3$, giving $\omega_0 = 2.41$.

Transfer functions of odd order that present simultaneous conditions on the amplitude and phase are given by [9.4]

$$S_{12}(s) = \frac{\mathrm{Ev}[R_n(-s)R_{n+1}(s)]}{R_n(s)R_{n+1}(s)}$$

where Ev stands for the even part and

$$R_n(s) = H_n(s)$$

$$R_{n+1}(s) = 2H_{n+1}(s) - H_n(s)$$

The amplitude response also has a full transmission point at $\omega = \omega_0$ where $\omega_0^2 = \alpha(2n - 1 - \alpha)$, and the phase characteristic will be the same as that of $H_n(s)$.

This transfer function provides results very similar to those shown in Figures 9.5 and 9.6. The main difference is that these transfer functions have an odd degree $N = 2n + 1$.

9.3.3 Synthesis of Non-Minimum-Phase Functions with Simultaneous Conditions on the Amplitude and Phase

In this section the synthesis of the even-order transfer functions is provided [9.5]. In this case the transfer functions are of the type [9.7]

$$S_{12}(s) = \frac{\mathrm{Re}[D_r(s)D_r(-s)]}{D_r(s)D_r^*(s)}$$

$$D_r(s) = H_r(s) + j\frac{(2r-3-\alpha)s}{(2r-3)(2r-1-\alpha)}H_{r-1}(s)$$

$$D_r^*(s) = H_r(s) - j\frac{(2r-3-\alpha)s}{(2r-3)(2r-1-\alpha)}H_{r-1}(s)$$

This transfer function can be realized under the form of the electrical network shown in Figure 9.7. This network is composed of shunt capacitors and admittance inverters of value 1 and coupling admittance inverters of value K_i. These have for the $ABCD$ matrix

$$\begin{pmatrix} 0 & j/K_i \\ jK_i & 0 \end{pmatrix}$$

Since the network is symmetrical, it can be defined entirely from its even-mode circuit according to the Barlett bisection theorem. The even- and odd-mode circuits of this network are shown in Figure 9.8. The even-mode circuit is made of shunt susceptances $C_i s + jK_i$ and the odd-mode circuit is made of shunt susceptances $C_i s - jK_i$.

From the Barlett bisection theorem, $S_{12}(s)$ can be expressed as a function of the admittance $Y_e(s)$ of the even-mode circuit and the admittance $Y_o(s)$ of the

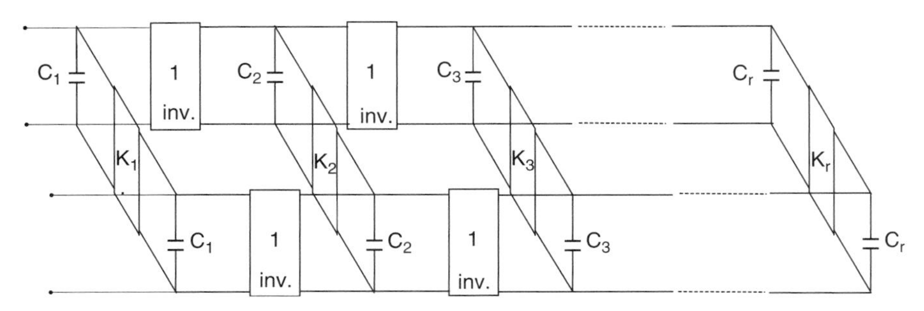

Figure 9.7 Realization of the non-minimum-phase transfer function.

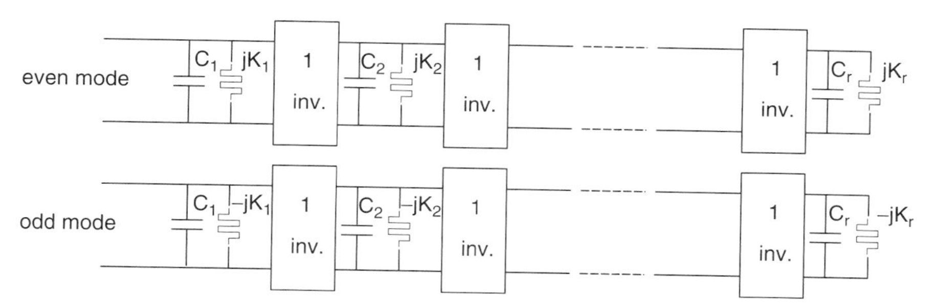

Figure 9.8 Even- and odd-mode circuits of the network of Figure 9.7.

odd-mode function as

$$S_{12}(s) = \frac{Y_e(s) - Y_o(s)}{[1 + Y_e(s)][1 + Y_o(s)]}$$

In addition, since $Y_e(s) = Y(s)$ and $Y_o(s) = Y^*(s)$, $S_{12}(s)$ can be expressed as

$$S_{12}(s) = \frac{Y(s) - Y^*(s)}{[1 + Y(s)][1 + Y^*(s)]}$$

By introducing the polynomial

$$F_r(s) = (1 - j)D_r(s) = [E_1(s) + O_1(s)] + j[E_2(s) + O_2(s)]$$

where E stands for even part and O stands for odd part, $S_{12}(s)$ can be expressed as

$$S_{12}(s) = \frac{\text{Im}[F_r(s)F_r(-s)]}{F_r(s)F_r^*(s)}$$

where Im stands for the imaginary part. In this situation, $Y(s)$ can be expressed as

$$Y(s) = \frac{E_1(s) + jO_2(s)}{O_1(s) + jE_2(s)} \quad \text{for } r \text{ even} \quad \text{and} \quad Y(s) = \frac{O_1(s) + jE_2(s)}{E_1(s) + jO_2(s)} \quad \text{for } r \text{ odd}$$

The elements of the even-mode circuit are then obtained from $Y(s)$ as follows. First the capacitor is extracted using

$$C_1 = \left.\frac{Y(s)}{s}\right|_{s \to \infty}$$

Then the remaining admittance is computed:

$$Y'(s) = Y(s) - C_1 s$$

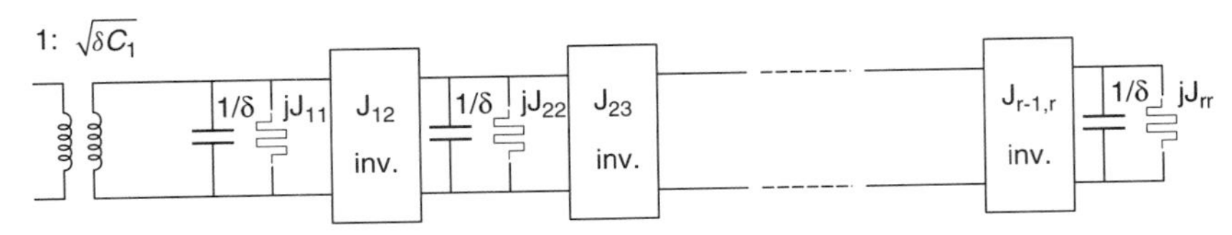

Figure 9.9 Even-mode prototype circuit.

The shunt reactance is extracted using

$$jK_1 = Y'(s)|_{s \to \infty}$$

The remaining admittance is then

$$Y''(s) = Y'(s) - jK_1$$

The admittance inverter of value 1 is extracted, leading to the remaining admittance:

$$Y'''(s) = \frac{1}{Y''(s)}$$

which is of the same form as $Y(s)$ but of a lower order. The procedure is then repeated to extract all the capacitors C_i and shunt reactances K_i of the even-mode circuit of Figure 9.8.

As will be seen in the next section, the even-mode equivalent circuit of the generalized interdigital circuit will not be easily comparable with the even-mode prototype circuit of Figure 9.8. Therefore, the even-mode prototype circuit of Figure 9.8 is modified using the techniques of Chapter 2 to make all the capacitors equal to $1/\delta$. This has the effect of introducing an ideal transformer at the input of the prototype and scaling the shunt reactances and admittance inverters as shown in Figure 9.9, with

$$J_{k,k+1} = \frac{1}{\delta\sqrt{C_k C_{k+1}}} \qquad k = 1, 2, \ldots, r-1$$

$$J_{kk} = \frac{K_k}{\delta C_k} \qquad k = 1, 2, \ldots, r$$

It will be seen later that δ will play an important role in frequency transformation, making it possible to use a low-pass prototype rather than a bandpass prototype in the design method.

9.4 DESIGN METHOD

9.4.1 Even-Mode Equivalent Circuit

Since the structure of Figure 9.4 is symmetrical, according to the Barlett bisection theorem, it can be defined entirely by its even-mode structure. To define the even-mode equivalent circuit, an open-circuit plane following the axis of symmetry of the structure must be introduced. From Figure 9.4, the axis of symmetry occurs in the middle of the unit elements (u.e.'s) Y_{kk}. The u.e. Y_{kk} corresponds to a line of length θ and characteristic admittance Y_{kk} as shown in Figure 9.10. The open-circuit plane of the symmetry occurs halfway through this line at $\theta/2$. The admittance seen at the input of an open line of length $\theta/2$ is given by $Y = jY_{kk}\tan(\theta/2)$. The last element in Figure 9.10 shows how the $\theta/2$ line will be represented in future equivalent circuits.

Figure 9.11 shows the even-mode equivalent circuit of the generalized interdigital filter. It is important to note that even-mode circuit prototypes with desired filtering characteristics are commonly provided using admittance inverters rather than unit elements. Therefore, one must transform this equivalent circuit so that it is based on admittance inverters rather than on unit elements.

From the $ABCD$ matrix identity below, a unit element can be modeled as an admittance inverter J placed between two admittances Y:

$$\begin{pmatrix} \cos\theta & j\dfrac{\sin\theta}{Y_{k,k+1}} \\ jY_{k,k+1}\sin\theta & \cos\theta \end{pmatrix} = \begin{pmatrix} 1 & 0 \\ -jY_{k,k+1}\cot\theta & 1 \end{pmatrix} \begin{pmatrix} 0 & j\dfrac{\sin\theta}{Y_{k,k+1}} \\ j\dfrac{Y_{k,k+1}}{\sin\theta} & 0 \end{pmatrix} \begin{pmatrix} 1 & 0 \\ -jY_{k,k+1}\cot\theta & 1 \end{pmatrix}$$

Y_kk
u.e.
≡
θ
Y_kk
and
θ/2
θ/2
Y_kk
Y_kk
open-circuit plane
θ/2
Y_kk
Y = jY_kk tan(θ/2)
≡
Y_kk
Y = jY_kk tan(θ/2)
circuit representation

Figure 9.10 Equivalent circuit of the transverse u.e. Y_{kk}.

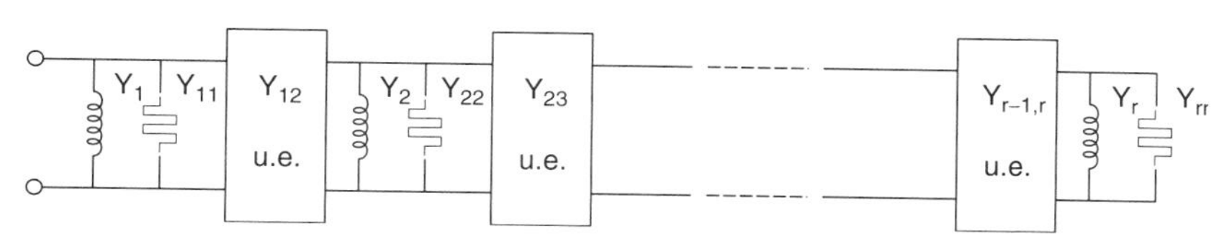

Figure 9.11 Even-mode equivalent circuit with unit elements.

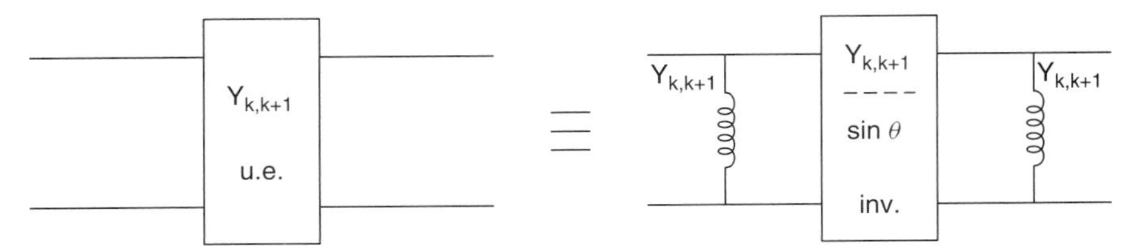

Figure 9.12 Transforming a unit element into an admittance inverter.

where $J = Y_{k,k+1}/\sin\theta$ and $Y = -jY_{k,k+1}\cot\theta$. For the equivalent-circuit representations, the unit element will be replaced by the admittance inverter and two admittances as shown in Figure 9.12. One notes that through θ, the admittance inverter is dependent on frequency. From the *ABCD* matrix identity

$$\begin{pmatrix} 0 & j\dfrac{\sin\theta}{Y_{k,k+1}} \\ j\dfrac{Y_{k,k+1}}{\sin\theta} & 0 \end{pmatrix}$$
$$= \begin{pmatrix} \sqrt{\sin\theta} & 0 \\ 0 & \dfrac{1}{\sqrt{\sin\theta}} \end{pmatrix} \begin{pmatrix} 0 & j\dfrac{1}{Y_{k,k+1}} \\ jY_{k,k+1} & 0 \end{pmatrix} \begin{pmatrix} \dfrac{1}{\sqrt{\sin\theta}} & 0 \\ 0 & \sqrt{\sin\theta} \end{pmatrix}$$

this frequency dependence can be removed from the admittance inverter. This matrix identity means that the admittance inverter $Y_{k,k+1}/\sin\theta$ can be modeled as a frequency-independent admittance inverter placed between two ideal transformers as shown in Figure 9.13.

The effect of the two transformers can be shifted to the admittances surrounding the admittance inverter. This is done by multiplying all the admittances by the factor $n^2 = \sin\theta$. Also note that the effect of the transformer at the input of the network will be neglected, since for $\lambda/4$ lines, $\theta \approx \pi/2$ and $\sqrt{\sin\theta} \approx 1$. The final even-mode equivalent circuit of the generalized interdigital filter is given in

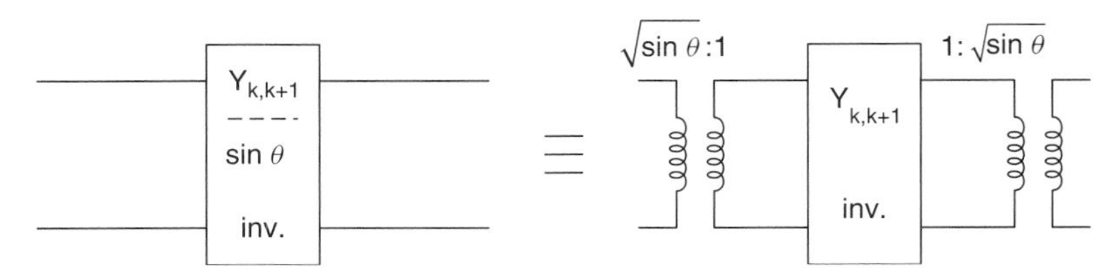

Figure 9.13 Removing the frequency dependence from the admittance inverter.

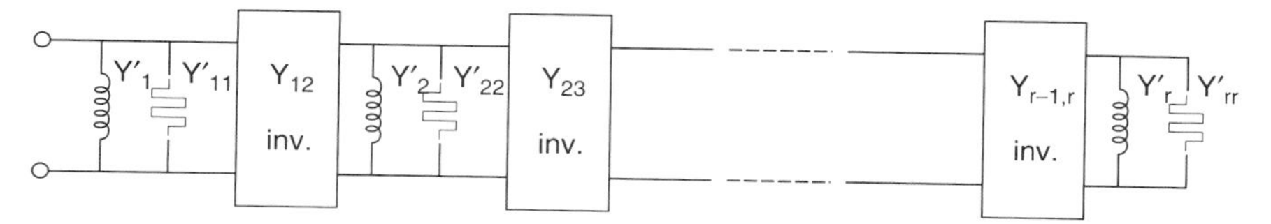

Figure 9.14 Even-mode equivalent circuit with admittance inverters.

Figure 9.14, where

$$Y_1' = (Y_1 + Y_{12}) \sin\theta \qquad Y_r' = (Y_r + Y_{r-1,r}) \sin\theta$$
$$Y_{11}' = Y_{11} \sin\theta \qquad Y_{rr}' = Y_{rr} \sin\theta$$
$$Y_k' = (Y_{k-1,k} + Y_k + Y_{k,k+1}) \sin\theta$$
$$Y_{kk}' = Y_{kk} \sin\theta$$

The admittance inverters are simply $Y_{k,k+1}$ and the total admittances are given by

$$-j(Y_1 + Y_{12}) \cot\theta \sin\theta + jY_{11} \tan\frac{\theta}{2} \sin\theta$$
$$= -j(Y_1 + Y_{11} + Y_{12}) \cos\theta + jY_{11}$$
$$-j(Y_r + Y_{r-1,r}) \cot\theta \sin\theta + jY_{rr} \tan\frac{\theta}{2} \sin\theta$$
$$= -j(Y_r + Y_{rr} + Y_{r-1,r}) \cos\theta + jY_{rr}$$
$$-j(Y_{k-1,k} + Y_k + Y_{k,k+1}) \cot\theta \sin\theta + jY_{kk} \tan\frac{\theta}{2} \sin\theta$$
$$= -j(Y_{k-1,k} + Y_k + Y_{k,k+1} + Y_{kk}) \cos\theta + jY_{kk}$$

The even-mode equivalent circuit of Figure 9.14 is then compared to the even-mode prototype circuit of Figure 9.9. By matching the elements of Figures 9.9 and 9.14, we have

$$Y_{k,k+1} = J_{k,k+1}$$
$$-j(Y_1 + Y_{11} + Y_{12}) \cos\theta + jY_{11} = j\frac{1}{\delta}\omega + jJ_{11}$$
$$-j(Y_r + Y_{rr} + Y_{r-1,r}) \cos\theta + jY_{rr} = j\frac{1}{\delta}\omega + jJ_{rr}$$
$$-j(Y_{k-1,k} + Y_k + Y_{k,k+1} + Y_{kk}) \cos\theta + jY_{kk} = j\frac{1}{\delta}\omega + jJ_{kk}$$

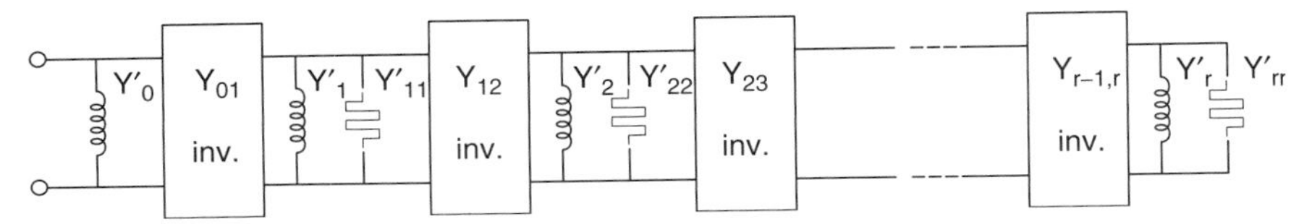

Figure 9.15 Even-mode equivalent circuit with the additional input line.

By using the frequency transformation $\omega \to -\delta \cos\theta$ and identifying according to the sign of j, these equations reduce to

$$Y_{k,k+1} = J_{k,k+1}$$

$$Y_1 + Y_{11} + Y_{12} = 1 \qquad Y_{11} = J_{11}$$

$$Y_r + Y_{rr} + Y_{r-1,r} = 1 \qquad Y_{rr} = J_{rr}$$

$$Y_{k-1,k} + Y_k + Y_{k,k+1} + Y_{kk} = 1 \qquad Y_{kk} = J_{kk}$$

Finally, we need to match the input/outputs of the two circuits. The prototype circuit starts with an ideal transformer of ratio $1 : \sqrt{\delta C_1}$, and there is nothing left in the equivalent circuit to match this. We use the technique for narrowband interdigital filters described in Chapter 7. It consists of adding two nonresonant lines, one at the input, the other at the output of the generalized interdigital filter. The even-mode equivalent circuit of the generalized interdigital filter with these additional lines is shown in Figure 9.15.

In summary:

$$Y_{k,k+1} = J_{k,k+1} = \frac{1}{\delta\sqrt{C_k C_{k+1}}} \qquad k = 1, 2, \ldots, r-1$$

$$Y_{kk} = J_{kk} = \frac{K_k}{\delta C_k} \qquad k = 1, 2, \ldots, r$$

$$Y_k = 1 - Y_{k-1,k} - Y_{k,k+1} - Y_{kk} \qquad k = 1, 2, \ldots, r-1$$

with

$$Y_{01} = \frac{1}{\sqrt{\delta C_1}}$$

$$Y_0 = 1 - \frac{1}{\sqrt{\delta C_1}}$$

$$Y_r = 1 - Y_{r-1,r} - Y_{rr}$$

9.4.2 Frequency Transformation

The prototype elements are provided for a low-pass filter with a −3-dB cutoff frequency of Ω. Through the use of the frequency transformation, it will be possible to associate this −3-dB cutoff frequency with the upper and lower −3-dB cutoff frequencies f_1 and f_2 of the target bandpass filter with center frequency f_0. First, let's express the frequency dependence of θ:

$$\theta = \beta\frac{\lambda_0}{4} = \frac{\omega}{c}\frac{\lambda_0}{4} = \frac{2\pi f}{4c}\frac{c}{f_0} = \frac{\pi}{2}\frac{f}{f_0}$$

Then the association of Ω with f_1 and f_2 is as follows:

$$-\delta\cos\left(\frac{\pi}{2}\frac{f_1}{f_0}\right) = -\Omega$$

$$-\delta\cos\left(\frac{\pi}{2}\frac{f_2}{f_0}\right) = \Omega$$

Since f_0 is taken as the center of f_1 and f_2, $f_0 = (f_1 + f_2)/2$ and

$$-\delta\cos\left(\frac{\pi}{2}\frac{2f_1}{f_1+f_2}\right) = -\delta\sin\left[\frac{\pi}{2}\left(1-\frac{2f_1}{f_1+f_2}\right)\right] = -\delta\sin\left(\frac{\pi}{2}\frac{f_2-f_1}{f_1+f_2}\right) = -\Omega$$

$$-\delta\cos\left(\frac{\pi}{2}\frac{2f_2}{f_1+f_2}\right) = -\delta\sin\left[\frac{\pi}{2}\left(1-\frac{2f_2}{f_1+f_2}\right)\right] = -\delta\sin\left(\frac{\pi}{2}\frac{f_1-f_2}{f_1+f_2}\right) = \Omega$$

which results in the same formula:

$$\delta = \Omega/\sin\left(\frac{\pi}{2}\frac{f_2-f_1}{f_1+f_2}\right)$$

9.4.3 Physical Parameters of the Interdigital Structure

The design technique proposed in the preceding sections does not provide the final physical properties of the lines, such as the width of the lines, the spacing between the lines, and the width of the slots. However, as will be shown, knowledge of $Y_{k,k+1}$, Y_k, and Y_{kk} will be sufficient for defining the physical properties. The technique relies on relating the $Y_{k,k+1}$, Y_k, and Y_{kk} with the capacitances defined in Figure 9.16, where

t is the thickness of a bar
b is the distance between the two main ground planes
W_k is the width of bar k
$S_{k,k+1}$ is the distance between bars k and $k+1$
C_{pk} is the parallel-plate capacitance of bar k
C'_f is the fringing capacitance for an isolated bar (0 or $n+1$)
$(C'_{fe})_{k,k+1}$ is the even-mode fringing capacitance for bars k and $k+1$
$(C'_{fo})_{k,k+1}$ is the odd-mode fringing capacitance for bars k and $k+1$

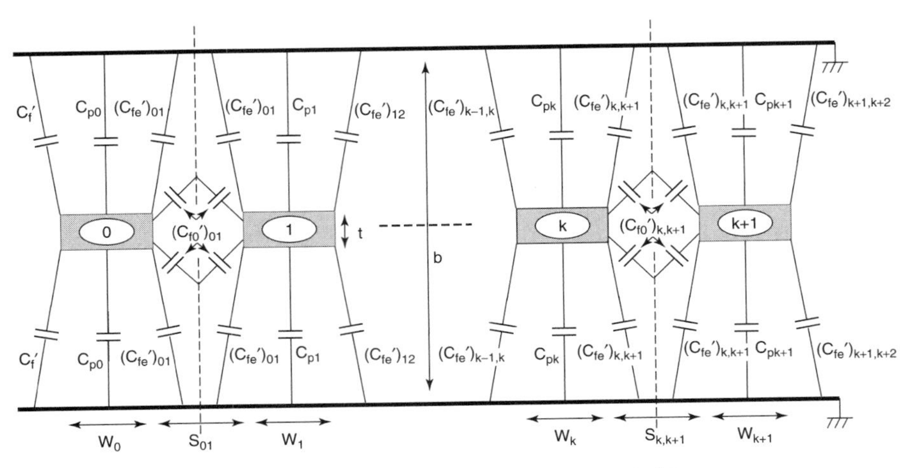

Figure 9.16 Definition of the capacitances in coupled bar structures.

Getsinger [9.8], computed the various capacitances of Figure 9.16 and provided a set of curves relating the capacitances, the distance between ground planes b, the thickness of the bars t, and the distance S between consecutive bars. One curve provides C'_{fe}/ε and $\Delta C/\varepsilon = C'_{fo}/\varepsilon - C'_{fe}/\varepsilon$ as functions of the spacing between bars S/b with the bar thickness t/b as a parameter. Another curve provides the fringing capacitance C'_f/ε of an isolated bar as a function of bar thickness t/b. In these curves, the ratios C/ε of static capacitance per unit length to the permittivity of the dielectric of the medium are related to the characteristic impedance Z_c of a lossless transmission line by

$$Z_c\sqrt{\varepsilon_r} = \frac{\eta}{C\varepsilon} \quad \text{or} \quad \frac{C}{\varepsilon} = \frac{\eta Y_c}{\sqrt{\varepsilon_r}}$$

where η is the impedance of free space per square meter ($\eta = 376.7\ \Omega/\text{m}^2$) and ε_r is the relative permittivity of the dielectric medium. These curves allow us to find the distance between bars $S_{k,k+1}$ when the $\Delta C_{k,k+1}/\varepsilon$ are given. In addition, the parallel-plate capacitance will not be directly available, and the additional curves will be used to compute its final value. Once the parallel-plate capacitance is known, the width of the bars can be computed using the relation

$$\frac{C_{pk}}{\varepsilon} = 2\frac{W_k/b}{1 - t/b} \quad \text{or} \quad \frac{W_k}{b} = \frac{1}{2}\left(1 - \frac{t}{b}\right)\frac{C_{pk}}{\varepsilon}$$

To relate the capacitances in Figure 9.16 to the interdigital structure, the capacitance model of Figure 9.17 is used. Comparing Figures 9.16 and 9.17 provides the following relations between the two sets of capacitances:

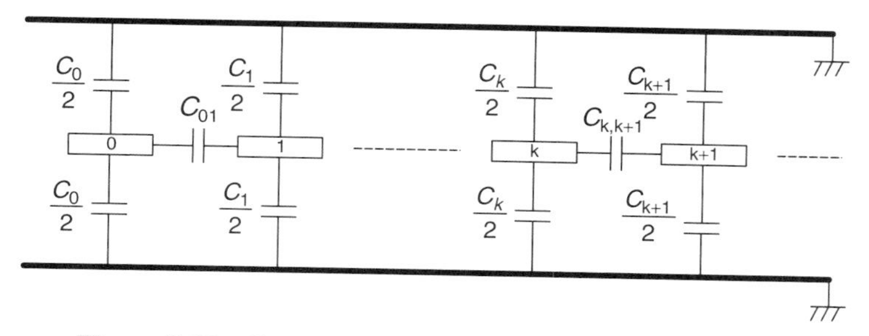

Figure 9.17 Capacitance model of the interdigital structure.

$$C_{k,k+1} = [C'_{fo} - C'_{fe}]_{k,k+1} = \Delta C_{k,k+1}$$
$$C_k = 2[C_{pk} + (C'_{fe})_{k,k+1} + (C'_{fe})_{k-1,k}]$$
$$C_0 = 2[C_{p0} + (C'_{fe})_{01} + C'_f]$$
$$C_n = 2[C_{pn} + (C'_{fe})_{n-1,n} + C'_f]$$

and

$$\frac{\Delta C_{k,k+1}}{\varepsilon} = \frac{C_{k,k+1}}{\varepsilon}$$
$$\frac{C_{pk}}{\varepsilon} = \frac{1}{2}\frac{C_k}{\varepsilon} - \frac{(C'_{fe})_{k,k+1}}{\varepsilon} - \frac{(C'_{fe})_{k-1,k}}{\varepsilon}$$
$$\frac{C_{p0}}{\varepsilon} = \frac{1}{2}\frac{C_0}{\varepsilon} - \frac{(C'_{fe})_{01}}{\varepsilon} - \frac{C'_f}{\varepsilon}$$
$$\frac{C_{pn}}{\varepsilon} = \frac{1}{2}\frac{C_n}{\varepsilon} - \frac{(C'_{fe})_{n-1,n}}{\varepsilon} - \frac{C'_f}{\varepsilon}$$

The capacitances of Figure 9.16 are computed from the $Y_{k,k+1}$ and Y_k of the equivalent circuit:

$$\frac{C_{k,k+1}}{\varepsilon} = \frac{\eta Y_c}{\sqrt{\varepsilon_r}} Y_{k,k+1} \qquad k = 1, 2, \ldots, r-1$$
$$\frac{C_k}{\varepsilon} = \frac{\eta Y_c}{\sqrt{\varepsilon_r}} Y_k \qquad k = 2, 3, \ldots, r-1$$
$$\frac{C_{01}}{\varepsilon} = \frac{\eta Y_c}{\sqrt{\varepsilon_r}} Y_{01} \qquad \frac{C_1}{\varepsilon} = \frac{\eta Y_c}{\sqrt{\varepsilon_r}} Y_1$$
$$\frac{C_0}{\varepsilon} = \frac{\eta Y_c}{\sqrt{\varepsilon_r}} Y_0 \qquad \frac{C_r}{\varepsilon} = \frac{\eta Y_c}{\sqrt{\varepsilon_r}} Y_r$$

where Y_c is the desired characteristic admittance at the input/output of the structure (e.g., $Y_C = 1/50\ \Omega$) and assuming that $Y_{k,k+1}$ and Y_k are provided for 1-Ω terminations. At this stage the various capacitance ratios are computed from the $Y_{k,k+1}$ and Y_k of the equivalent circuit. Then the $\Delta C_{k,k+1}/\varepsilon$ are used with the Getsinger curves to find the $S_{k,k+1}/b$ distances for a fixed t/b ratio. Note that the t/b ratio is decided on at the very start. Knowledge of $S_{k,k+1}/b$ then provides the capacitances $(C'_{fe})_{k,k+1}/\varepsilon$. Using only t/b, the capacitances C'_f/ε can be found using the other curve. At this stage the parallel-plate capacitances C_{pk}/ε can finally be computed, from which the widths W_k of the bars can be defined using the equation.

To design the generalized interdigital filter, additional steps are needed to account for the slot coupling between the two sets of lines. Shimizu and Jones [9.9] provide an equation relating the coupling between two lines through a slot as shown in Figure 9.18. This coupling can be expressed as

$$\frac{V_2}{V_1} = \frac{\sqrt{\varepsilon_r}\, Z_0 \pi^2}{1920\,(K\,\{\mathrm{sech}[(\pi/2)(W/b)]\})^2} \left(\frac{d}{b}\right)^2 e^{-\pi(\tau/d)}$$

where $K(m)$ is the complete elliptic integral of the first kind and $\mathrm{sech}\, x = 1/\cosh x$. Therefore, if the coupling and the width of the bar are given, one could define the value of the slot width d. Malherbe [9.10] provides additional results to define the equation for the coupling between two transverse lines k, k. These are summarized below.

$$KK = \frac{\sqrt{\varepsilon_r} Z_0 \pi^2}{1920\,(K\,\{\mathrm{sech}[(\pi/2)(W_k/b)]\})^2} \left(\frac{d_k}{b}\right)^2 e^{-\pi(\tau/d_k)}$$

$$\sqrt{\varepsilon_r} Z_0 = \frac{\eta}{\sqrt{C_k/\varepsilon(C_k/\varepsilon + 2C_{kk}/\varepsilon)}} \qquad \text{with} \quad \frac{C_{kk}}{\varepsilon} = \frac{\eta Y_c}{\sqrt{\varepsilon_r}} Y_{kk} \quad k = 1, 2, \ldots, r$$

$$KK = \frac{C_{kk}/\varepsilon}{C_k/\varepsilon + C_{kk}/\varepsilon}$$

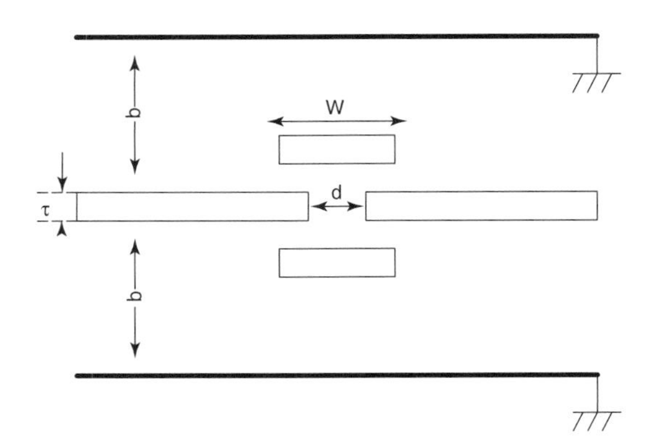

Figure 9.18 Slot coupling.

Unfortunately, introduction of the slot has the effect of reducing the parallel-plate capacitance C_{pk} and introducing an additional fringing capacitance C'_{fk}. The reduction in parallel-plate capacitance will be equal to

$$\frac{\Delta C_{pk}}{\varepsilon} = 2\frac{d_k/b}{1 - t/b}$$

and the additional fringing capacitance is given by

$$\frac{C'_{fk}}{\varepsilon} = \left.\frac{C'_{fe}}{\varepsilon}\right|_{d_k}$$

and is computed from the Getsinger C'_{fe}/ε curve for $S = d_k$.

Since the new value of parallel-plate capacitance C_{pk} is now smaller than initially, the corresponding bar width W_k will be different as well. Therefore, an iterative process must be used until the values of the various elements no longer change between iterations.

9.5 DESIGN EXAMPLE

The filter is to have a center frequency $f_0 = 2.7$ GHz with a bandwidth of 54 MHz. The amplitude and phase function is of order $N = 2r = 2 \times 3 = 6$, with the parameter $\alpha = 0.5$. The magnitude and delay of the low-pass prototype function are shown in Figure 9.19. The magnitude is equal to -3 dB at $\omega = \Omega = 2.66$ rad/s. The synthesis of the low-pass prototype leads to the following element values for the even-mode circuit of Figure 9.8:

$$\begin{aligned} C_1 &= 0.2500 \qquad & C_2 &= 0.6169 \qquad & C_3 &= 0.7880 \\ K_1 &= 0.03130 \qquad & K_2 &= 0.2261 \qquad & K_3 &= 0.7947 \end{aligned}$$

The distance between plates is chosen as $b = 10$ mm and the thickness of the lines as $t = 2$ mm, $\varepsilon_r = 1$ for the air, the termination admittance $Y_c = 1/50\ \Omega$, and $\eta = 376.7\ \Omega/\text{m}^2$.

The length of the lines is given by

$$l = \frac{\lambda_0}{4} = \frac{c/\sqrt{\varepsilon_r}}{4f_0} = \frac{3 \times 10^8/1}{4 \times 2.7 \times 10^9} = 27.76 \text{ mm}$$

The frequency transformation is such that

$$\delta = \Omega/\sin\left(\frac{\pi}{2}\frac{f_2 - f_1}{f_1 + f_2}\right) = 2.66/\sin\left(\frac{\pi}{2}\frac{54}{2 \times 2700}\right) = 169.3478$$

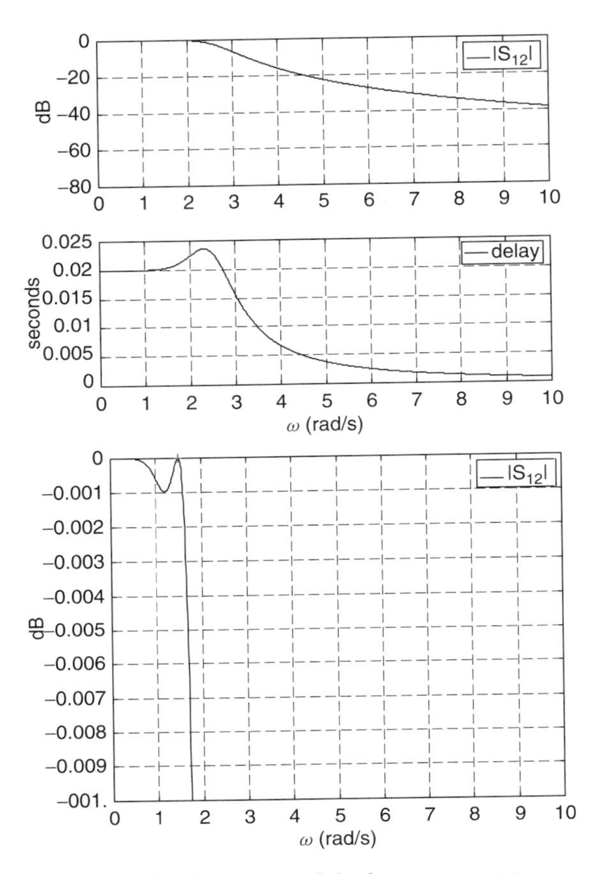

Figure 9.19 Responses of the low-pass prototype.

The admittances $Y_{k,k+1}$, Y_{kk}, and Y_k are then computed using

$$Y_{k,k+1} = J_{k,k+1} = \frac{1}{\delta\sqrt{C_k C_{k+1}}} \qquad k = 1, 2, \ldots, r-1$$

$$Y_{kk} = J_{kk} = \frac{K_k}{\delta C_k} \qquad k = 1, 2, \ldots, r$$

$$Y_k = 1 - Y_{k-1,k} - Y_{k,k+1} - Y_{kk} \qquad k = 1, 2, \ldots r-1$$

with

$$Y_{01} = \frac{1}{\sqrt{\delta C_1}}$$

$$Y_0 = 1 - \frac{1}{\sqrt{\delta C_1}}$$

$$Y_r = 1 - Y_{r-1,r} - Y_{rr}$$

$$\begin{array}{llll} Y_0 = 0.8463 & Y_1 = 0.7005 & Y_2 = 0.9743 & Y_3 = 0.9856 \\ Y_{01} = 0.1537 & Y_{11} = 0.0007 & Y_{12} = 0.0150 & \\ Y_{22} = 0.0022 & Y_{23} = 0.0085 & Y_{33} = 0.060 & \end{array}$$

The capacitance ratios are then computed using

$$\frac{C}{\varepsilon} = \frac{\eta Y_c}{\sqrt{\varepsilon_r}} Y$$

$$\begin{array}{llll} C_0/\varepsilon = 6.3761 & C_1/\varepsilon = 5.2273 & C_2/\varepsilon = 7.3406 & C_3/\varepsilon = 7.4253 \\ C_{01}/\varepsilon = 1.1579 & C_{11}/\varepsilon = 0.0056 & C_{12}/\varepsilon = 0.1133 & \\ C_{22}/\varepsilon = 0.0163 & C_{23}/\varepsilon = 0.0638 & C_{33}/\varepsilon = 0.0449 & \end{array}$$

Since $\Delta C_{k,k+1}/\varepsilon = C_{k,k+1}/\varepsilon$, using the Getsinger plots of $\Delta C/\varepsilon$ versus S/b for $t/b = 2$ mm/10 mm $= 0.2$, one can find the spacing between lines $S_{k,k+1}$. The values are summarized as follows:

$$\begin{array}{lll} C_{01}/\varepsilon = 1.1579 & C_{12}/\varepsilon = 0.1133 & C_{23}/\varepsilon = 0.0638 \\ S_{01}/b = 0.28 & S_{12}/b = 0.99 & S_{23}/b = 1.16 \end{array}$$

At this stage the parallel-plate capacitances are computed using

$$\frac{C_{pk}}{\varepsilon} = \frac{1}{2}\frac{C_k}{\varepsilon} - \frac{(C'_{fe})_{k,k+1}}{\varepsilon} - \frac{(C'_{fe})_{k-1,k}}{\varepsilon} \quad \text{and} \quad \begin{aligned} \frac{C_{p0}}{\varepsilon} &= \frac{1}{2}\frac{C_0}{\varepsilon} - \frac{(C'_{fe})_{01}}{\varepsilon} - \frac{C'_f}{\varepsilon} \\ \frac{C_{pr}}{\varepsilon} &= \frac{1}{2}\frac{C_r}{\varepsilon} - \frac{(C'_{fe})_{r-1,r}}{\varepsilon} - \frac{C'_f}{\varepsilon} \end{aligned}$$

The $(C'_{fe})_{k,k+1}/\varepsilon$ are computed using the Getsinger plots of C'_{fe}/ε versus S/b for $t/b = 0.2$:

$$\begin{array}{lll} S_{01}/b = 0.28 & S_{12}/b = 0.99 & S_{23}/b = 0.0638 \\ \left(C'_{fe}\right)_{01}/\varepsilon = 0.31 & \left(C'_{fe}\right)_{12}/\varepsilon = 0.63 & \left(C'_{fe}\right)_{23}/\varepsilon = 0.65 \end{array}$$

and the fringing capacitance for an isolated bar is provided by the plot of C'_f versus $t/b = 0.2$. In this case,

$$\frac{C'_f}{\varepsilon} = 0.69$$

Then the width of the lines is given by

$$\frac{W_k}{b} = \frac{1}{2}\left(1 - \frac{t}{b}\right)\frac{C_{pk}}{\varepsilon}$$

The values of the parallel-plate capacitances and the line widths are summarized as follows:

$$C_{p0}/\varepsilon = 2.188 \qquad C_{p1}/\varepsilon = 1.6736 \qquad C_{p2}/\varepsilon = 2.3903 \qquad C_{p3}/\varepsilon = 2.3727$$
$$W_0 = 0.8752 \qquad W_1 = 0.6795 \qquad W_2 = 0.9561 \qquad W_3 = 0.9491$$

The width of the slots is then computed using

$$KK = \frac{\sqrt{\varepsilon_r} Z_0 \pi^2}{1920\,(K\,\{\text{sech}[(\pi/2)(W_k/b)]\})^2} \left(\frac{d_k}{b}\right)^2 e^{-\pi(\tau/d_k)}$$

$$\sqrt{\varepsilon_r} Z_0 = \frac{\eta}{\sqrt{C_k/\varepsilon(C_k/\varepsilon + 2C_{kk}/\varepsilon)}} \quad \text{with} \quad \frac{C_{kk}}{\varepsilon} = \frac{\eta Y_c}{\sqrt{\varepsilon_r}} Y_{kk} \qquad k = 1, 2, \ldots, r$$

$$KK = \frac{C_{kk}/\varepsilon}{C_k/\varepsilon + C_{kk}/\varepsilon}$$

This is solved by plotting the function for several values of d and checking where it intersects with the desired coupling factor KK defined from C_k/ε and C_{kk}/ε. The procedure starts with using the initial value of W_k. The results at the first iteration are provided below.

$$W_1 = 6.795 \text{ mm} \qquad W_2 = 9.561 \text{ mm} \qquad W_3 = 9.491 \text{ mm}$$
$$KK_1 = 0.0011 \qquad KK_2 = 0.0022 \qquad KK_3 = 0.060$$
$$d_1 = 2.85 \text{ mm} \qquad d_2 = 3.61 \text{ mm} \qquad d_3 = 4.82 \text{ mm}$$

The procedure is then to reduce or augment the various capacitances by the factors presented in the preceding section. This leads to new values of W_k, and the d_k are recomputed. This is repeated until the values no longer change between iterations. In the end, the following values were found:

$$W_0 = 8.752 \text{ mm} \qquad W_1 = 6.877 \text{ mm} \qquad W_2 = 9.838 \text{ mm} \qquad W_3 = 1.004 \text{ mm}$$
$$d_1 = 2.86 \text{ mm} \qquad d_2 = 3.61 \text{ mm} \qquad d_3 = 4.80 \text{ mm}$$

One can see that the process converges very rapidly.

9.6 REALIZATIONS AND MEASURED PERFORMANCE

In constructing the generalized filter of the example described in Section 9.5, it was found that to avoid capacitive effects, it was necessary to reduce the $\lambda/4$ lengths. The reduced lengths were such that $L = l - \Delta l$. Laboratory experiments show that $\Delta l = 5.26$ mm provided good results. The center frequency measured was at first shifted from the center frequency expected. It was, however, corrected easily using adjusting screws. The magnitude and delay expected and measured

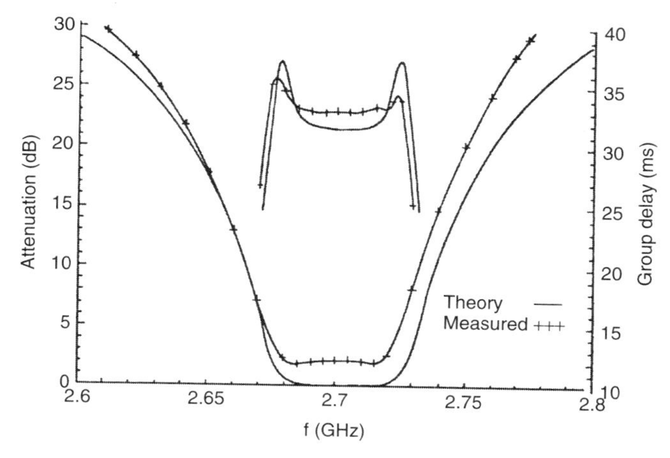

Figure 9.20 Theoretical and experimental responses.

TABLE 9.2. Filter Performance Expected and Measured

$N = 6 \alpha = 0.5$[a]	Theory	Measured
f_0 (GHz)	2.70	2.70
BW (MHz)	54	50
τ_0 (ns)	31.36	32.94
$\Delta\tau$ (ns)	3.7	2
VSWR	1.03	1.15
Losses (dB)	0	2

[a] τ_0 is the group delay at the center frequency and $\Delta\tau$ is the variation of the group delay.

are shown in Figure 9.20, and a summary of the filter performance is given in Table 9.2.

It can be seen that the experimental group delay in the passband is slightly greater than its theoretical value (32.94 ns instead of 31.36 ns). This is due to the additional electrical length introduced by the input and output lines. This difference has little significance since only phase linearity variation is important. It is interesting to note that the magnitude measured shows better selectivity at high frequencies. The filter realized is shown in Figure 9.21. Figure 9.22 shows the two third-order interdigital structures and the coupling plane, and Figure 9.23 provides a close-up of one of the interdigital structures.

9.7 CONCLUSIONS

Filters with simultaneous conditions on the amplitude and phase require non-minimum-phase structures. The design technique is provided for non-minimum-phase

Figure 9.21 Sixth-order generalized interdigital filter.

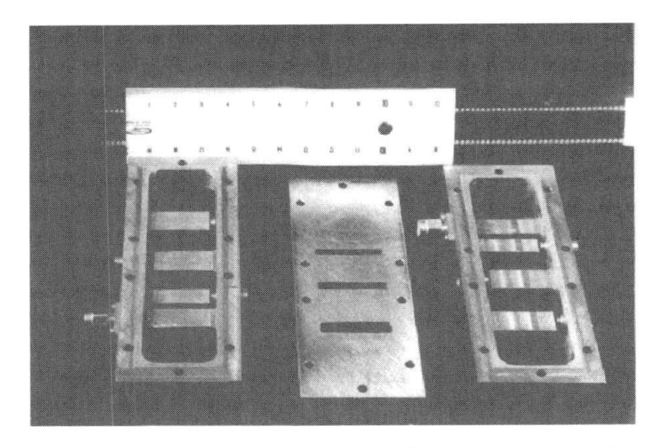

Figure 9.22 Coupling plane and the two third-order interdigital structures.

Figure 9.23 Close-up of one of the interdigital structures.

symmetrical responses using a generalized interdigital structure. Asymmetrical responses with simultaneous conditions on the amplitude and phase are more difficult to design. The techniques of this chapter for linear-phase filters can be extended by finding polynomials $H_n^m(s)$ which give a group delay of the form

$$T(s) = \frac{1}{2}\left[\frac{H_n^{m\prime}(s)}{H_n^m(s)} + \frac{H_n^{m\prime}(-s)}{H_n^m(-s)}\right] = 1 + K\frac{s^{2(n-m)}(s^2+\omega_0^2)^m}{H_n^m(s)H_n^m(-s)}$$

Such a filter will provide a delay with $n-m$ conditions at the origin $s=0$ and m conditions around ω_0. This would give a filter with a linear phase in two frequency bands that would be very interesting for multiplexers applications [9.11].

Acknowledgments

This work was the result of an exchange with the government of Brazil.

REFERENCES

[9.1] J. D. Rhodes, *Theory of Electrical Filters*, Wiley, New York, 1976.

[9.2] J. D. Rhodes, The theory of generalized interdigital linear phase networks, *IEEE Trans. Circuit Theory*, vol. 16, no. 3, pp. 280–288, 1969.

[9.3] P. Jarry and H. Abdalla, Jr., Arbitrary phase and group delay polynomials, presented at the IEEE International Symposium on Circuits and Systems, ISCAS'82, Rome, May 10–12, 1982.

[9.4] H. Abdalla, Jr. and P. Jarry, Linear phase odd degree filters with maximum flatness amplitude characteristics, *Proc. European Conference on Circuit Theory and Design*, ECCTD'85, pp. 686–688, Sept. 1985.

[9.5] J. D. Rhodes, A low-pass prototype network for microwave linear phase filters, *IEEE Trans. Microwave Theory Tech.*, vol. 18, no. 6, pp. 290–301, 1970.

[9.6] H. Abdalla, Jr. and P. Jarry, Microwave linear phase filters, *Proc. 14th European Microwave Conference*, EMC'84, Leuven, Belgium, pp. 358–363, Sept. 1984.

[9.7] H. Abdalla Jr. and P. Jarry, *Transfer functions at non minimum phase with one finite transmission zero*, *Proc. 4th Brazilian Telecommunications Symposium*, pp. 54–57, University of Rio de Ganeiro, Sept. 3–5, 1986.

[9.8] W. J. Getsinger, Coupled rectangular bars between parallel plates, *IRE Trans. Microwave Theory Tech.*, vol. 10, no. 1, pp. 65–72, 1962.

[9.9] J. K. Shimizu and E. M. T. Jones, Coupled-transmission line directional couplers, *IRE Trans. Microwave Theory Tech.*, vol. 6, pp. 403–406, Oct. 1958.

[9.10] J. A. Malherbe, Microwave Transmission Line Filters, Artech House, Dedham, MA, 1976.

[9.11] H. Abdalla, Jr. and P. Jarry, Linear phase filters with two bands, *Proc. 1st Brazilian Telecommunications Symposium*, University of Rio de Ganeiro, pp. 51–56, Sept. 5–9, 1983.

Selected Bibliography

Rhodes, J. D., The generalized direct-coupled cavity linear phase filter, *IEEE Trans. Microwave Theory Tech.*, vol. 18, pp. 308–313, Sept. 1970.

Atia, A. E., and A. E. Williams, Non-minimum-phase optimum-amplitude bandpass waveguide filters, *IEEE Trans. Microwave Theory Tech.*, vol. 22, pp. 425–431, Apr. 1974.

Levy, R., Linear phase filters and the demise of external equalization, *Microwave Sys. News*, vol. 6, pp. 65–74, Aug.–Sept. 1976.

10

TEMPERATURE-STABLE NARROWBAND MONOMODE TE_{011} LINEAR-PHASE FILTERS

10.1 INTRODUCTION

In this chapter we provide a design method for monomode TE_{011} cylindrical cavity filters that have an equiripple amplitude response in the passband and improved phase linearity. The technique is based on a non-minimum-phase function and a low-pass cross-coupled prototype circuit where the additional degrees

Advanced Design Techniques and Realizations of Microwave and RF Filters,
By Pierre Jarry and Jacques Beneat

of freedom have been used to improve the phase linearity. The design method consists of matching the coupling graphs and coupling matrices of prototype circuit and TE_{011} filter. Practical techniques for canceling the effects of other modes in the cavities are also provided. The method is presented for a sixth-order filter but can be extended to other orders. A first realization of the sixth-order TE_{011} filter using copper provides the desired magnitude and group delay responses but suffers from significant response variations when temperature is varied. For satellite applications, a TE_{011} filter made with Invar and a thin layer of copper shows that the responses can remain constant even when the temperature is varied from -10 to $+50^{\circ}$C.

10.2 TE_{011} FILTERS

TE_{011} filters of degree 5 and 6 are shown in Figure 10.1. They consist of cylindrical TE_{011} resonant mode cavities that are coupled through small rectangular apertures on the side of the cylinders. The design of these filters will involve the use of graphs. The graphs representing these filters are shown in Figure 10.2.

10.3 LOW-PASS PROTOTYPE

10.3.1 Amplitude

A sixth-order generalized Chebyshev function is given by [10.1]

$$|S_{12}(j\omega)|^2 = \frac{1}{1 + \varepsilon^2 R_6^2(\omega)}$$

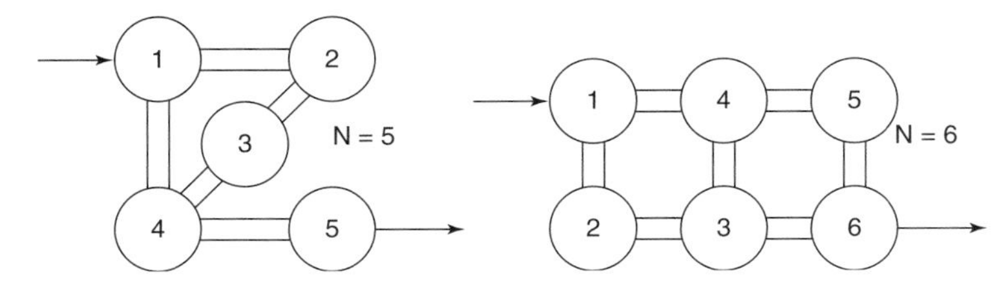

Figure 10.1 Top view of TE_{011} filters.

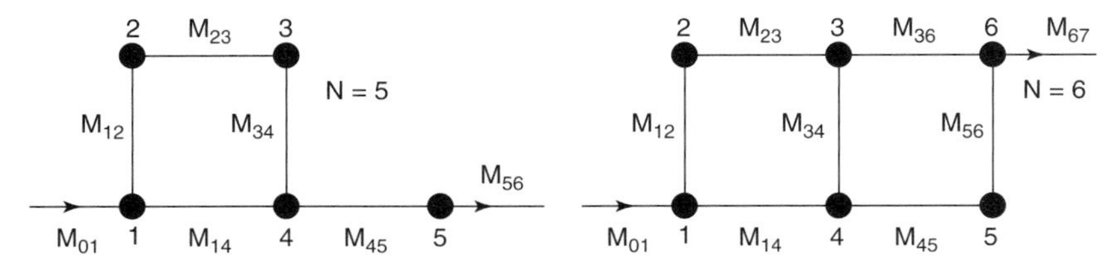

Figure 10.2 Coupling graphs.

where ε provides the ripple in the passband and $R_6(\omega)$ is given by

$$R_6(\omega) = \frac{1}{2}\left[\frac{F_3(\omega)}{F_3(-\omega)} + \frac{F_3(-\omega)}{F_3(\omega)}\right] = \frac{1}{2}\frac{F_3^2(\omega) + F_3^2(-\omega)}{F_3(\omega)F_3(-\omega)}$$

where

$$F_3(\omega) = \left(a_2\omega + \sqrt{\omega^2 - 1}\right)\left(\omega_f\sqrt{\omega^2 - 1} + \omega\sqrt{\omega_f^2 - 1}\right)\left(\omega + \sqrt{\omega^2 - 1}\right)$$

In this function, ω_f provides a transmission zero, and the parameter a_2 can be used to shape the group delay. The transfer function $S_{12}(s)$ then becomes

$$|S_{12}(j\omega)|^2 = \frac{[F_3(\omega)F_3(-\omega)]^2}{[F_3(\omega)F_3(-\omega)]^2 + (\varepsilon^2/4)[F_3^2(\omega) + F_3^2(-\omega)]^2}$$

The transfer function $S_{12}(s)$ will be given by

$$|S_{12}(s)|^2 = A\frac{F_3(s/j)F_3(-s/j)}{\prod_{k=1}^{6}(s + p_k)}$$

where the poles $p_k = \sigma_k + j\omega_k$ are given by computing the roots of the denominator of $|S_{12}(j\omega)|^2$ and selecting those that are in the left half-plane. Note that the term $\sqrt{\omega^2 - 1}$ will appear in powers of 2 in the denominator of $|S_{12}(j\omega)|^2$, so the roots can be found using the polynomial root-finding function `roots` in MATLAB. Similar observations are also found regarding the $\sqrt{\omega^2 - 1}$ terms in the numerator.

10.3.2 Delay

The group delay of the transfer function depends only on the poles and is expressed as

$$T_g(\omega) = \sum_{k=1}^{6}\frac{\sigma_k}{\sigma_k^2 + (\omega + \omega_k)^2}$$

An example of this transfer function for a −20-dB return loss is shown in Figure 10.3 for $a_2 = 2$ and $\omega_f = 2$ rad/s. The parameter a_2 can be used to improve the group delay [10.2]. However, there is no analytical method to define the optimum value of a_2. This can be done using an optimization method such as the Newton–Raphson technique. Table 10.1 provides the value of a_2 for different values of ε and where the optimization is run on 50% or 80% of the passband of the filter. The transmission zero was pushed toward infinity for better group delay results ($\omega_f \rightarrow \infty$).

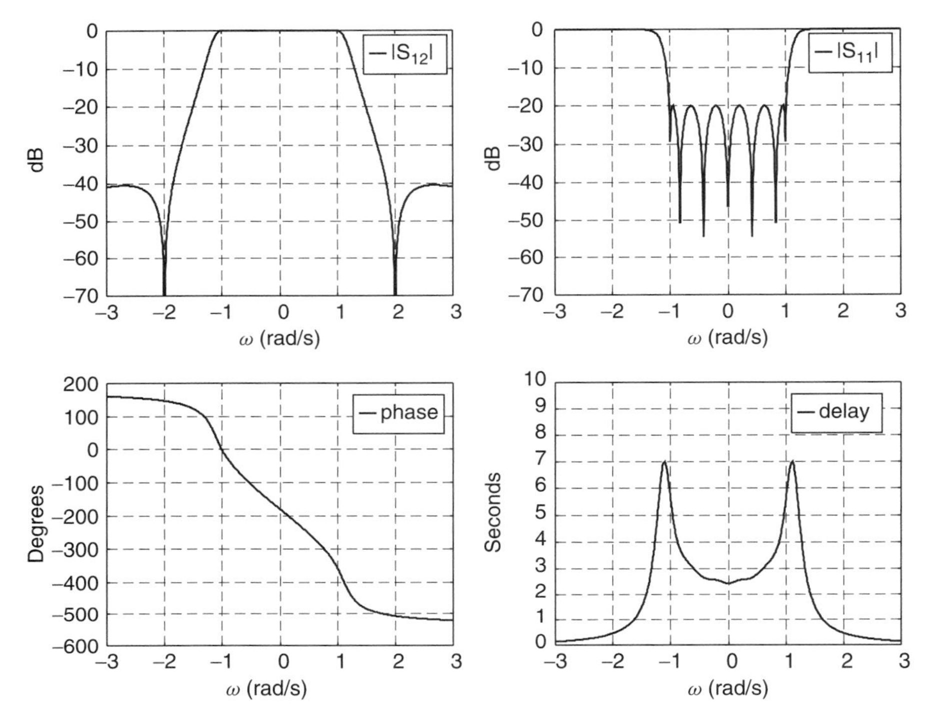

Figure 10.3 Generalized Chebyshev response ($a_2 = 2$ and $\omega_f = 2$ rad/s).

TABLE 10.1. Values of the Parameter a_2 for Optimized Group Delay

ε (dB)	50% of Passband	80% of Passband
0.001	1.2018	1.2473
0.004	1.2456	1.3161
0.01	1.2786	1.3722
0.04	1.3356	1.4771
0.1	1.3806	1.5614

10.3.3 Synthesis of the Low-Pass Prototype

The transfer function can be realized using the sixth-order non-minimum-phase low-pass prototype circuit using admittance inverters as shown in Figure 10.4. Compared to minimum-phase prototype circuits, there are two additional non-minimum couplings through K_{16} and K_{25}.

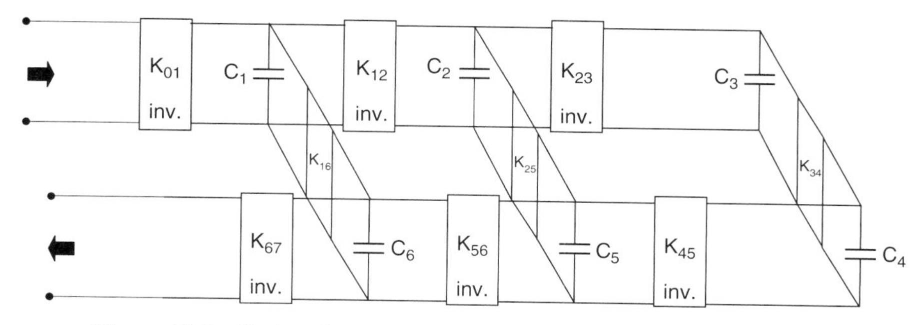

Figure 10.4 Sixth-order non-minimum-phase low-pass prototype circuit.

The system is kept symmetrical, so that

$$\begin{array}{l} C_1 = C_6 \\ C_2 = C_5 \\ C_3 = C_4 \end{array} \quad \text{and} \quad \begin{array}{ll} K_{01} & = K_{12} = K_{23} = K_{45} = K_{56} = K_{67} = 1 \\ K_{34} & \neq 1,\ K_{16} \neq 1,\ K_{25} \neq 1 \end{array}$$

It can be shown that the values of the capacitors and remaining inverters are given by

$$\begin{aligned} C_1 &= \frac{1}{d_2} \\ C_2 &= \frac{d_2}{\chi} \\ C_3 &= \frac{\chi}{d_0 + n_0 K_{25} - d_0 K_{16} K_{25}} \end{aligned} \quad \text{and} \quad \begin{aligned} K_{16} &= \frac{n_2 - C_1 d_1}{d_2} \\ K_{25} &= \frac{d_1 - C_2 n_0 + C_2 d_0 K_{16}}{\chi} \\ K_{34} &= \frac{n_0 - d_0 K_{16}}{d_0 + n_0 K_{25} - d_0 K_{16} K_{25}} \end{aligned}$$

where $\chi = n_1 - d_0 C_1 + d_1 K_{16}$. The various quantities are computed from the poles of the transfer function sorted by $\sigma_1 \leq \sigma_2 \leq \sigma_3$ and $|\omega_1| \geq |\omega_2| \geq |\omega_3|$.

$$\begin{aligned} n_0 &= \sigma_1\sigma_2\omega_3 - \sigma_1\omega_2\sigma_3 \\ &\quad + \omega_1\sigma_2\sigma_3 + \omega_1\omega_2\omega_3 \\ n_1 &= \sigma_1\sigma_2 + \sigma_2\sigma_3 + \sigma_1\sigma_3 \\ &\quad + \omega_1\omega_2 + \omega_2\omega_3 - \omega_1\omega_3 \\ n_2 &= \omega_1 - \omega_2 + \omega_3 \end{aligned} \qquad \begin{aligned} d_0 &= \sigma_1\sigma_2\sigma_3 + \sigma_1\omega_2\omega_3 - \omega_1\sigma_2\omega_3 \\ &\quad + \omega_1\omega_2\sigma_3 \\ d_1 &= \omega_1(\sigma_2 + \sigma_3) - \omega_2(\sigma_1 + \sigma_3) \\ &\quad + \omega_3(\sigma_1 + \sigma_2) \\ d_2 &= \sigma_1 + \sigma_2 + \sigma_3 \end{aligned}$$

Due to the symmetry, the Barlett bisection theorem can be used and the prototype circuit can be defined entirely by its even-mode circuit given in Figure 10.5. The coupling graph of the prototype circuit of Figure 10.4 is given in Figure 10.6,

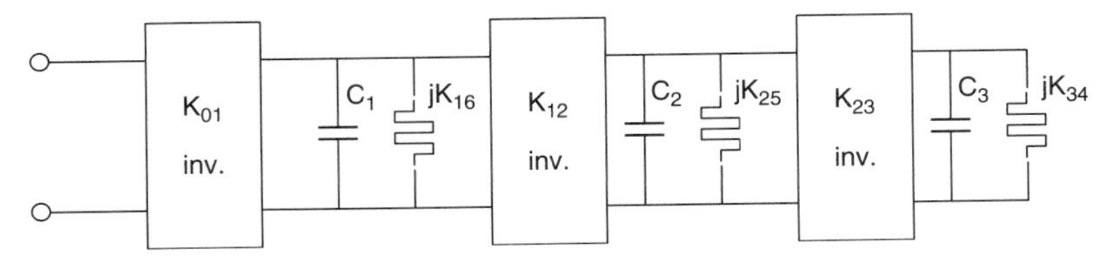

Figure 10.5 Sixth-degree even-mode prototype circuit.

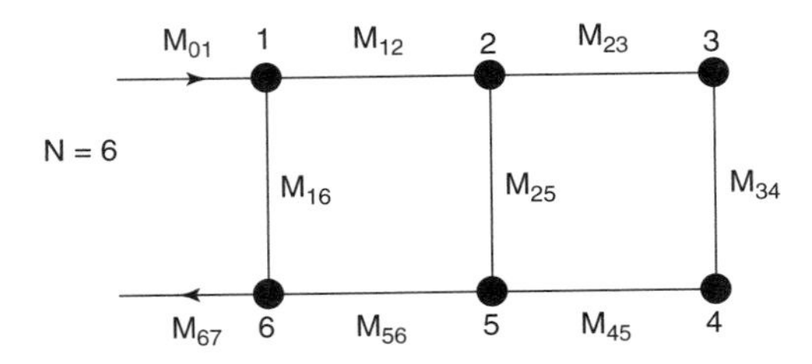

Figure 10.6 Coupling graph of the prototype circuit.

where

$$M_{01} = M_{67} = \frac{K_{01}}{\sqrt{C_1}} \qquad M_{34} = \frac{K_{34}}{C_3}$$

$$M_{12} = M_{56} = \frac{K_{12}}{\sqrt{C_1 C_2}} \quad \text{and} \quad M_{16} = \frac{K_{16}}{C_1}$$

$$M_{23} = M_{45} = \frac{K_{23}}{\sqrt{C_2 C_3}} \qquad M_{25} = \frac{K_{25}}{C_2}$$

10.4 DESIGN METHOD

10.4.1 Matching the Coupling

From Figure 10.5 we can define the coupling values M for the coupling graph of the prototype circuit [10.3]. Unfortunately, the coupling graph of the TE$_{011}$ filter has a different form from the coupling graph of the prototype circuit, as shown in Figure 10.7. This difference is also seen in the coupling matrices of

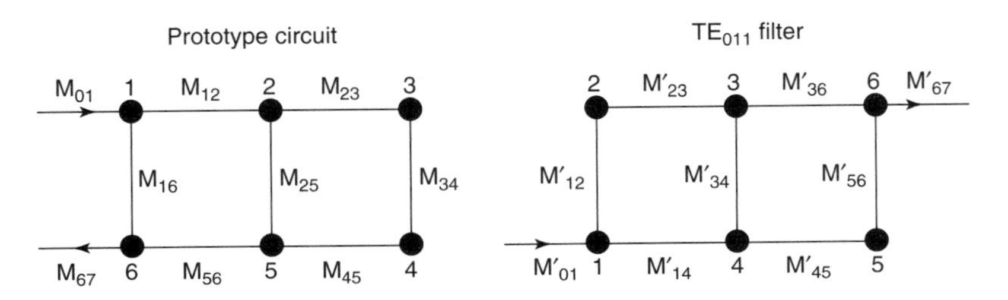

Figure 10.7 Two different coupling graphs.

the following two configurations:

$$[M] = \begin{pmatrix} 0 & M_{01} & 0 & 0 & 0 & 0 & 0 & 0 \\ M_{01} & 0 & M_{12} & 0 & 0 & 0 & M_{16} & 0 \\ 0 & M_{12} & 0 & M_{23} & 0 & M_{25} & 0 & 0 \\ 0 & 0 & M_{23} & 0 & M_{34} & 0 & 0 & 0 \\ 0 & 0 & 0 & M_{34} & 0 & M_{45} & 0 & 0 \\ 0 & 0 & M_{25} & 0 & M_{45} & 0 & M_{56} & 0 \\ 0 & M_{16} & 0 & 0 & 0 & M_{56} & 0 & M_{67} \\ 0 & 0 & 0 & 0 & 0 & 0 & M_{67} & 0 \end{pmatrix}$$

$$[M'] = \begin{pmatrix} 0 & M'_{01} & 0 & 0 & 0 & 0 & 0 & 0 \\ M'_{01} & 0 & M'_{12} & 0 & M'_{14} & 0 & 0 & 0 \\ 0 & M'_{12} & 0 & M'_{23} & 0 & 0 & 0 & 0 \\ 0 & 0 & M'_{23} & 0 & M'_{34} & 0 & M'_{36} & 0 \\ 0 & M'_{14} & 0 & M'_{34} & 0 & M'_{45} & 0 & 0 \\ 0 & 0 & 0 & 0 & M'_{45} & 0 & M'_{56} & 0 \\ 0 & 0 & 0 & M'_{36} & 0 & M'_{56} & 0 & M'_{67} \\ 0 & 0 & 0 & 0 & 0 & 0 & M'_{67} & 0 \end{pmatrix}$$

The main difference occurs for M_{16}, M_{25}, and M'_{14}, M'_{36}. To find suitable values for all M' couplings, one can use a method based on a rotation matrix R such that

$$[M'] = [R]^{\mathrm{T}}[M][R]$$

The matrix R should be such that it preserves the properties of the initial matrix M. One such matrix is given in terms of a pivot $[x, y]$ and a rotation angle θ by

$$\begin{aligned} R_{kk} &= 1 & k \neq x, k \neq y \\ R_{xx} &= R_{yy} = C = \cos\theta & \\ R_{xy} &= -R_{yx} = S = \sin\theta & \\ R_{jk} &= 0 & j \neq x, k \neq y \end{aligned} \qquad [R] = \begin{bmatrix} 1 & 0 & 0 & 0 & 0 & 0 \\ 0 & C & 0 & 0 & S & 0 \\ 0 & 0 & 1 & 0 & 0 & 0 \\ 0 & 0 & 0 & 1 & 0 & 0 \\ 0 & -S & 0 & 0 & C & 0 \\ 0 & 0 & 0 & 0 & 0 & 1 \end{bmatrix}$$

In the example provided, the pivot was chosen as [2,5]. It should be noted that one rotation is not enough to cause the two coupling matrices to be identical. The procedure should then be repeated using several consecutives but different rotations. This can be tedious, in particular finding the correct angle for each rotation. In the case of symmetrical networks, one could apply the rotation technique on the even-mode coupling matrices. In this case the matrices reduce to 3×3 matrices, and only one rotation will be necessary.

For the even-mode representations, the two networks should have a given symmetry. For the prototype circuit, the symmetry is clear from Section 10.4.

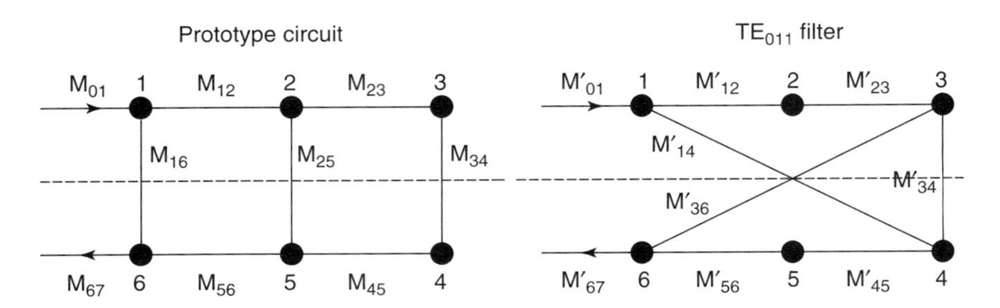

Figure 10.8 Symmetry for even-mode networks.

For the TE_{011} filter [10.4] the network is first redrawn to be more comparable to the prototype; then the symmetry chosen is indicated in Figure 10.8. From Figure 10.8, the two even-mode coupling matrices M_e and M'_e can be defined such that

$$[M_e] = \begin{bmatrix} M_{e11} & M_{e12} & 0 \\ M_{e12} & M_{e22} & M_{e23} \\ 0 & M_{e23} & M_{e33} \end{bmatrix} \quad \text{and} \quad [M'_e] = \begin{bmatrix} 0 & M'_{e12} & M'_{e13} \\ M'_{e12} & 0 & M'_{e23} \\ M'_{e13} & M'_{e23} & M'_{e33} \end{bmatrix}$$

where

$$\begin{aligned} M_{e11} &= M_{16} \\ M_{e12} &= M_{12} \\ M_{e22} &= M_{25} \\ M_{e23} &= M_{23} \\ M_{e33} &= M_{34} \end{aligned} \quad \text{and} \quad \begin{aligned} M'_{e12} &= M'_{12} \\ M'_{e13} &= M'_{14} \\ M'_{e23} &= M'_{23} \\ M'_{e33} &= M'_{34} \end{aligned}$$

Note that the TE_{011} filter will not be able to produce the coupling $M_{e11} = M_{16}$, so that in the design, one should have prototypes where $M_{16} = 0$. In this case there is only one rotation matrix (pivot = [2,3]):

$$[M'_e] = [R]^{\mathrm{T}}[M_e][R] \quad \text{with} \quad [R] = \begin{bmatrix} 1 & 0 & 0 \\ 0 & C & S \\ 0 & -S & C \end{bmatrix}$$

Analytically, it can be shown that the resulting element M'_e [2,2] is given by

$$M'_e\ [2,3] = C^2 M_{e22} - 2SC M_{e23} + S^2 M_{e33}$$

Since this element must be equal to zero, it can be used to define the angle of rotation θ:

$$M_{e33}(\tan\theta)^2 - 2M_{e23}\tan\theta + M_{e22} = 0$$

10.4.2 Selecting the Cavities

In cylindrical cavities, the resonating frequency of a $TE_{n,m,l}$ resonant mode is given by

$$f_{n,m,l} = \frac{c}{2\pi}\sqrt{\left(\frac{x'_{n,m}}{a}\right)^2 + \left(\frac{l\pi}{d}\right)^2}$$

and the resonating frequency of a $TM_{n,m,l}$ resonant mode is given by

$$f_{n,m,l} = \frac{c}{2\pi}\sqrt{\left(\frac{x_{n,m}}{a}\right)^2 + \left(\frac{l\pi}{d}\right)^2}$$

where $x_{n,m}$ is the mth root of the nth Bessel function $[J_n(x) = 0]$ and $x'_{n,m}$ is the mth root of the derivative of the nth Bessel function $[J'_n(x) = 0]$.

The unloaded Q factor for a mode $TE_{n,m,l}$ is given by

$$Q_0 = \frac{1}{2\pi}\frac{\lambda_0}{\delta_S}\frac{\left[1 - (n/x'_{n,m})^2\right]\sqrt{(x'_{n,m})^2 + (l\pi a/d)^2}}{\left[(x'_{n,m})^2 + (2a/d)(l\pi a/d)^2 + (1 - 2a/d)(nl\pi a/x'_{n,m}d)^2\right]}$$

where

$$\delta_S = \sqrt{\frac{2}{\mu_0\mu_s\sigma_s\omega}}$$

is the penetration depth and σ_s and μ_s are the conductivity and permittivity of the metal. The lowest frequency of resonance occurs for the TE_{111} mode, but the TE_{011} mode provides higher Q factors. The TE_{011} mode gives very low loss but must be used in a single-mode fashion. This leads to rather bulky filters compared to dual-mode structures [10.4].

It is also important to check the distribution of the different modes in the cavity. For example, when comparing the distribution of the modes with the Q factor, the ratio $D/L = 1.39$ between diameter and length of the cavity provides a good compromise. To define the best means of coupling, the electric and magnetic fields for the TE_{011} mode are shown in Figure 10.9. For the best coupling efficiency, one should place the couplings at the maximums of the electric field (electric coupling by probe or antenna) or at the maximums of the magnetic field (magnetic coupling by iris). From Figure 10.9 we decide to place the input/output probes and the iris at the half length and height of the cavity.

In addition, the TE_{011} is degenerate with the TM_{111} mode, and special means are necessary to cancel the effects of this mode. The introduction of a plunger that does not make contact with the walls of the cavity degrades the Q factor of the TM_{111} mode and lowers its resonant frequency. There are two methods to further reduce the resonant frequency of the TM_{111} mode: placing a thin dielectric layer on the plunger or placing a set of plugs on the plunger. The second method is illustrated in Figure 10.10.

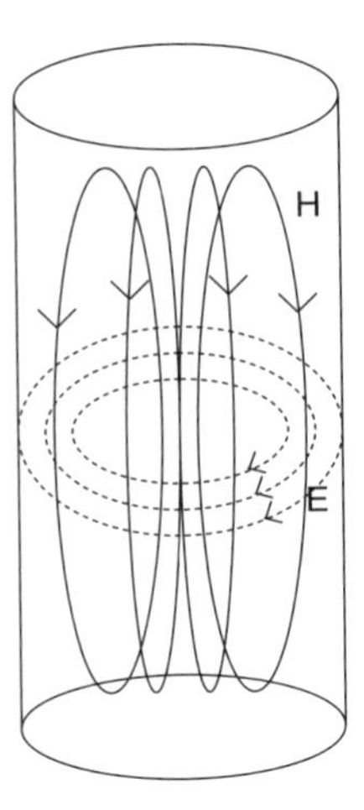

Figure 10.9 Electric and magnetic waves for the TE_{011} resonant mode.

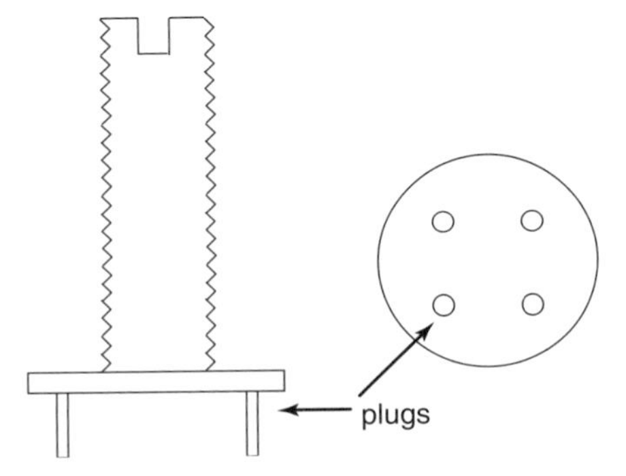

Figure 10.10 TE_{011} tuning plunger with four plugs to eliminate theTM_{111} mode.

10.4.3 Defining the Coupling

The coupling of the cavities are made through a rectangular iris as shown in Figure 10.11. The static magnetic polarizability of the aperture for zero thickness is given [10.5] by

$$P = 0.061l^3 + 0.197hl^2$$

The magnetic polarizability in the case of a thickness t is given by

$$P' = \frac{P}{1 - (2l/\lambda)^2} \times 10^{-\left(1.346\, t/l \sqrt{1-(2l/\lambda)^2}\right)}$$

For the TE_{011} mode, the denormalized coupling coefficients are given in terms of the magnetic polarizability above by

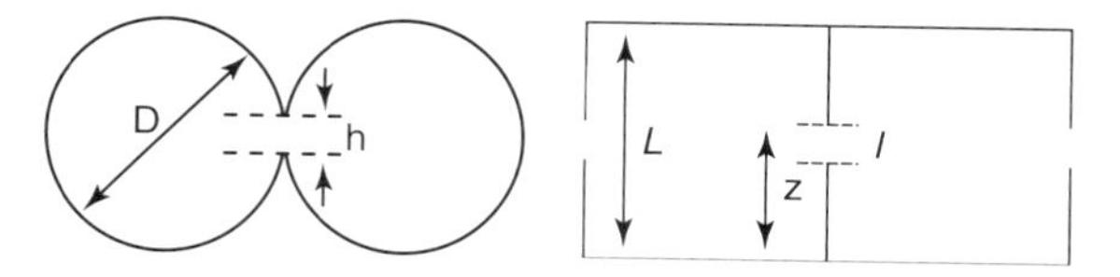

Figure 10.11 Top and side views of the coupling between two cylindrical cavities.

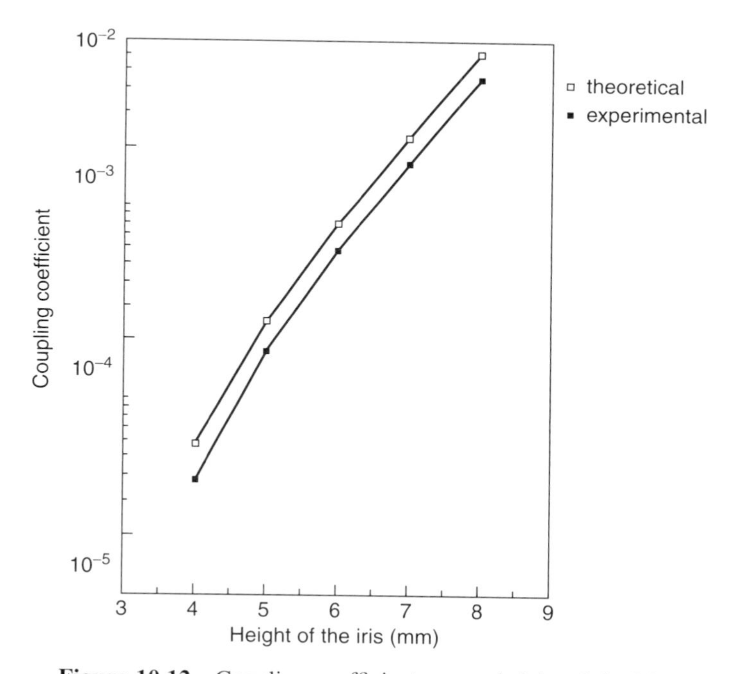

Figure 10.12 Coupling coefficient versus height of the iris.

$$P' = \frac{LD^4M''}{3.78\lambda_0 \sin^2(\pi z/L)}$$

where z is taken as the midheight of the cavity (e.g., $z = L/2$). These formulas are, however, approximate [10.6], so it is useful to conduct laboratory experiments to better define the actual coupling. Figure 10.12 shows the theoretical and measured denormalized coupling coefficient M'' versus the height l of the iris when the thickness is fixed at $t = 5$ mm, the width at $h = 4$ mm, and the cavity dimensions at $D = 29.28$ mm and $L = 21.06$ mm, providing a resonant frequency of $f_0 = 14{,}375$ MHz.

The last stage in the design consists of defining the denormalized coupling coefficients M'' in terms of the normalized coupling coefficients M' from

Section 10.4.1. These are given simply by

$$M''_{jk} = \frac{\Delta\omega}{\omega_0} M'_{jk}$$

where $\Delta\omega$ is the bandwidth and ω_0 is the center frequency of the desired bandpass filter.

10.5 DESIGN EXAMPLE

The first step consists of defining the elements of the even-mode prototype of Figure 10.5. Selection of the low-pass prototype is that of the combined amplitude phase function with ripple in the band of $\varepsilon = 0.004$ dB, providing about -30 dB return loss [10.6]. The parameter $a_2 = 1.3161$ is selected for optimization of the group delay over 80% of the passband and $\omega_f \to \infty$. In this case the roots of $[F_3(\omega)F_3(-\omega)]^2 + (\varepsilon^2/4)[F_3^2(\omega) + F_3^2(-\omega)]^2$, the denominator of $|S_{12}(j\omega)|^2$, are given in Table 10.2.

For the case where $\sigma_1 \leq \sigma_2 \leq \sigma_3$ and $|\omega_1| \geq |\omega_2| \geq |\omega_3|$, we select

$$\begin{aligned} \sigma_1 &= 0.2530 & \qquad \omega_1 &= 1.2503 \\ \sigma_2 &= 0.5927 \quad \text{and} & \omega_2 &= 0.8000 \\ \sigma_3 &= 0.6451 & \omega_3 &= 0.2447 \end{aligned}$$

Applying the synthesis equations of Section 10.3, we find that

$$\begin{aligned} C_1 &= 0.6708 & \qquad K_{16} &\approx 0 \\ C_2 &= 1.2674 \quad \text{and} & K_{25} &= 0.2031 \\ C_3 &= 1.5939 & K_{34} &= 0.8524 \end{aligned}$$

TABLE 10.2

Roots in ω	Roots in s	Selected Roots in s
-1.2503 + 0.2530i	0.2530 - 1.2503i	-0.2530 + 1.2503i
-1.2503 - 0.2530i	0.2530 - 1.2503i	-0.2530 - 1.2503i
1.2503 + 0.2530i	-0.2530 + 1.2503i	-0.5927 + 0.8000i
1.2503 - 0.2530i	0.2530 + 1.2503i	-0.5927 - 0.8000i
-0.8000 + 0.5927i	-0.5927 - 0.8000i	-0.6451 + 0.2447i
-0.8000 - 0.5927i	0.5927 - 0.8000i	-0.6451 - 0.2447i
0.8000 + 0.5927i	-0.5927 + 0.8000i	
0.8000 - 0.5927i	0.5927 + 0.8000i	
-0.2447 + 0.6451i	-0.6451 - 0.2447i	
-0.2447 - 0.6451i	0.6451 - 0.2447i	
0.2447 + 0.6451i	-0.6451 + 0.2447i	
0.2447 - 0.6451i	0.6451 + 0.2447i	

The responses of the low-pass cross-coupled prototype circuit of Figure 10.4 using 1-Ω terminations are given in Figure 10.13. The normalized coupling coefficients of the prototype circuit are then given by

$$\begin{array}{lcl} M_{01} = M_{67} = 1.221 & & M_{16} = 0 \\ M_{12} = M_{56} = 1.0845 & \text{and} & M_{25} = 0.1603 \\ M_{23} = M_{45} = 0.7036 & & M_{34} = 0.5348 \end{array}$$

Solving the condition

$$M_{e33}(\tan\theta)^2 - 2M_{e23}\tan\theta + M_{e22} = 0$$

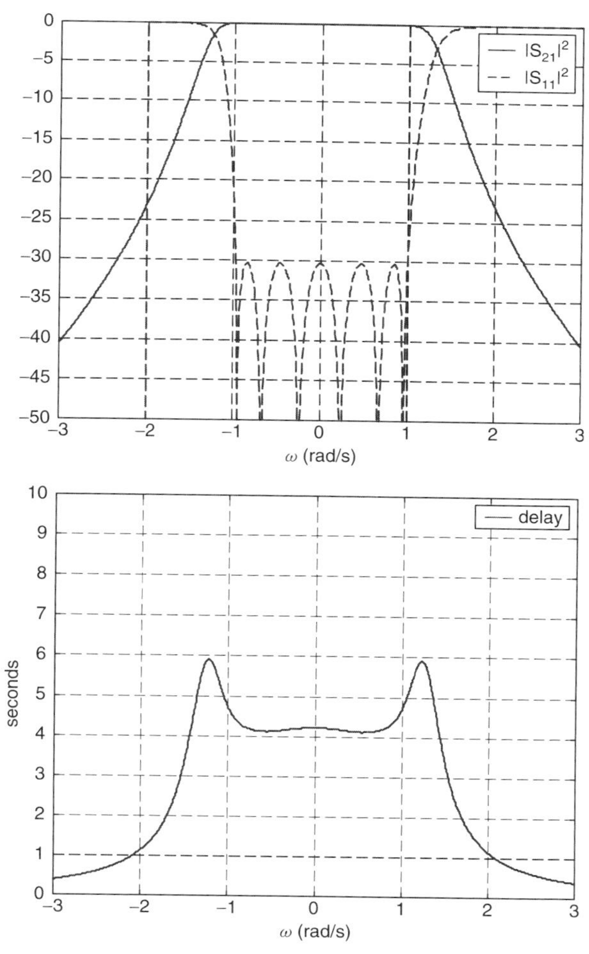

Figure 10.13 Simulations of the low-pass cross-coupled prototype circuit.

leads to two possible angles, $\theta = 0.1187$ rad and $\theta = 1.1919$ rad. Selecting the smallest angle value, we have for

$$[M_e'] = [R]^{\mathrm{T}}[M_e][R]$$

the following matrices:

$$[M_e] = \begin{bmatrix} 0 & 1.0845 & 0 \\ 1.0845 & 0.1602 & 0.7036 \\ 0 & 0.7036 & 0.5348 \end{bmatrix} \quad \text{and} \quad [M_e'] = \begin{bmatrix} 0 & 1.0769 & 0.1284 \\ 1.0769 & 0 & 0.6398 \\ 0.1284 & 0.6398 & 0.6950 \end{bmatrix}$$

where

$$[R] = \begin{bmatrix} 1 & 0 & 0 \\ 0 & 0.9930 & 0.1184 \\ 0 & -0.1184 & 0.9930 \end{bmatrix}$$

so that the normalized couplings for the TE_{011} filter are given by

$$\begin{aligned} M_{01}' &= M_{67}' = 1.221 \\ M_{12}' &= M_{56}' = 1.0769 \\ M_{23}' &= M_{45}' = 0.6398 \end{aligned} \quad \text{and} \quad \begin{aligned} M_{14}' &= M_{36}' = 0.1284 \\ M_{34}' &= 0.6950 \end{aligned}$$

For a filter centered at $f_0 = 14{,}375$ MHz with a bandwidth of $\Delta f = 15$ MHz, the denormalized coupling coefficients are

$$\begin{aligned} M_{01}'' &= M_{67}'' = 10.19 \times 10^{-4} \\ M_{12}'' &= M_{56}'' = 9.00 \times 10^{-4} \\ M_{23}'' &= M_{45}'' = 5.34 \times 10^{-4} \end{aligned} \quad \text{and} \quad \begin{aligned} M_{14}'' &= M_{36}'' = 1.07 \times 10^{-4} \\ M_{34}'' &= 5.80 \times 10^{-4} \end{aligned}$$

Then the couplings from the graph are identified with those of the physical structure as shown in Figure 10.14. Plots similar to Figure 10.12 are then used to define the height l of the iris to produce the desired M_{jk}'' coupling coefficient.

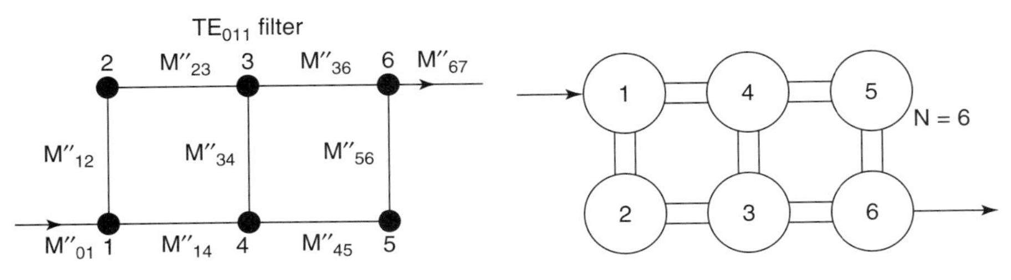

Figure 10.14 Coupling correspondences between the graph and the physical structure.

10.6 REALIZATIONS AND MEASURED PERFORMANCE

10.6.1 Amplitude and Phase Performance

This filter was specified with a center frequency of 14.375 GHz, a 15 MHz bandwidth, and about a −30-dB return loss and phase linearity in the passband. This filter corresponds to the design example of Section 10.5 and was realized using copper [10.6]. The measured forward and return coefficients are shown in Figure 10.15. Figure 10.16 shows the measured group delay and spurious response. Agreement with the expected response is close, but there is an insertion loss of about 2.4 dB, corresponding to a narrowband (1%) and a Q value of about 14,500. The filter constructed is shown in Figures 10.17 and 10.18. Figure 10.18 provides a better view of the six cavities, the non-minimum-phase couplings, and the plugs that eliminate the TM_{111} mode [10.5,10.6].

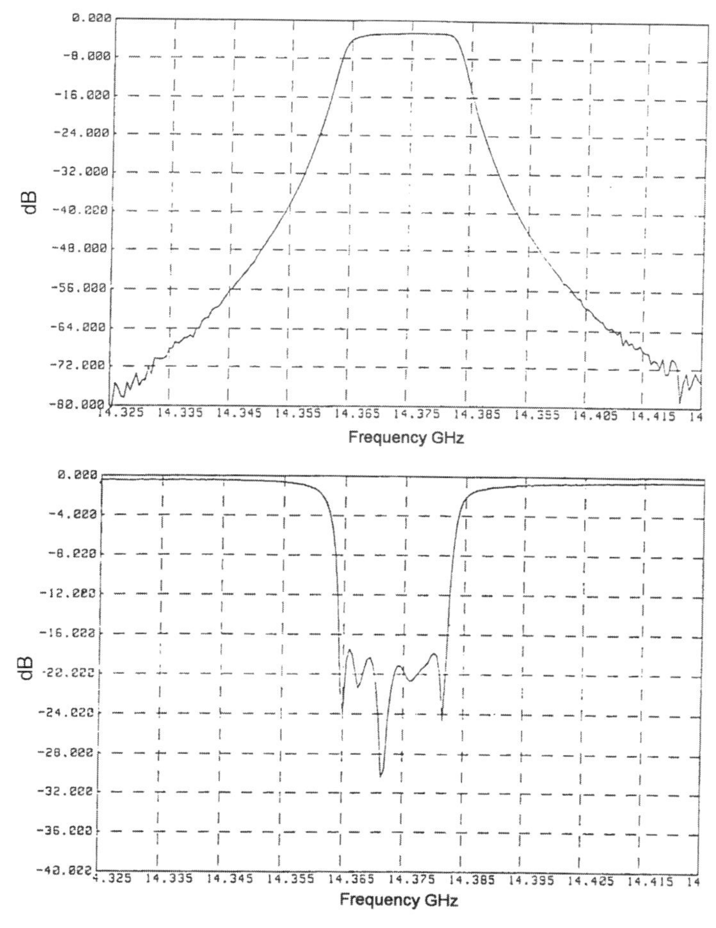

Figure 10.15 Forward and return coefficients.

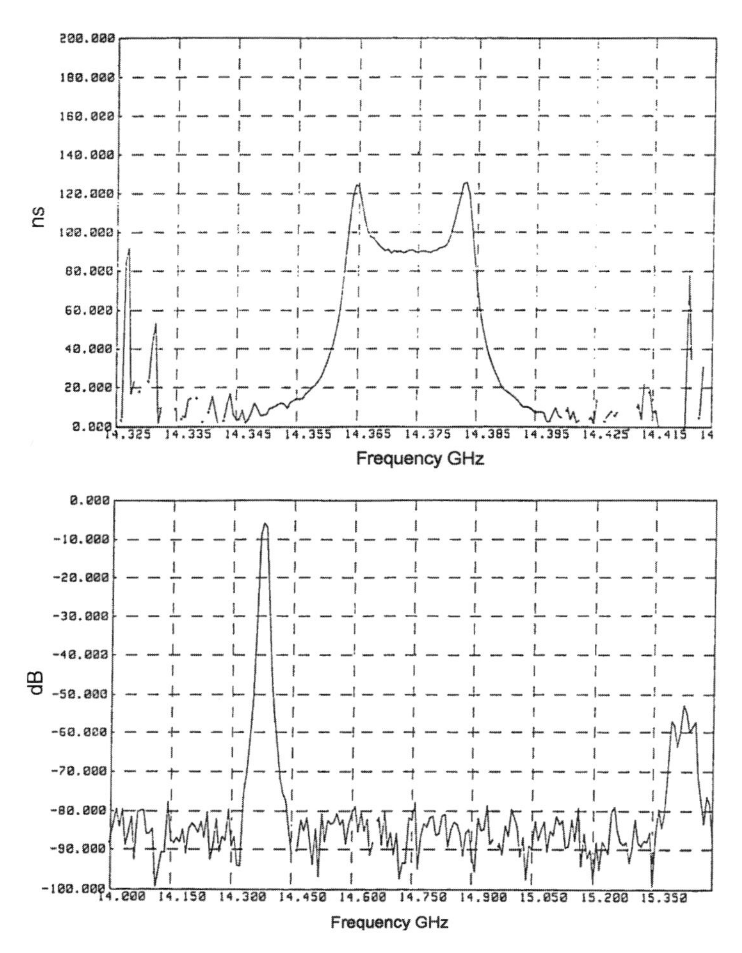

Figure 10.16 Group delay and spurious responses.

10.6.2 Temperature Performance

For satellite applications the responses of the filter must remain the same even when temperature varies from -10 to $+50^{\circ}$C. The first realization of the TE_{011} filter used copper as the main construction material. Figure 10.19 shows the magnitude and group delay of the filter measured as the temperature is varied. There are very substantial variations in both the magnitude and the group delay.

Another filter was then constructed using Invar, and a thin layer of copper was placed on the surfaces of the filter to ensure good conductivity. The magnitude and group delay measured for this filter are shown in Figure 10.20. The stability of the responses against temperature is nearly perfect.

Figure 10.17 Realization of the sixth-degree TE_{011} filter.

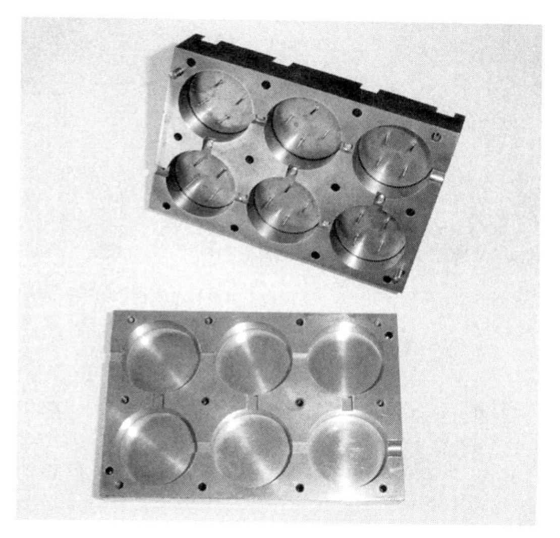

Figure 10.18 Internal view of the filter showing the couplings and plugs.

10.7 CONCLUSIONS

The additional degrees of freedom of non-minimum-phase functions can be used to improve the phase linearity of filters. However, the prototype circuit will have additional couplings that are not always possible to obtain depending on the

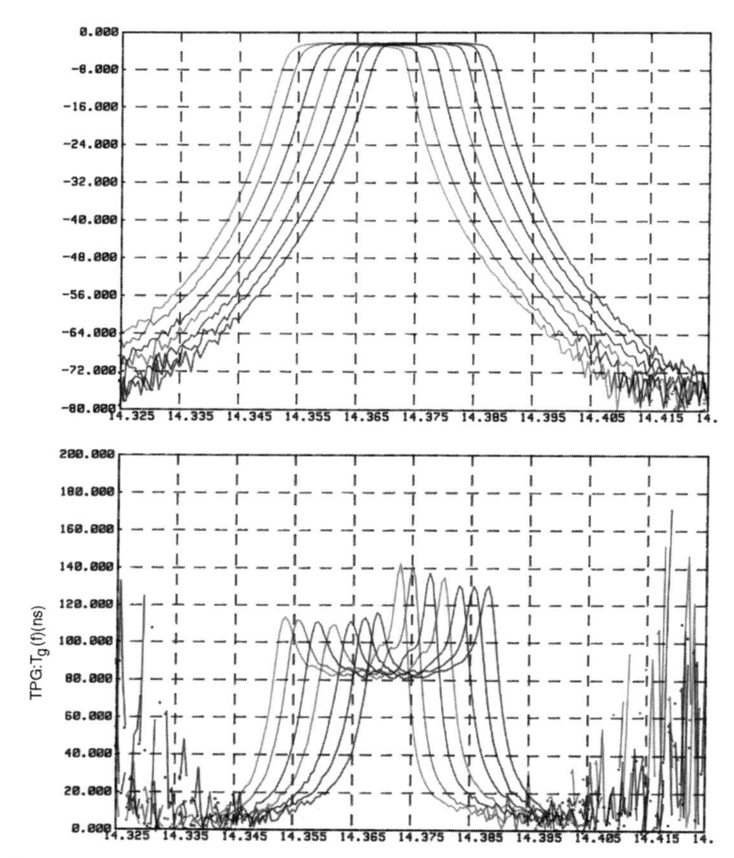

Figure 10.19 Amplitude and group delay responses of the copper filter for temperatures varying from −10 to +50°C.

structure used. A TE_{011} filter based on cylindrical cavity resonators with additional side coupling can be used to achieve these additional couplings. Another consideration for filters intended for satellite applications is how well they maintain their responses over wide temperature changes. Using Invar with a thin layer of copper provides filters that are not only stable in temperature but also have lighter weights.

Acknowledgments

This work was supported by TRT (Philips).

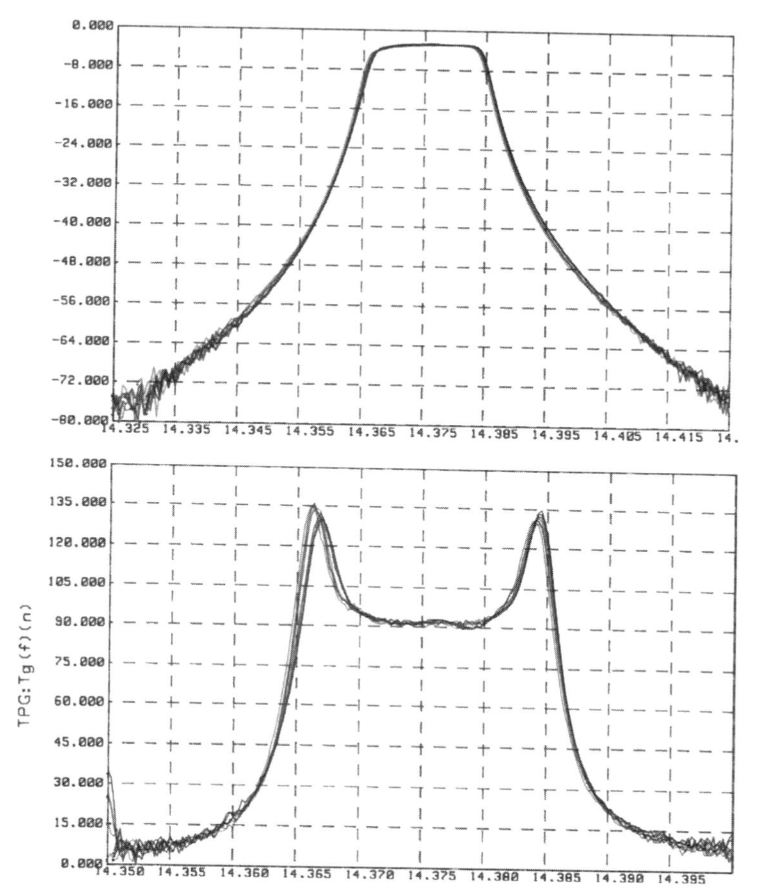

Figure 10.20 Amplitude and group delay responses of the Invar filter for temperatures varying from −10 to +50°C.

REFERENCES

[10.1] I. H. Zabalawi, A Generalized Chebyshev channelizing filter, *Int. J. Circuit Theory Appl.*, vol. 13, no. 1, pp. 37–45, 1985.

[10.2] T. T. Ha, *Solid-Stage Microwave Amplifier Design*, Wiley-Interscience, New York, 1981.

[10.3] J. D. Rhodes and I. H. Zabalawi, Synthesis of symmetrical dual mode in-line prototype networks, *Int. J. Circuit Theory Appl.*, vol. 8, no. 2, pp. 145–160, 1980.

[10.4] A. E. Atia and A. E. Williams, General TE_{011} mode waveguide bandpass filters, *IEEE Trans. Microwave Theory Tech.*, vol. 24, pp. 640–648, Oct. 1976.

[10.5] G. L. Matthaei, L. Young, and E. M. T. Jones, *Microwave Filters, Impedance-Matching Networks and Coupling Structures*, McGraw-Hill, New York; 1964.

[10.6] J. F. Favennec, P. Jarry, and A. Blanchard, Synthesis and realisations of a very narrow-band linear phase microwave filter, *Microwave Nat. Days*, Grenoble, France, pp. 145–146, Mar. 20–22, 1991.

Selected Bibliography

Atia, A. E. and A. E. Williams, Temperature compensation of TE_{011} mode circular cavities, *IEEE Trans. Microwave Theory Tech.*, vol. 24, pp. 668–669, Oct. 1976.

Cloete, J. H., Tables of non-minimum phase even degree low-pass prototype networks for the design of microwave linear-phase filters, *IEEE Trans. Microwave Theory Tech.*, vol. 27, pp. 123–128, Feb. 1979.

Franti, L. F. and G. M. Paganuzzi, Pseudo-elliptical phase-equalized filter allowing good optimization for both phase and amplitude response, *IEE Proc.*, vol. 127, pt. H, no. 6, pp. 342–345, 1980.

Jokela, K. T., Narrow-band stripline or microstrip filters with transmission zeros at real and imaginary frequencies, *IEEE Trans. Microwave Theory Tech.*, vol. 28, pp. 542–547, June 1980.

Levy, R., Filters with single transmission zeros at real or imaginary frequencies, *IEEE Trans. Microwave Theory Tech.*, vol. 24, no. 4, pp. 172–181, 1976.

Rhodes, J. D. and R. J. Cameron, General extracted poles synthesis technique with applications to low-loss TE_{011} mode filters, *IEEE Trans. Microwave Theory Tech.*, vol. 28, no. 9, pp. 1018–1028, 1980.

Rhodes, J. D. and I. H. Zabalawi, Selective linear phase filters possessing a pair of j-axis transmission zeros, *Int. J. Circuit Theory Appl.*, vol. 10, pp. 251–263, 1982.

PART IV

NON-MINIMUM-PHASE ASYMMETRICAL RESPONSE FILTERS

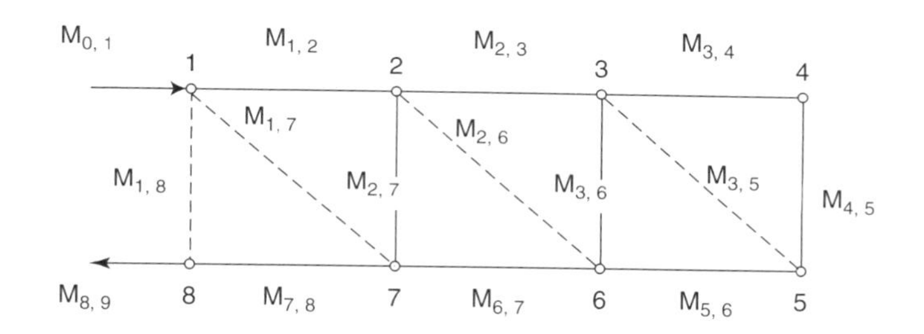

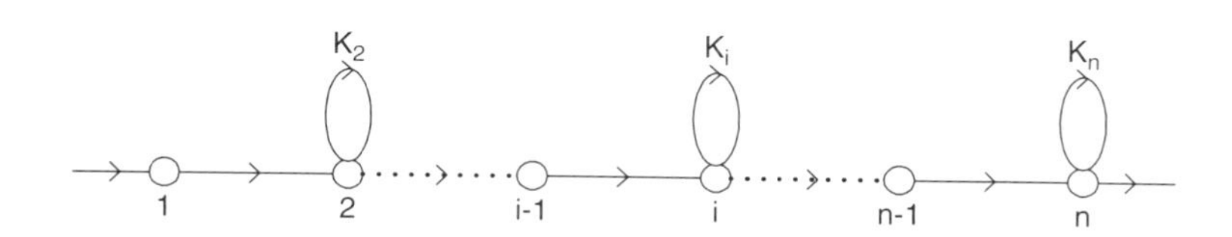

11

ASYMMETRICAL CAPACITIVE-GAP COUPLED LINE FILTERS

11.1 INTRODUCTION

Designing filters with asymmetrical responses is rather complex. The low-pass prototype transfer function is complex valued and the low-pass prototype circuits are not physically realizable. In this chapter, asymmetrical generalized Chebyshev functions with an arbitrary number of transmission zeros are described in detail. The function is synthesized using an in-line low-pass prototype circuit. The extraction procedures for the in-line structure are simpler than that of the more traditional cross-coupled structure. A realizable bandpass structure is deduced from the low-pass structure through the use of a frequency transformation.

Advanced Design Techniques and Realizations of Microwave and RF Filters,
By Pierre Jarry and Jacques Beneat

A filter topology called capacitive gap–parallel coupled line (CGCL) is shown to be suitable for producing asymmetrical generalized Chebyshev functions. In addition, since suspended substrate structure (SSS) filters present several advantages, such as a high Q factor, wide bandwidths, and good temperature stability, the CGCL filter is realized using SSS. Realization of the filter shows that upper stopband rejection is improved, and compared to traditional capacitive gap filters it uses less space and is easy to manufacture. It is also found that it seems difficult to place transmission zeros on the left side of the passband.

11.2 CAPACITIVE-GAP COUPLED LINE FILTERS

A capacitive-gap coupled line filter is depicted in Figure 11.1. The structure starts with transmissions lines (1, 3, 5, ...) of length approximately equal to $\lambda/2$, separated by gaps as in the capacitive-gap filter. Additional lines of length approximately equal to $\lambda/4$ are introduced (2, 4, ...) to provide additional coupling that can create zeros of transmission. The widths of the lines, the gaps, and the distances to the coupled lines are not generally equal when realizing an asymmetrical general Chebyshev response.

11.3 SYNTHESIS OF LOW-PASS ASYMMETRICAL GENERALIZED CHEBYSHEV FILTERS

The low-pass transmission function $|S_{21}(j\omega)|^2$ for the case of a symmetrical or asymmetrical generalized Chebyshev response is given by Cameron [11.1]:

$$|S_{21}(j\omega)|^2 = \frac{1}{1 + \varepsilon^2 F_n^2(\omega)}$$

where ε provides Chebyshev-type ripple in the passband and the function $F_n(\omega)$ is defined as

$$F_n(\omega) = \cosh\left(\sum_{k=1}^{n} \cosh^{-1} \frac{\omega - 1/\omega_k}{1 - \omega\omega_k}\right)$$

where ω_k represents an arbitrary zero of transmission. Note that ω_k can be chosen finite or infinite.

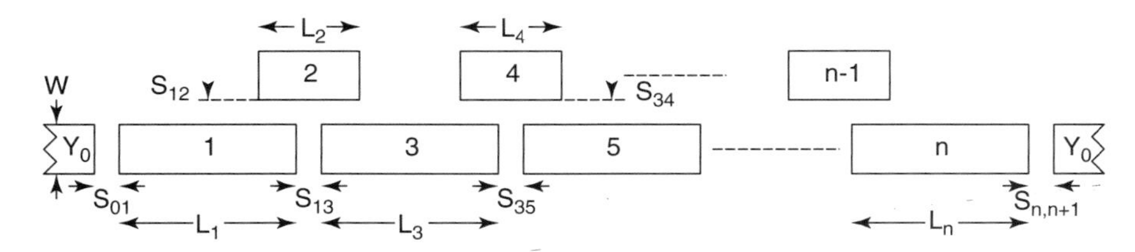

Figure 11.1 Capacitive-gap coupled line filter.

The function $F_n(\omega)$ can be given under the form of the ratio of two real-valued coefficients polynomials in ω:

$$F_n(\omega) = \frac{P_n(\omega)}{D_n(\omega)}$$

where the denominator polynomial is simply

$$D_n(\omega) = \prod_{k=1}^{n} \left(1 - \frac{\omega}{\omega_k}\right)$$

Means for computing the numerator were first proposed by Cameron [11.1]. However, Amari [11.2] provides a simple way to compute this polynomial:

$$P_{n+1}(\omega) = P_n(\omega)\left[(1+\alpha)\omega - \left(\frac{1}{\omega_{n+1}} + \frac{\alpha}{\omega_n}\right)\right] - \alpha P_{n-1}\left(1 - \frac{\omega}{\omega_n}\right)^2$$

with $\alpha = \sqrt{1-(1/\omega_{n+1})^2/1-(1/\omega_n)^2}$. The recursion starts with $P_0(\omega) = 1$ and $P_1(\omega) = \omega - 1\omega_1$.

For example, for the set of frequencies [3.37, ∞, ∞] with $\varepsilon = 0.1$ corresponding to a -20-dB return loss, we have

$$P_3(\omega) = 3.9099\omega^3 - 0.5935\omega^2 - 2.9099\omega + 0.2967$$

$$D_3(\omega) = -0.2967\omega + 1.0000$$

The main goal, however, is to define the transfer function $S_{21}(s)$. This is done by expressing

$$|S_{21}(j\omega)|^2 = \frac{D_n^2(\omega)}{D_n^2(\omega) + \varepsilon^2 P_n^2(\omega)}$$

The transfer function is then given by

$$S_{21}(s) = \frac{D_n(s/j)}{Q_n(s)}$$

The polynomial coefficients of the numerator in s are given by dividing the polynomial coefficients of $D_n(\omega)$, by j^k, where k represents the power of ω associated with the coefficient.

For the denominator polynomial, one has to compute the roots in ω of $D_n^2(\omega) + \varepsilon^2 P_n^2(\omega)$ by using the MATLAB `roots.m` function, for example; these roots are then multiplied by j to form the roots in s, and one then selects roots that have a real part that is negative for stability purposes. One should note that for the symmetrical case, the roots follow the traditional root and its complex conjugate, while for the asymmetrical case, one can have roots without their complex

TABLE 11.1

Roots in ω	Roots in s	Selected Roots in s
-1.3787 + 0.7432i	-0.7432 - 1.3787i	-0.7432 - 1.3787i
-1.3787 - 0.7432i	0.7432 - 1.3787i	-1.1719 + 0.2415i
0.2415 + 1.1719i	-1.1719 + 0.2415i	-0.4286 + 1.2891i
0.2415 - 1.1719i	1.1719 + 0.2415i	
1.2891 + 0.4286i	-0.4286 + 1.2891i	
1.2891 - 0.4286i	0.4286 + 1.2891i	

conjugate. This means that for the asymmetrical case the transfer function $S_{21}(s)$ will be the ratio of two complex-valued coefficient polynomials in s.

In our case, we have the results shown in Table 11.1. The polynomial $Q_n(s)$ can be constructed from the selected roots in s using the MATLAB function `poly.m`, for example. The transfer function $S_{21}(s)$ is given by the complex-coefficient polynomials

$$D_3(s) = j0.2967s + 1$$

$$Q_3(s) = s^3 + (2.3438 - j0.1519)s^2 + (3.4908 - j0.5455)s + (2.3673 - j0.9372)$$

The return loss transfer function $S_{11}(s)$ for a lossless system can be found by

$$|S_{11}(j\omega)|^2 = 1 - |S_{21}(j\omega)|^2 = 1 - \frac{D_n^2(\omega)}{D_n^2(\omega) + \varepsilon^2 P_n^2(\omega)} = \frac{\varepsilon^2 P_n^2(\omega)}{D_n^2(\omega) + \varepsilon^2 P_n^2(\omega)}$$

Then $S_{11}(s)$ is given as the ratio of two polynomials:

$$S_{11}(s) = \frac{\varepsilon P_n(s/j)}{Q_n(s)}$$

The denominator polynomial is the same as that of $S_{21}(s)$, and the polynomial coefficients of the numerator in s are given by dividing the polynomial coefficients of $P_n(\omega)$ by j^k, where k represents the power of ω associated with the coefficient. We obtain for our case

$$\varepsilon P_3(s) = j0.3929s^3 + 0.0597s^2 + j0.2924s + 0.0299$$

Note that $Q_3(s)$ needs to be scaled such that $\max|S_{11}(j\omega)| = \max|S_{21}(j\omega)| = 1$. This can be done by plotting $|S_{11}(j\omega)|$ and $|S_{21}(j\omega)|$ and finding the maximum value. In the case above, we find $\max|S_{11}(j\omega)| = \max|S_{21}(j\omega)| = 0.392944$, so that we multiply $Q_3(s)$ by 0.392944:

$$Q_3(s) = (0.3929 + j0.0000)s^3 + (0.9210 - j0.0597)s^2 + (1.3717 - j0.2143)s + (0.9302 - j0.3683)$$

The responses of the example are then given in Figure 11.2.

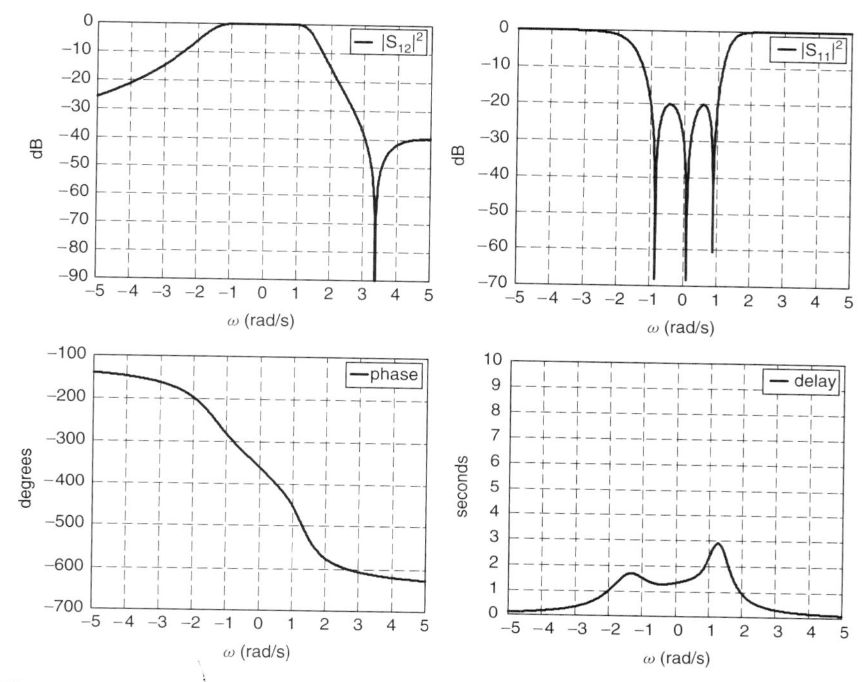

Figure 11.2 Reponses for an asymmetrical generalized Chebyshev transfer function.

11.3.1 In-Line Network

Potential prototype circuits that can realize generalized Chebyshev functions are shown in Figures 11.3, 11.4 and 11.5. The technique for defining the elements of cross-coupled network of Figure 11.3 was given by Cameron [11.3] and is based on a successive extraction of *ABCD* matrices defined from $S_{21}(s)$ and

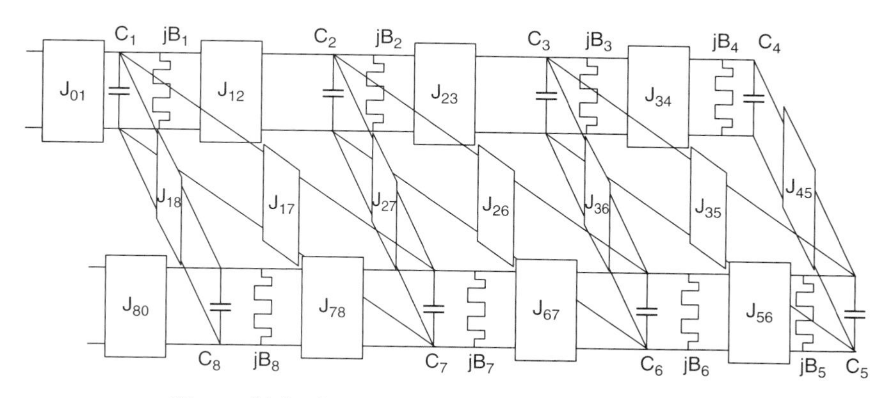

Figure 11.3 Low-pass cross-coupled ladder network.

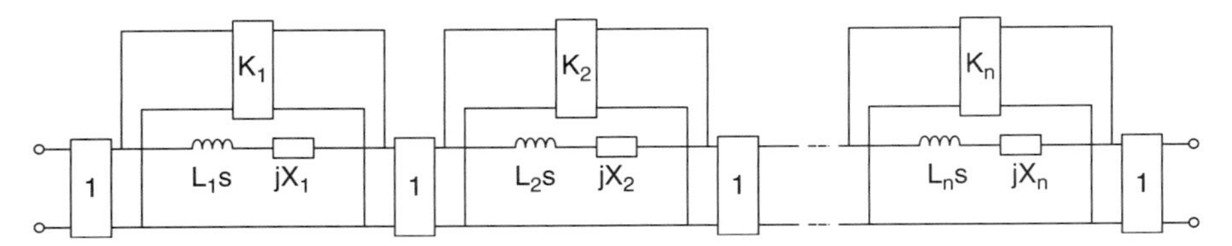

Figure 11.4 Low-pass in-line impedance inverter network.

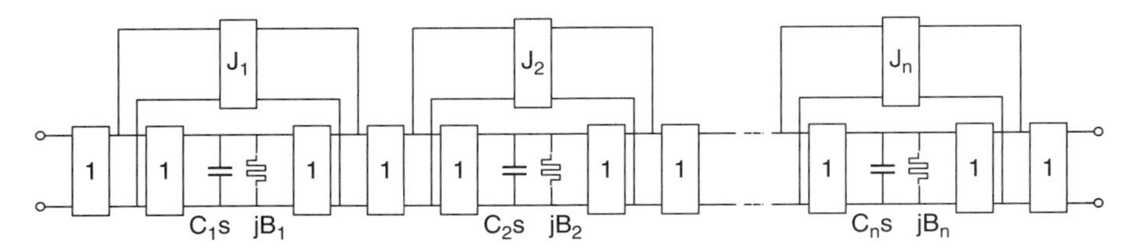

Figure 11.5 Low-pass in-line admittance inverter network.

$S_{11}(s)$. This technique is rather complex, as the extraction goes back and forth between input and output. This technique is described in more detail in Chapter 13. On the other hand, the technique for defining the elements of the in-line structures of Figures 11.4 and 11.5 relies on successive extractions from the input impedance or input admittance [11.4]. The network based on impedance inverters of Figure 11.4 is described in Chapter 12, while we present here the network based on admittance inverters of Figure 11.5. Note that the structures of Figures 11.4 and 11.5 provide the same response when taking $L_k = C_k$, $X_k = B_k$, and $K_k = -1/J_k$. This is shown in Appendix 4.

11.3.2 Analysis of the In-Line Network

The in-line network consists of admittance inverters cascaded with parallel resonators (C_k,B_k). The susceptances B_k are very important for the asymmetrical aspect of the response of the filter [11.5]. The fundamental element is shown in Figure 11.6. It can be shown using even- and odd-mode characterizations that the transfer function of the fundamental element is such that

$$S_{21}(s) = \frac{Y_e(s) - Y_o(s)}{[1 + Y_e(s)][1 + Y_o(s)]} = \frac{1 + jJ(Cs + jB)}{1 - (jJ)\{1 + (1 + jJ)[(C/2)s + j(B/2)]\}}$$

where $Y_e(s)$ and $Y_o(s)$ are the even- and odd-mode admittances. One can easily check that $S_{21}(j\omega)$ is equal to zero when

$$1 - J(C\omega + B) = 0$$

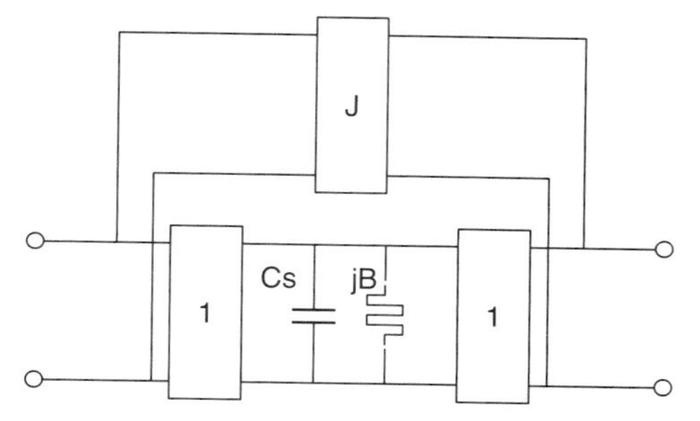

Figure 11.6 Fundamental element of the prototype circuit of Figure 11.5.

Therefore, the fundamental element produces a zero of transmission at $\omega = \omega_Z$ such that

$$\omega_z = \frac{1/J - B}{C}$$

Note that ω_Z can be positive or negative, depending on the values and signs of J and B. This means that the fundamental element can produce zeros of transmission on the left or right side of the passband.

For the purposes of extracting a fundamental element of Figure 11.6 from the input admittance $Y_{\text{in}}(s)$, it will be necessary to define an intermediary circuit. This intermediary circuit is based on the two identities shown in Figure 11.7. First, the J inverter is replaced by a corresponding admittance Π network. Then a parallel admittance placed after an inverter can be brought in front of the inverter as a series impedance. The parallel admittance and the series impedance will have the same value if the inverter is unity.

The resulting equivalent circuit of the fundamental element is shown in Figure 11.8. One should note, however, that Cs and jB are treated as impedances while $-jJ$ is treated as an admittance. Also note that to move the effect of the

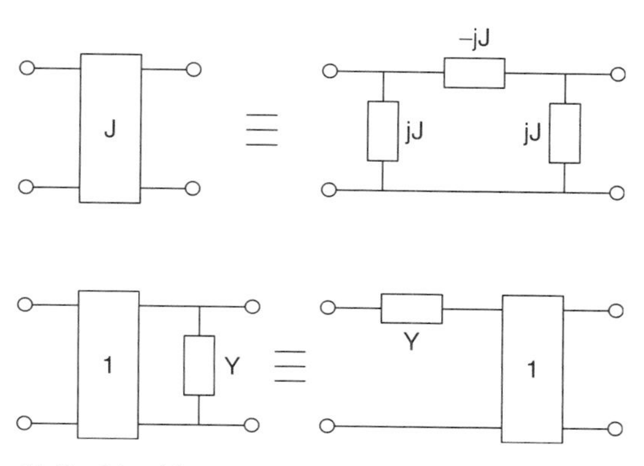

Figure 11.7 Identities used to construct the intermediary circuit.

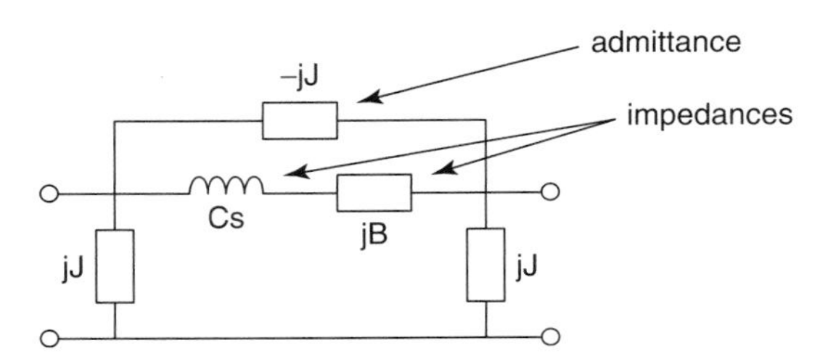

Figure 11.8 Equivalent circuit of the fundamental element.

two cascaded unity admittance inverters on the loads, the admittance inverters J used in Figure 11.5 should be taken as $-J$, found from the synthesis method presented below.

The structure of the intermediary circuit is shown in Figure 11.9. The series impedances seen between the two parallel admittances jJ for the circuits of Figures 11.8 and 11.9 are given by

$$Z(s) = \frac{1}{[1/(Cs + jB)] - jJ} \quad \text{and} \quad Z(s) = \frac{1}{C's + jB'} + jX'$$

The two impedances and therefore the two circuits are the same if

$$J = \frac{1}{X'}$$
$$C = C'X'^2$$
$$B = (X'B' - 1)X'$$

One notes that the impedance from Figure 11.9 becomes infinity when

$$jC'\omega + jB' = 0$$

When this impedance reaches infinity, the circuit produces the zero of transmission. Therefore, we have

$$C'\omega_Z + B' = 0 \quad \text{or} \quad B' = -\omega_Z C'$$

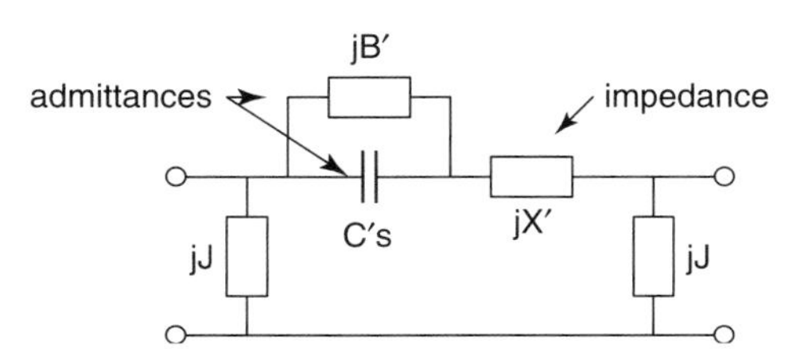

Figure 11.9 Intermediary circuit for extraction purposes.

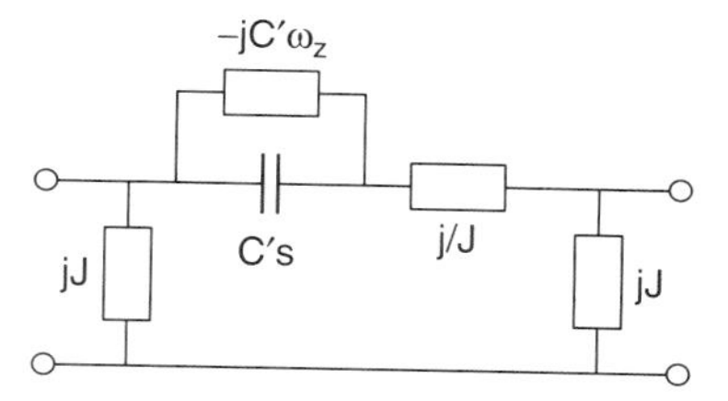

Figure 11.10 Intermediary circuit of the fundamental element for extraction purposes.

We can then express the conditions for the two circuits to be equivalent as

$$J = \frac{1}{X'}$$

$$C = \frac{C'}{J^2}$$

$$B = \left(-\frac{1}{J}\omega_z C' - 1\right)\frac{1}{J} = -\frac{J + \omega_Z C'}{J^2}$$

and the intermediary circuit can be drawn directly using J, the capacitance C', and the zero of transmission ω_z as shown in Figure 11.10. The extraction procedure will consist of extracting J and the capacitance C'. The capacitance C and the susceptances B of the fundamental element can be then be found from J and C'.

11.3.3 Synthesis of the In-Line Network

The structure of Figure 11.5 can produce as many zeros of transmission as there are fundamental elements. One can, however, decide to not include all the fundamental elements. For example, Figure 11.11 shows the structure that can realize a third-order generalized Chebyshev function with one zero of transmission [11.6].

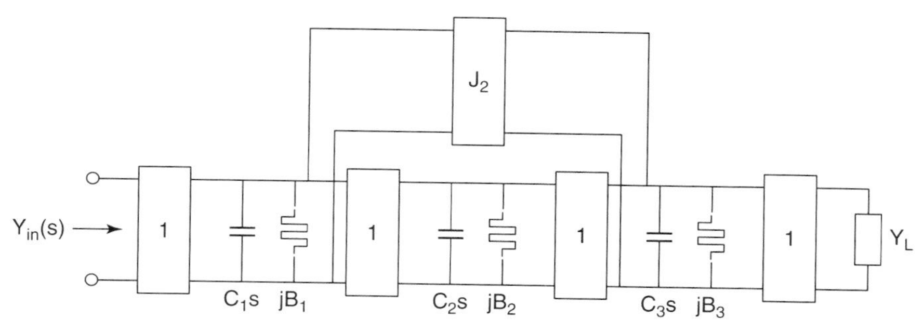

Figure 11.11 Prototype circuit for a third-order generalized Chebyshev function with one zero of transmission.

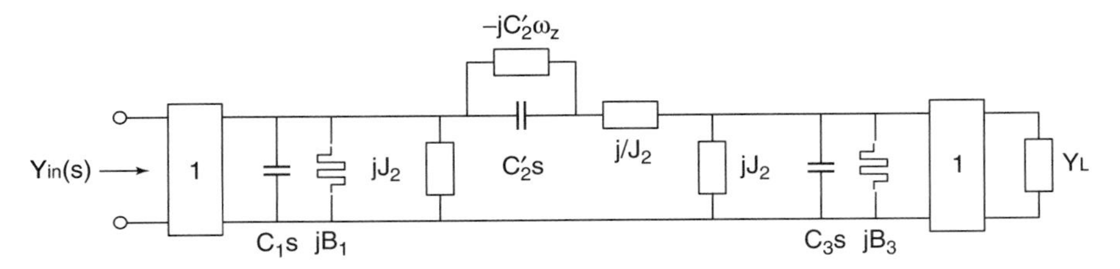

Figure 11.12 Intermediary circuit used for the extraction technique.

The extraction technique consists of replacing a fundamental element by the intermediary circuit as shown in Figure 11.12. The input admittance $Y_{\text{in}}(s)$ of the network in Figure 11.12 is given by

$$Y_{\text{in}}(s) = \cfrac{1}{C_1 s + jB_1 + jJ_2 + \cfrac{1}{\cfrac{j}{J_2} + \cfrac{1}{C_2'(s - j\omega_Z)} + \cfrac{1}{C_3 s + jB_3 + jJ_2 + 1/Y_L}}}$$

The input impedance for the generalized Chebychev function is defined in terms of $S_{11}(s)$ as

$$Z_{\text{in}}(s) = Z_L \frac{1 + S_{11}(s)}{1 - S_{11}(s)}$$

Note that for the procedure to start properly, the degree of numerator and denominator polynomials should be different by 1. Therefore, it might be necessary to perform some scaling of the polynomials $P(s)$ and $Q(s)$ to match their highest-degree coefficients. In the case of complex-valued coefficients, one can also scale a polynomial by $\pm j$ without affecting the magnitude response.

The unity admittance inverter is extracted first, leaving the new input admittance as

$$Y(s) = C_1 s + jB_1 + jJ_2 + \cfrac{1}{\cfrac{j}{J_2} + \cfrac{1}{C_2'(s - j\omega_Z)} + \cfrac{1}{C_3 s + jB_3 + jJ_2 + 1/Y_L}}$$

The capacitance C_1 is extracted using

$$C_1 = \left.\frac{Y(s)}{s}\right|_{s \to \infty}$$

The input admittance becomes

$$Y_1(s) = Y(s) - C_1 s$$

The susceptance B_1 is extracted using

$$jB_1 = Y_1(s)|_{s\to\infty}$$

The input admittance becomes

$$Y_2(s) = Y_1(s) - jB_1$$

The admittance inverter J_2 is extracted using

$$jJ_2 = Y_2(s)|_{s\to j\omega_Z}$$

At this stage the input impedance $Z_2'(s)$ is computed as

$$Z_2'(s) = \frac{1}{Y_2(s) - jJ_2} - \frac{j}{J_2}$$

The capacitance C_2' of the intermediary circuit is then extracted using

$$\frac{1}{C_2'} = (s - j\omega_Z)Z_2'(s)|_{s\to j\omega_z}$$

from which the capacitance C_2 and susceptance B_2 are computed using

$$C_2 = \frac{C_2'}{J_2^2} \quad \text{and} \quad B_2 = -\frac{J_2 + \omega_Z C_2'}{J_2^2}$$

The input impedances becomes

$$Z_3(s) = Z_2'(s) - \frac{1}{C_2'(s - j\omega_Z)} \quad \text{and} \quad Y_3(s) = \frac{1}{Z_3(s)}$$

The capacitance C_3 is then extracted using

$$C_3 = \left.\frac{Y_3(s)}{s}\right|_{s\to\infty}$$

The input admittance becomes

$$Y_4(s) = Y_3(s) - C_3 s - jJ_2$$

The susceptance B_3 and the final load can be found knowing that

$$Y_4(s) = jB_3 + \frac{1}{Y_L}$$

The procedure can be generalized using the flowchart shown in Figure 11.13.

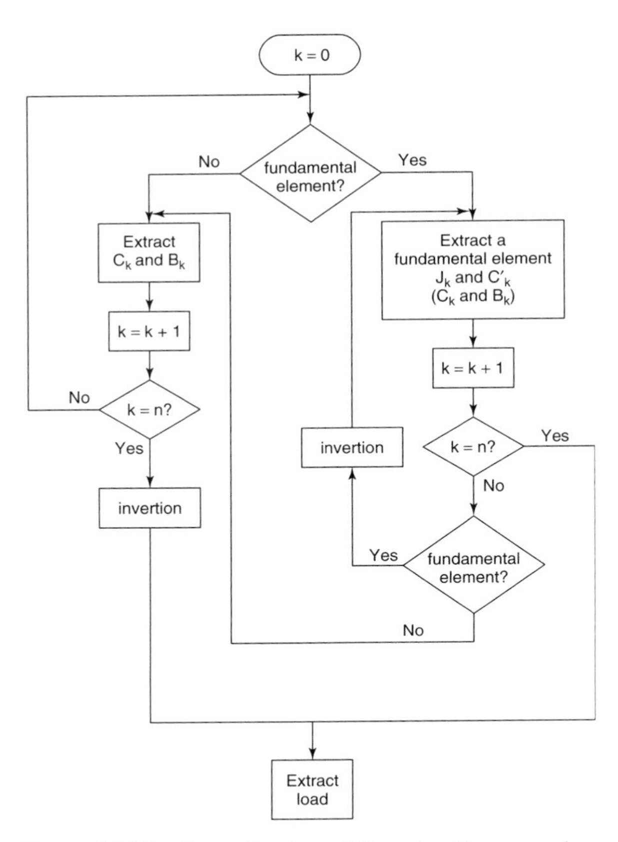

Figure 11.13 Generalization of the extraction procedure.

11.3.4 Frequency Transformation

The low-pass prototype circuit can be modified to produce a bandpass response as shown in Figure 11.14. This circuit is obtained using the low pass-to-bandpass transformation technique:

$$s \to \alpha\left(\frac{s}{\omega_0} + \frac{\omega_0}{s}\right) \quad \text{with} \quad \begin{aligned} \omega_0 &= \sqrt{\omega_1 \omega_2} \\ \alpha &= \frac{\omega_0}{\omega_2 - \omega_1} \end{aligned}$$

The admittance inverters are independent of frequency, so they remain the same. We then match the admittances and their derivatives such that

$$C\left[\alpha\left(\frac{\omega}{\omega_0} - \frac{\omega_0}{\omega}\right)\right] + B = C_a\omega - \frac{1}{L_a\omega} \quad \text{and} \quad C\left[\alpha\left(\frac{1}{\omega_0} + \frac{\omega_0}{\omega^2}\right)\right] = C_a + \frac{1}{L_a\omega^2}$$

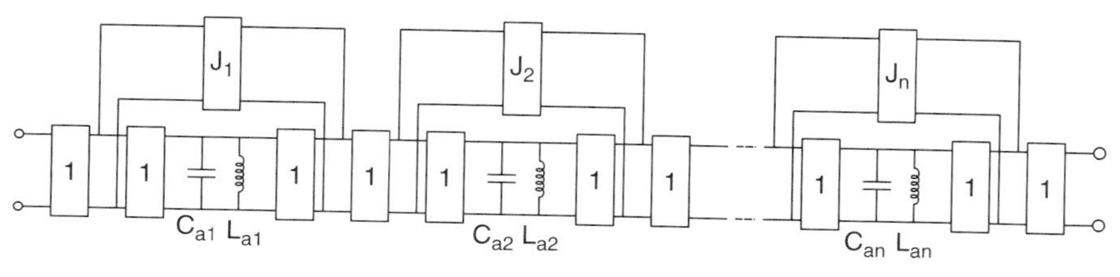

Figure 11.14 Bandpass in-line admittance inverter network.

Solving the two equations at $\omega = \omega_0$, we find that

$$B = C_a\omega_0 - \frac{1}{L_a\omega_0} \quad \text{and} \quad \frac{2\alpha}{\omega_0}C = C_a + \frac{1}{L_a\omega_0^2}$$

To define the bandpass elements, we will use the relations

$$C_{ak} = \frac{1}{2}\left(\frac{2\alpha C_k + B_k}{\omega_0}\right) \quad \text{and} \quad L_{ak} = \frac{2}{\omega_0(2\alpha C_k - B_k)}$$

The results are valid as long as $\omega/\omega_0 + \omega_0/\omega = 2$. This means that ω should be close to ω_0, and the design technique will work properly for narrowband applications.

11.4 DESIGN METHOD

Once the bandpass prototype circuit has been defined, the main challenge is to define the physical parameters of the capacitive-gap coupled line filter of Figure 11.1. First, we provide an equivalent circuit of the CGCL structure, in particular an equivalent circuit that can be compared with the fundamental element of the prototype circuit. Malherbe [11.7], shows how a pair of coupled lines can be modeled as the circuit in Figure 11.15. When port 1 is left on an open

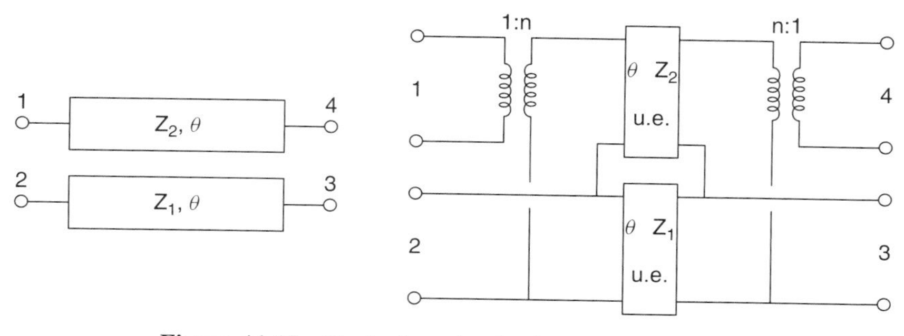

Figure 11.15 Equivalent circuit of two coupled lines.

circuit, the unit element becomes a distributed admittance, and this configuration has the equivalent circuit shown in Figure 11.16. By cascading this circuit and its reverse, the equivalent circuit of the capacitive-gap coupled line can now be defined and is shown in Figure 11.17.

Note that in the equivalent circuit, the gap has been modeled as an admittance inverter. The effect of the two transformers is placed onto the impedance Z using the identity shown in Figure 11.18. By replacing the gaps by admittance inverters and terminating the structure by lines with characteristic admittances equal to Y_1, we have the equivalent circuit of Figure 11.19.

The design technique then consists of identifying the equivalent circuit of the CGCL filter and that of the bandpass prototype as shown in Figure 11.20. In the design method the input and output inverters of the equivalent circuit will be taken as the unity inverters of the prototype. The first unit element of length $2\theta_0 \approx \pi$ will be identified with the resonant cell (L_{a1}, C_{a1}), and the last unit element of length $2\theta_0$ will be identified with the resonant cell (L_{a3}, C_{a3}). The

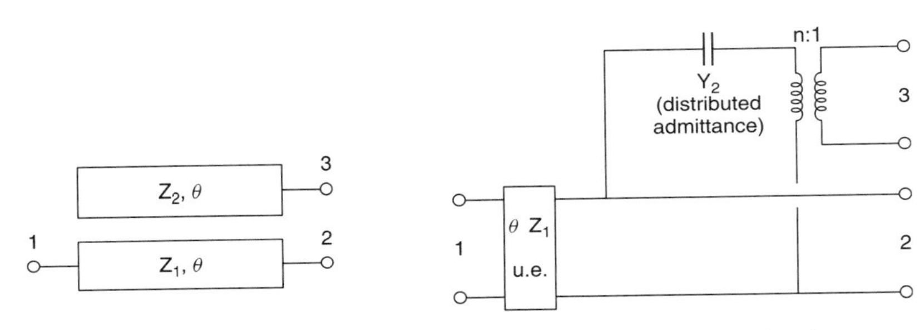

Figure 11.16 Equivalent circuit when one port is open-circuit.

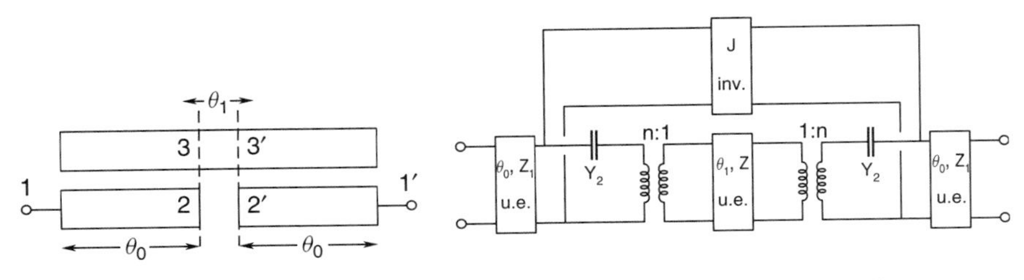

Figure 11.17 Equivalent circuit of a capacitive gap and coupled lines.

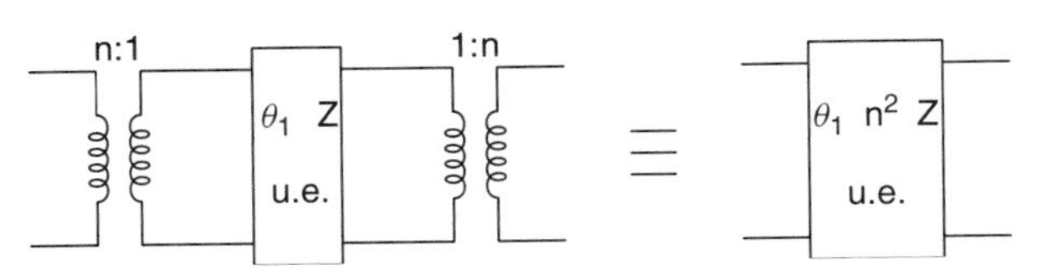

Figure 11.18 Moving the effect of the transformers on the impedance.

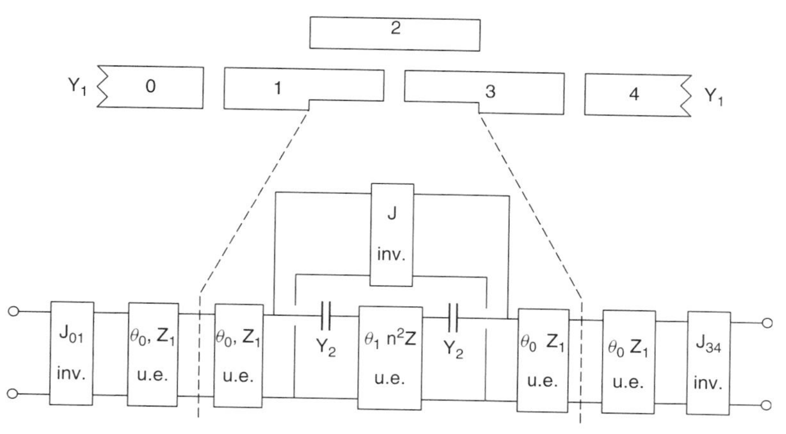

Figure 11.19 Equivalent circuit of a CGCL filter of order 3 and one zero of transmission.

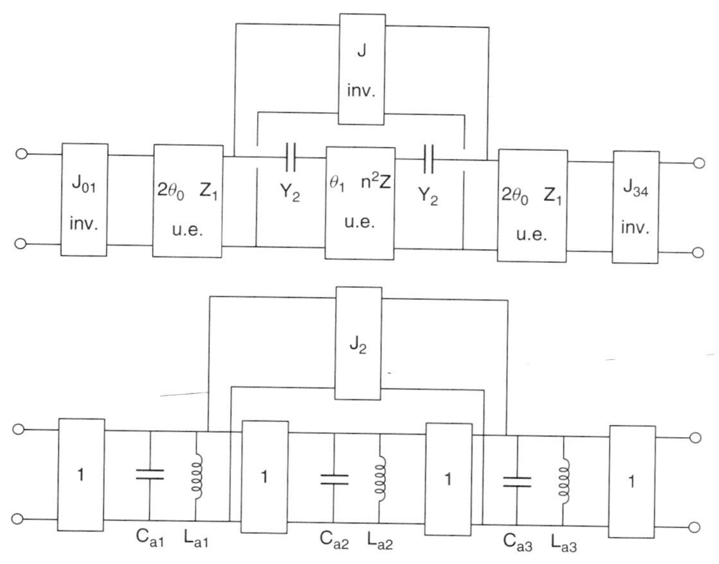

Figure 11.20 Equivalent circuit and bandpass prototype circuit for a third-order generalized Chebyshev response with one transmission zero.

unity inverters will be achieved using gaps (as modeled in Chapter 5), and the central element will be compared with the fundamental element (J_2, L_{a2}, C_{a2}) of the prototype.

Identification with the Fundamental Element The technique consists of taking $J = J_2$ and identifying the circuits shown in Figure 11.21. The *ABCD* matrix of

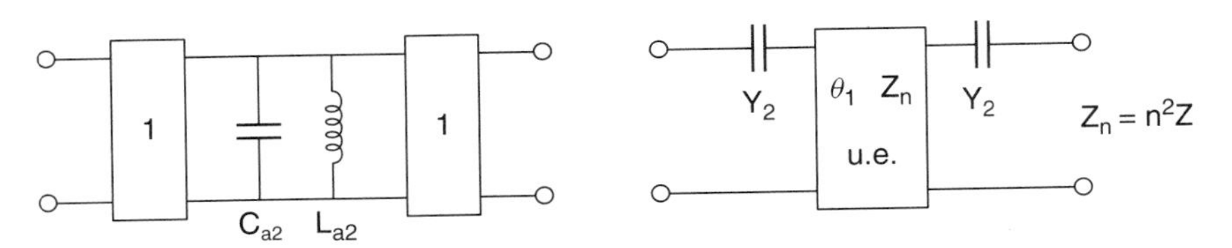

Figure 11.21 Identification with the fundamental element.

each circuit should be made equal:

$$(-1)\begin{pmatrix} 1 & Y_{a2} \\ 0 & 1 \end{pmatrix} = \begin{pmatrix} \dfrac{\sin\theta_1}{Z_n}\left(\dfrac{Z_2}{\tan\theta} + \dfrac{Z_n}{\tan\theta_1}\right) & \dfrac{j\sin\theta_1}{Z_n}\left(Z_n^2 - \dfrac{Z_2^2}{\tan^2\theta} - \dfrac{2Z_2Z_n}{\tan\theta\tan\theta_1}\right) \\ \dfrac{j\sin\theta_1}{Z_n} & \dfrac{\sin\theta_1}{Z_n}\left(\dfrac{Z_2}{\tan\theta} + \dfrac{Z_n}{\tan\theta_1}\right) \end{pmatrix}$$

where $Y_{a2} = jC_{a2}\omega - 1/jL_{a2}\omega$. For the C parameters of the two $ABCD$ matrices to be identical, the length θ_1 must be zero. Clearly, this is not practical, so we will let θ_1 be very small but not zero. In this case, the matrix of the equivalent circuit is approximately

$$\begin{pmatrix} A & B \\ C & D \end{pmatrix} \approx \begin{pmatrix} 1 & -j\dfrac{Z_2}{\tan\theta} \\ 0 & 1 \end{pmatrix}$$

Therefore, the two matrices will be identical if $jC_{a2}\omega - 1/jL_{a2}\omega$ and $-j(Z_2/\tan\theta)$ are identical.

To define Z_2, we decide to match the slope of these two admittances such that

$$\left.\frac{\theta_0}{2}\frac{dB_\theta}{d\theta}\right|_{\theta=\theta_0=\pi/2} = \left.\frac{\omega_0}{2}\frac{dB_\omega}{d\omega}\right|_{\omega=\omega_0}$$

with

$$B_\theta = -\frac{2Z_2}{\tan\theta} \quad \text{and} \quad B_\omega = C_{a2}\omega - \frac{1}{L_{a2}\omega}$$

This gives us

$$Z_2 = \frac{C_{a2}\omega_0 + 1/L_{a2}\omega_0}{\pi}$$

Note that the prototype elements are computed for 1-Ω terminations, so this value of Z_2 is normalized to 1-Ω. Its value will need to be scaled to account for traditional 50-Ω terminations.

In the case of suspended substrate stripline for two coupled lines Z_1 and Z_2, we have the following relations:

$$n^2 = \frac{Z_2}{Z_1} + 1$$

$$\frac{C_{11}}{\varepsilon_0} = \eta\sqrt{\varepsilon_{\text{eff}}}\frac{n(n-1)}{Z_2}$$

$$\frac{C_{12}}{\varepsilon_0} = \eta\sqrt{\varepsilon_{\text{eff}}}\frac{n}{Z_2}$$

where C_{11} and C_{12} are the parallel-plate and coupling capacitances and η is the impedance of free space per square meter ($\eta = 376.7\ \Omega/\text{m}^2$). The even- and odd-mode capacitances are given by

$$C_{oe} = C_{11}$$
$$C_{oo} = C_{11} + 2C_{12}$$

The associated even- and odd-mode impedances are

$$Z_{oe} = \frac{\eta\sqrt{\varepsilon_{\text{eff}}}\varepsilon_0}{C_{oe}} = \frac{Z_2}{n(n-1)}$$

$$Z_{oo} = \frac{\eta\sqrt{\varepsilon_{\text{eff}}}\varepsilon_0}{C_{oo}} = \frac{Z_2}{n(n+1)}$$

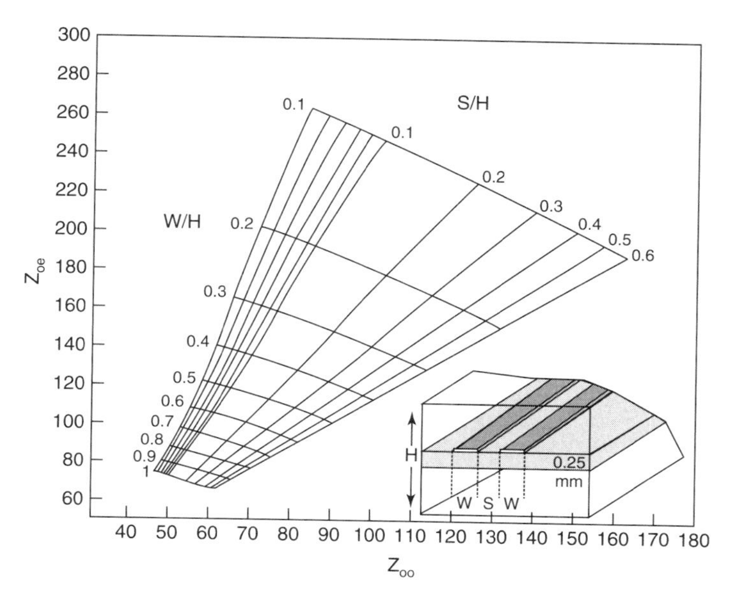

Figure 11.22 Z_{oe} versus Z_{oo} of coupled suspended substrate stripline.

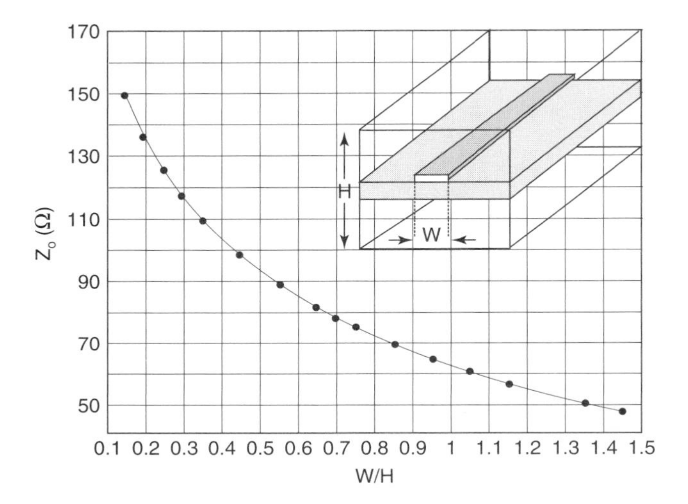

Figure 11.23 Characteristic impedance of a suspended substrate stripline.

One then uses graphs of Z_{oe} versus Z_{oo} with parameters W/H and S/H to define the values of the widths of the line W and the separation of the lines S when the height of the enclosure H is provided as well as the permittivity ε and the thickness t. One such graph is shown in Figure 11.22 for $\varepsilon_r = 2.22$ and $t = 0.25$ mm. Another useful graph for defining the physical parameters of the structure is that of the characteristic impedance of a line in suspended substrate stripline as shown in Figure 11.23.

11.5 DESIGN EXAMPLE

We wish to design a third-order generalized Chebyshev filter centered at 8500 MHz with a relative bandwidth of 400 MHz with one zero of transmission at 9200 MHz. The return loss is specified at -20 dB[11.8,11.9]. The first step is to define the location of the zero of transmission for the low-pass prototype. It is given by

$$\omega_Z = \alpha \left(\frac{\omega_{pz}}{\omega_0} - \frac{\omega_0}{\omega_{pz}} \right)$$

where ω_{pz} is the location of the bandpass zero (e.g., $2\pi \times 9200 \times 10^6$ rad/s). In this case we find that $\omega_z = 3.37$ rad/s.

The low-pass transfer function was provided in Section 11.3. However, when dealing with polynomials with complex coefficients, there is some additional scaling to be done to guarantee that the extraction technique can start properly. For example, note that in the case of S_{11}, we can multiply the polynomial $P(s)$ by $-j$ and still have the same magnitude characteristics. However, by doing so, the two polynomials $P(s)$ and $Q(s)$ now have the same coefficient in s^3, as seen below. Then after extracting the first admittance inverter, we start the extraction procedure with a rational function that can produce the first capacitor. The results of using the extraction procedure are given below.

```
P(s) = (0.3929+j0.0000)s^3 + (0.0000+j-0.0597)s^2
   + (0.2924+j0.0000)s^1 + (0.0000+j-0.0299)
Q(s) = (0.3929+j0.0000)s^3 + (0.9210+j-0.0597)s^2
   + (1.3717+j-0.2143)s^1 + (0.9302+j-0.3683)

Y(s):
Yn(s) = (0.7859+j0.0000)s^3 + (0.9210+j-0.1194)s^2
   + (1.6641+j-0.2143)s^1 + (0.9302+j-0.3981)
Yd(s) = (0.9210+j0.0000)s^2 + (1.0793+j-0.2143)s^1
   + (0.9302+j-0.3384)
```

The extraction procedure produces the following results.

```
C1 = 0.853328
B1 = 0.068953
J2 = -0.275203
C2 = 1.190100
B2 = -0.373153
C3 = 0.853328
B3 = 0.068953
load = 1.000000 + j-0.000000
```

Note that for the prototype circuit of Figure 11.5, one should take J_2 as + 0.275203, as mentioned in Section 13.3.2. The capacitor and inductor values of the bandpass prototype are given by

$$C_k = \frac{1}{2}\left(\frac{2\alpha C_k + B_k}{\omega_0}\right) \quad \text{and} \quad L_k = \frac{2}{\omega_0(2\alpha C_k - B_k)}$$

The simulation of the bandpass prototype circuit of Figure 11.14 using these values is shown in Figure 11.24. The characteristics of the suspended substrate stripline are provided in Figure 11.25.

The substrate is suspended symmetrically in the center of the enclosure. Selection of both depth and width of the channel is important because the dimensions must be sufficiently small to avoid propagation of waveguide modes within the channel that can produce unwanted coupling [11.10]. The layout of the filter is shown in Figure 11.26. There are, however, some symmetries in the prototype circuit so that $W_0 = W_4$, $L_1 = L_3$, and $S_{01} = S_{34}$. First, the input and output line width is taken as $W_0 = 2$ mm. Working with this width was necessary so that the entire filter can be integrated in the enclosure. From Figure 11.23 the input

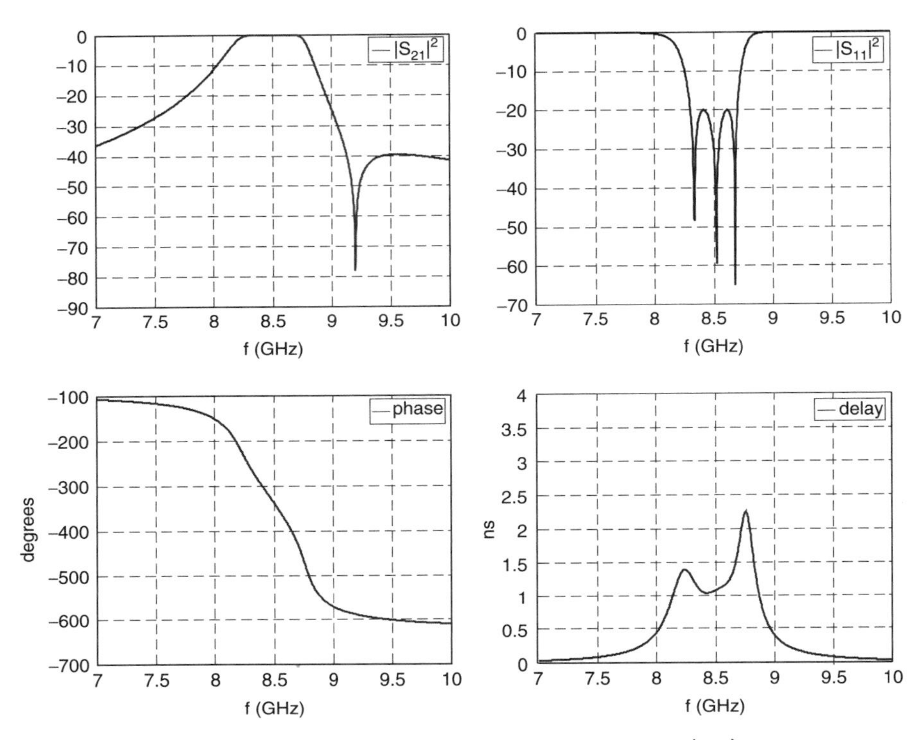

Figure 11.24 Bandpass generalized Chebyshev prototype circuit response.

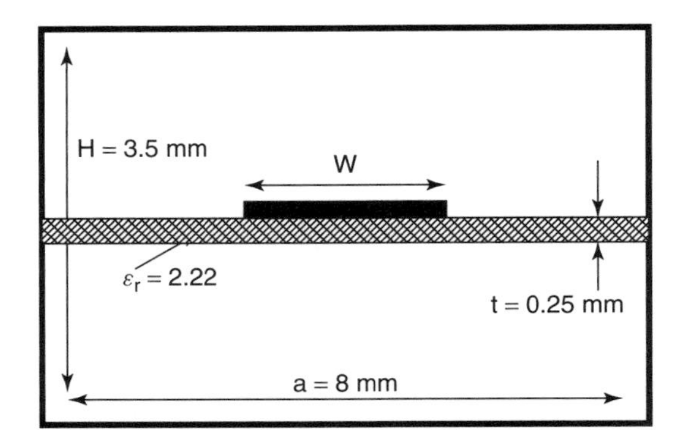

Figure 11.25 Choice of the suspended substrate stripline.

and output impedances would be equal to $Z_0 = 89.1\ \Omega$. To connect this filter to traditional 50-Ω equipment, an additional impedance transformer is needed, as explained later. The first line is also taken be $W_1 = 2$ mm and $Z_1 = 89.1\ \Omega$, except where it is being coupled with line 2, where it will have the width W_2.

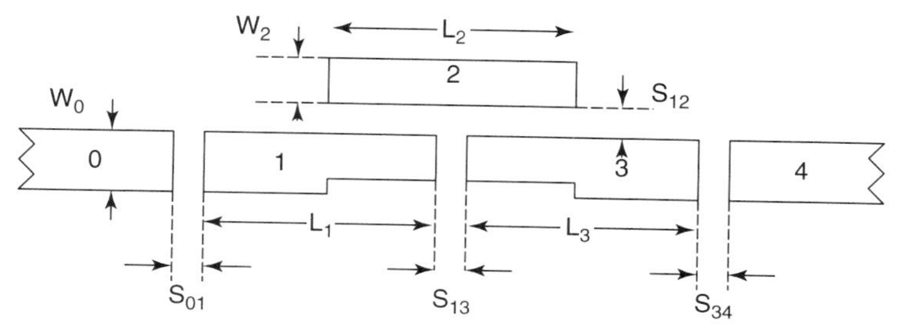

Figure 11.26 Layout of the CGCL filter.

As mentioned before, the gap S_{13} is to be very small. We decide on a value of $S_{13} = 0.2$ mm. From Chapter 5 on capacitive-gap filter design, we can compute the corresponding inverter value using

$$\frac{B_b}{Y_0} = \frac{H}{\lambda_g} \ln\left(\coth \frac{\pi S}{2H}\right) \quad \text{and} \quad \frac{B_b}{Y_0} = \frac{J/Y_0}{1 - \left(\dfrac{J}{Y_0}\right)^2}$$

This produces the normalized inverter value of $J/Y_0 = 0.2207$, which is different from the prototype normalized value of $J_2 = 0.2752$. To account for realizable values, it is good to provide some flexibility in the prototype circuit values. This is done by introducing scaling factors N_1 and N_2, which do not change the response of the prototype circuit, as shown in Figure 11.27. In this case we can use N_1 to match the inverter value of a gap of 0.2 mm:

$$N_1^2 J_2 = 0.2207 \quad \text{and} \quad N_1 = 0.895$$

The length L_1 is computed using the technique of Chapter 5 for the capacitive-gap filter. Once the gap $S = S_{13}$ is given, one can compute B_a, B_b, ϕ_1, and ϕ_2 to define the length to be extracted from the initial $\lambda_g/2$ line to create an inverter. In this case we find that $L_1 = 14.87$ mm.

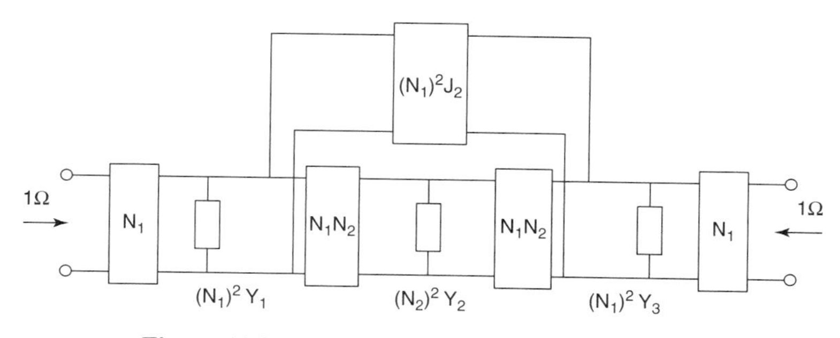

Figure 11.27 Modified bandpass prototype circuit.

The value of Z_2 denormalized to $Z_0 = 89.1\ \Omega$, which takes the prototype modification of Figure 11.27 into account, is given by

$$Z_2 = \frac{C_{a2}\omega_0 + 1/L_{a2}\omega_0}{\pi}\frac{Z_0}{N_1^2} = 1788.80\ \Omega$$

Using $Z_1 = 89.1\ \Omega$ and $Z_2 = 1788.80\ \Omega$, we find that the even and odd impedances are $Z_{oe} = 108.51\ \Omega$ and $Z_{oo} = 69.69\ \Omega$. From Figure 11.22, this provides $S_{12}/H = 0.15$ and $W_2/H = 0.56$. This gives $S_{12} = 0.525$ mm and $W_2 = 1.96$ mm.

The length L_2 would normally be taken as $\lambda_g/2$. However, since this line is open circuit at both ends, one should account for the fringing capacitances that change its effective electrical length. The fringing capacitances are shown in Figure 11.28, with

$$\frac{C_f}{\varepsilon_0} = \frac{C_{f1} + C_{f2}}{\varepsilon_0} = \Delta l\left(\frac{1}{(H-t)/2} + \frac{1}{(H-t)/2 + t/\varepsilon_r}\right)$$

The stringing capacitance for our structure is $C_f/\varepsilon_0 = 0.906$. This leads to $\Delta l = 0.76$ mm, and the final length L_2 is given by $L_2 = \lambda_g/2 - 2\Delta l = 15.04$ mm.

The value of the gap S_{01} is then computed to achieve an input admittance inverter of value N_1. Unfortunately, it is found that this gap is extremely small, on the order of 10 μm. Since this gap is used to match the input admittance inverter of the prototype, a $\lambda_g/4$ line is used instead, as shown in Figure 11.29. Indeed, the *ABCD* matrix of a quarter-wavelength line ($\theta = \pi\ /\ 2$) of impedance Z_{01} is the same as that of an inverter and can be matched to the admittance inverter of the prototype, as shown below. In this case, the denormalized impedance of this line is given by

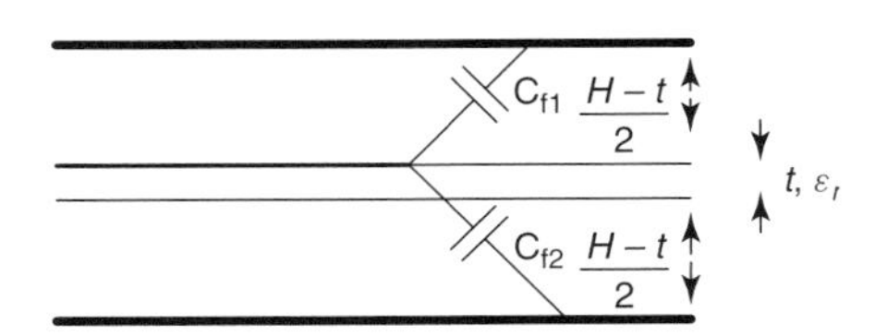

Figure 11.28 Fringing capacitances to take into account an open-circuit coupled line.

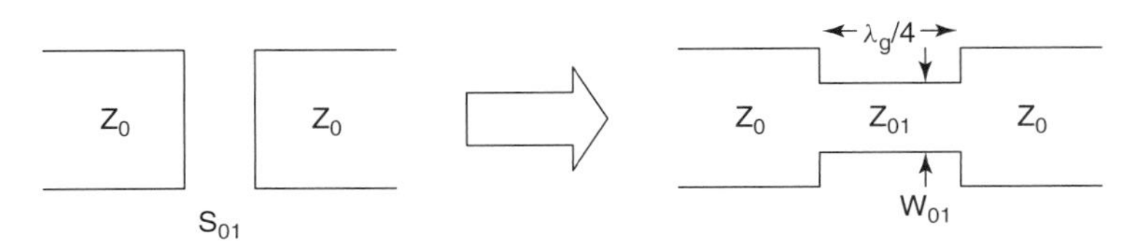

Figure 11.29 Replacing the gap by a $\lambda_g\ /\ 4$line.

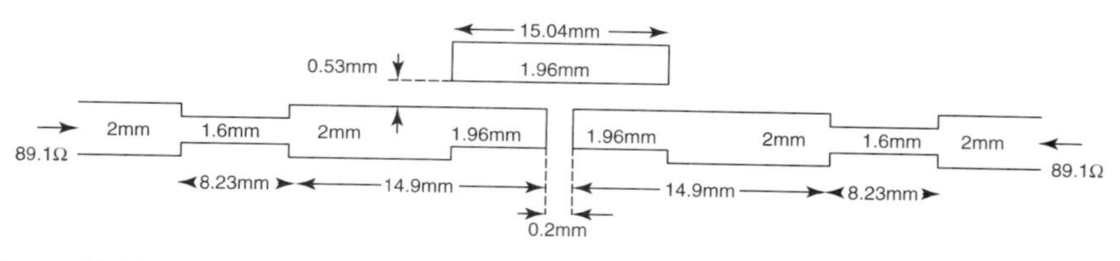

Figure 11.30 Third-order generalized Chebyshev SSS layout (for 89.1-Ω terminations).

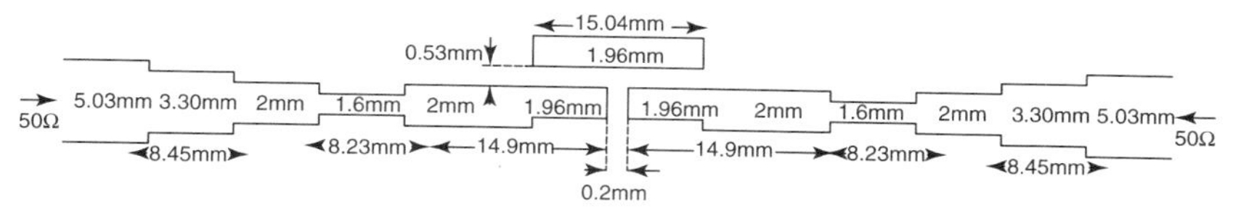

Figure 11.31 Third-order generalized Chebyshev SSS layout (for 50-Ω terminations).

$$\begin{pmatrix} \cos\theta & j\sin\theta\, Z_{01} \\ \dfrac{j\sin\theta}{Z_{01}} & \cos\theta \end{pmatrix}_{\theta\to\pi/2} = \begin{pmatrix} 0 & j\,Z_{01} \\ \dfrac{j}{Z_{01}} & 0 \end{pmatrix} = \begin{pmatrix} 0 & \dfrac{j}{N_1} \\ jN_1 & 0 \end{pmatrix}$$

$$Z_{01} = \frac{1}{N_1} Z_0 = 99.49\ \Omega$$

and from Figure 11.23, the width of this line should be $W_{01} = 1.60$ mm and the length $L_{01} = 8.23$ mm.

The suspended substrate stripline layout of the third-order generalized Chebyshev filter with one zero of transmission is shown in Figure 11.30. To adapt the filter to $Z_e = 50$-Ω loads, an impedance transformer is used. As shown before, a $\lambda_g/4$ line of impedance Z_i can be used as an inverter. The impedance is computed such that

$$Z_e = 50\ \Omega = \frac{Z_0}{Z_i^2}$$

This gives $Z_i = \sqrt{50 \times 89.1} = 66.74\ \Omega$. Using Figure 11.23, we define the width as $W_i = 3.29$ mm, and for $Z_e = 50\ \Omega$, $W_e = 5.03$ mm. The final layout for 50-Ω terminations is shown in Figure 11.31.

11.6 REALIZATION OF THE CGCL FILTER

The third-order filter of the design example was realized using suspended substrate stripline and is shown in Figure 11.32. The measured responses of the filter are shown in Figure 11.33 [11.9]. The transmission characteristic shows a transmission zero placed slightly above the desired frequency of 9200 MHz. Tuning

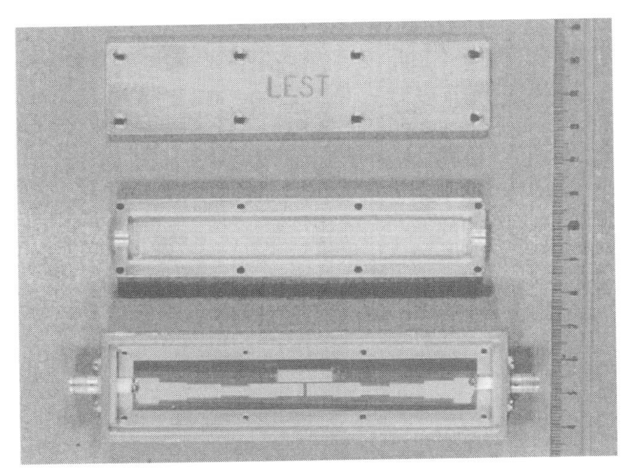

Figure 11.32 Third-order generalized Chebyshev CGCL filter.

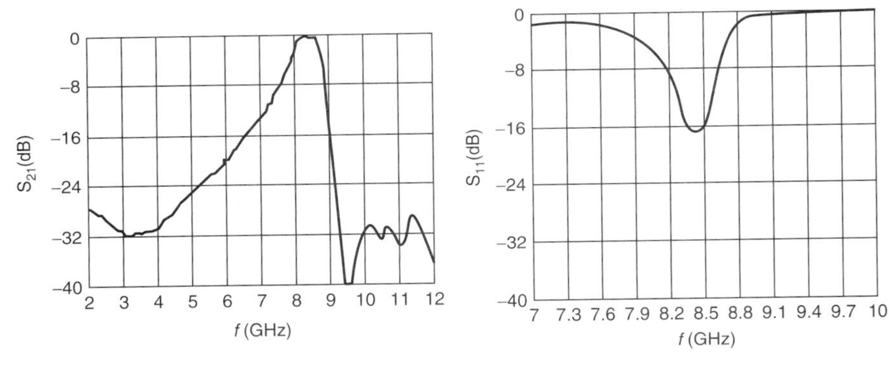

Figure 11.33 Forward and return coefficients of the third-order generalized Chebyshev CGCL filter.

the gap S_{13} allows as to place the zero at the correct frequency. This shows that the transmission zero is due to this coupling gap. The insertion loss is about 1 dB. Tuning screws were placed above the impedance transformers to improve the rejection in the lower stopband. The three resonances do not appear clearly in the return loss characteristic. However, the adaptation is acceptable and the return loss is about 16 dB.

11.7 CONCLUSIONS

In this chapter we provided a theory of a simple and original type of non-minimum-phase filter. The results confirm that asymmetrical responses are achievable using the CGCL topology implemented in SSS. SSS resonators and non-minimum-phase coupling have shown interesting results for the design of sophisticate microwave filters.

Acknowledgments

This work was carried out under contract with Thomson.

REFERENCES

[11.1] R. J. Cameron, Fast generation of Chebyshev filter prototypes with asymmetrically prescribed transmission zeros, *ESA J.*, vol. 6, pp. 83–95, 1982.

[11.2] S. Amari, Synthesis of cross-coupled resonator filters using an analytical gradient-based optimization technique, *IEEE Trans. Microwave Theory Tech.*, vol. 48, no. 9, pp. 1559–1564, 2000.

[11.3] R. J. Cameron, General prototype network-synthesis methods for microwave filters, *ESA J.*, vol. 6, pp. 192–206, 1982.

[11.4] E. Hanna, P. Jarry, E. Kerhervé, J. M. Pham, and D. Lo, Synthesis and design of a suspended substrate capacitive-gap parallel coupled-line (CGCL) bandpass filter with one transmission zero, *Proc. 10th IEEE International Conference on Electronic Circuits and Systems*, ICECS'03, University of Sharjah, United Arab Emirates, pp. 547–550, Dec. 14–17, 2003.

[11.5] E. Hanna, P. Jarry, E. Kerherve, and J. M. Pham, General prototype network method for suspended substrate microwave filters with asymmetrically prescribed transmission zeros: synthesis and realization, *Proc. IEEE Mediterranean Microwave Symposium*, MMS'04, Marseille, France, pp. 40–43, June 1–3 2004.

[11.6] E. Hanna, P. Jarry, E. Kerhervé, J. M. Pham, General prototype network with cross coupling synthesis method for microwave filters: application to suspended substrate stripline filters, Presented at the ESA-CNES International Wokshop on Microwave Filters, Toulouse, France, Sept. 13–15, 2004.

[11.7] J. A. G. Malherbe, Microwave Transmission Line Filters, Artech House, Dedham, MA, 1976.

[11.8] D. Lo, Hine Tong, and P. Jarry, Synthesis and compact realization of asymmetric microwave suspended substrate filters, *Proc. IEEE AP/MTTS*, University of Pennsylvania, Philadelphia, PA, pp. 41–44, Mar. 1981.

[11.9] E. Hanna, P. Jarry, E. Kerhervé, and J. M. Pham, A design approach for capacitive gap parallel-coupled line filters and fractal filters with the suspended substrate technology, *Research Signpost*, Special Volume on Microwave Filters and Amplifiers, pp. 49–71, ed. P. Jarry, 2005.

[11.10] P. Jarry, E. Kerhervé, O. Roquebrun, M. Guglieimi, and J. M. Pham, A new class of dual-mode microwave filter, *Proc. Asia Pacific Microwave Conference*, APMC'02, Kyoto, Japan, pp. 727–730, Nov. 19–22, 2002.

Selected Bibliography

Cohn, S. B., Parallel-coupled transmission-line-resonator filters, *IEEE Trans. Microwave Theory Tech.*, vol. 6, no. 2, pp. 223–231, 1958.

Golberg, D. E. *Genetics Algorithms in Search, Optimization and Machine Learning*, Addison-Wesley, Reading, MA, 1989.

Zaabad, A. H., Q. J. Zhang, and M. Nakhla, A neural network modelling approach to circuit optimisation and statistical design, *IEEE Trans. Microwave Theory Tech.*, vol. 43, pp. 1349–1358, June 1995.

12

ASYMMETRICAL DUAL-MODE TE_{102}/TE_{301} THICK IRIS RECTANGULAR IN-LINE WAVEGUIDE FILTERS WITH TRANSMISSION ZEROS

12.1 INTRODUCTION

When several frequency channels are present in a given frequency band, one needs high-selectivity filters to tune to a given channel. This requires filters with asymmetrical responses. In this chapter a novel design technique based on

Advanced Design Techniques and Realizations of Microwave and RF Filters,
By Pierre Jarry and Jacques Beneat

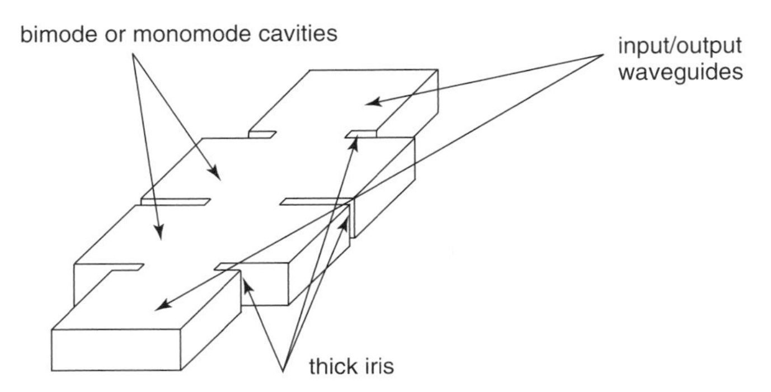

Figure 12.1 TE_{102}/TE_{301} filter.

an in-line prototype circuit using impedance inverters is presented. The filter topology to realize the asymmetrical response consists of rectangular monomode and bimode cavities coupled through thick irises. The design method is based on the use of a dedicated EM simulator for this structure and the use of progressive optimization steps. Realizations of these filters show that transmission zeros located on either the left or right of the passband are achievable. In addition, these filters are easy to manufacture and require nearly no tuning.

12.2 TE_{102}/TE_{301} FILTERS

The structure of the filter is shown in Figure 12.1. It consists of rectangular waveguides and cavities separated by centered thick irises. A bimode cavity (TE_{102}, TE_{301}) with a centered iris is capable of generating a transmission zero when the length and width are computed such that both resonant modes produce the same resonant frequency. The input and output waveguides are selected from standard values to operate using the TE_{10} mode. The monomode cavities operate using the TE_{101} resonant mode and are used to provide additional resonances (or poles) [12.1].

12.3 SYNTHESIS OF LOW-PASS ASYMMETRICAL GENERALIZED CHEBYSHEV FILTERS

The low-pass transmission function $|S_{21}(j\omega)|^2$ for the case of a symmetrical or asymmetrical generalized Chebyshev response is given by

$$|S_{21}(j\omega)|^2 = \frac{1}{1+\varepsilon^2 F_n^2(\omega)}$$

where ε provides a Chebyshev-type ripple in the passband and the function $F_n(\omega)$ is defined as

$$F_n(\omega) = \cosh\left(\sum_{k=1}^{n} \cosh^{-1}\frac{\omega - 1/\,\omega_k}{1-\omega\,\omega_k}\right)$$

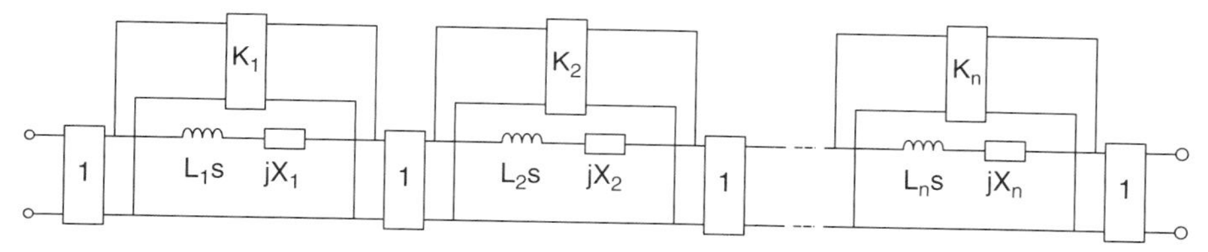

Figure 12.2 Low-pass in-line impedance inverter network.

where ω_k represents an arbitrary zero of transmission. Note that ω_k can be chosen finite or infinite. The transfer functions $S_{21}(s)$ and $S_{11}(s)$ can be obtained using the technique described in Chapter 11. These transfer functions can be realized using the in-line structure of Figure 12.2 [12.2].

12.3.1 Fundamental Element

It can be shown using even- and odd-mode characterizations that the transfer function of the fundamental element is such that [12.2]

$$S_{21}(s) = \frac{Y_e(s) - Y_o(s)}{[1 + Y_e(s)][1 + Y_o(s)]} = \frac{1 - (j/K)(Ls + jX)}{(1 + (j/K))\{1 + (1 - j/K)[(L/2)s + j(X/2)]\}}$$

where $Y_e(s)$ and $Y_o(s)$ are the even- and odd-mode admittances. One can easily check that $S_{21}(j\omega)$ is equal to zero when

$$1 + \frac{L\omega + X}{K} = 0$$

Therefore, the fundamental element produces a zero of transmission at $\omega = \omega_Z$ such that

$$\omega_z = -\frac{K + X}{L}$$

Note that ω_Z can be positive or negative, depending on the values and signs of K and X. This means that the fundamental element can produce zeros of transmission on the left or right side of the passband.

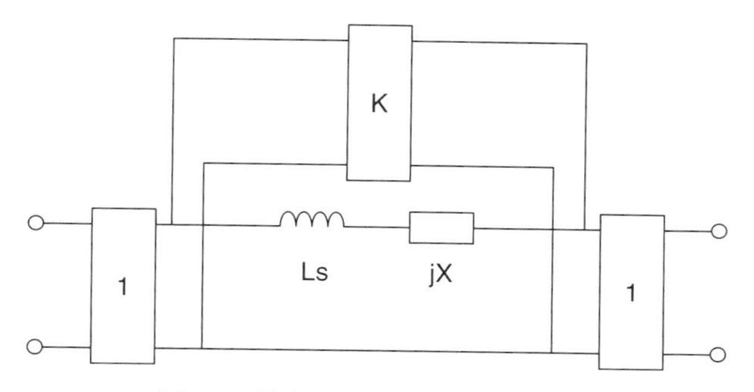

Figure 12.3 Fundamental element.

One should note that the fundamental element of Figure 12.3 will have the same $S_{21}(s)$ as that of the fundamental element using admittance inverters of Chapter 11 when taking $L = C$, $X = B$, and $K = -1/J$.

12.3.2 Analysis of the In-Line Network

As for the case of Chapter 11, it will be necessary to define an intermediary circuit equivalent to the fundamental element but that is easier to extract. This intermediary circuit is based on the two identities shown in Figure 12.4. First, the impedance inverter K is replaced by a corresponding admittance Π network. Then a parallel admittance placed after an inverter can be brought in front of the inverter as a series impedance. The parallel admittance and the series impedance will have the same value if the inverter is unity.

The resulting equivalent circuit of the fundamental element is shown in Figure 12.5. One should note, however, that Ls and jX are treated as impedances, whereas $-j/K$ is treated as an admittance. The structure of the intermediary circuit is shown in Figure 12.6. The series impedances seen between the two unity inverters for the circuits of Figures 12.5 and 12.6 are given by

$$Z(s) = \frac{1}{1/(Ls + jX) - j/K} \quad \text{and} \quad Z(s) = \frac{1}{C's + jB'} + jX'$$

The two impedances and therefore the two circuits are the same if

$$K = X'$$

$$L = C'X'^2$$

$$X = (X'B' - 1)X'$$

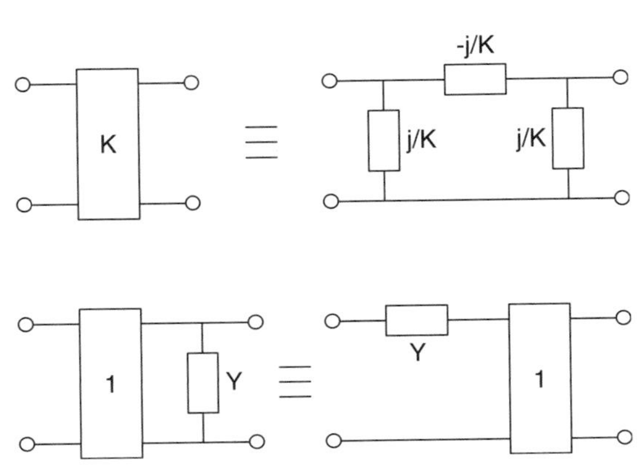

Figure 12.4 Identities used to construct the intermediary circuit.

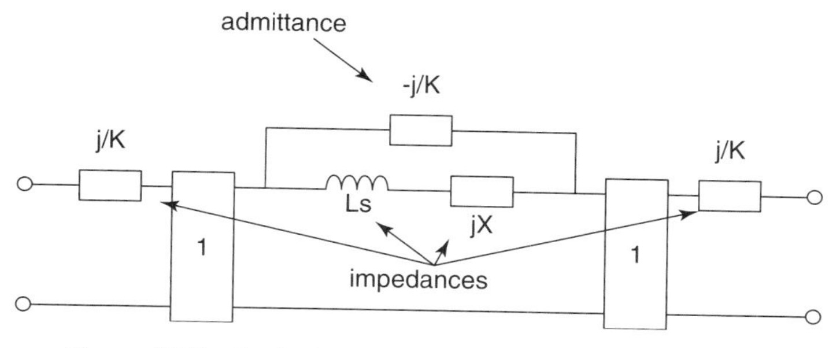

Figure 12.5 Equivalent circuit of the fundamental element.

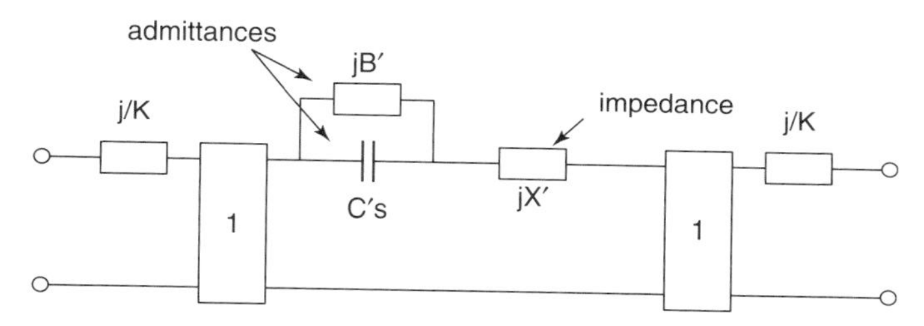

Figure 12.6 Intermediary circuit for extraction purposes.

One notes that the impedance from Figure 12.6 becomes infinity when

$$jC'\omega + jB' = 0$$

When this impedance reaches infinity, the circuit produces the zero of transmission. Therefore, we have

$$C'\omega_Z + B' = 0 \quad \text{or} \quad B' = -\omega_Z C'$$

We can then express the conditions for the two circuits to be equivalent as

$$\begin{aligned} K &= X' \\ L &= C'K^2 \\ X &= (-K\omega_z C' - 1)K = -\left(\frac{1}{K} + \omega_Z C'\right) K^2 \end{aligned}$$

and the intermediary circuit can be drawn directly using K, the capacitance C', and the zero of transmission ω_z, as shown in Figure 12.7. The extraction procedure will consist of extracting K and the capacitance C'. The capacitance C and the susceptances B of the fundamental element can then be found from K and C' [12.2].

The structure of Figure 12.2 can produce as many zeros of transmission as there are fundamental elements. One can, however, decide not to include all the

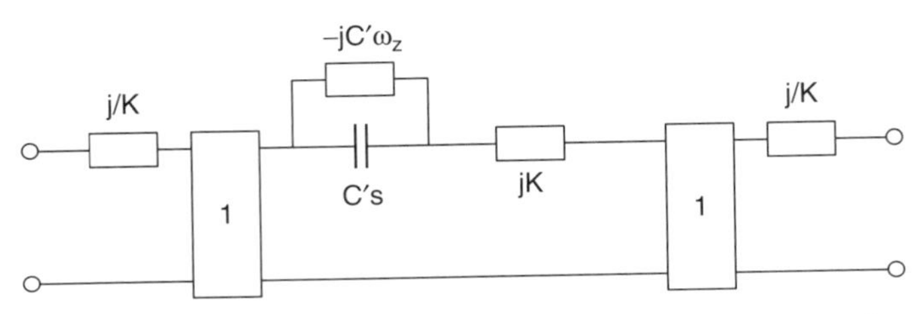

Figure 12.7 Intermediary circuit of the fundamental element for extraction purposes.

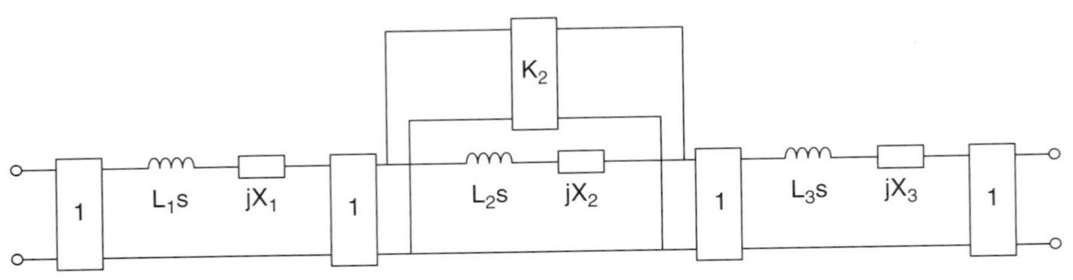

Figure 12.8 Prototype circuit for a third-order generalized Chebyshev function with one zero of transmission.

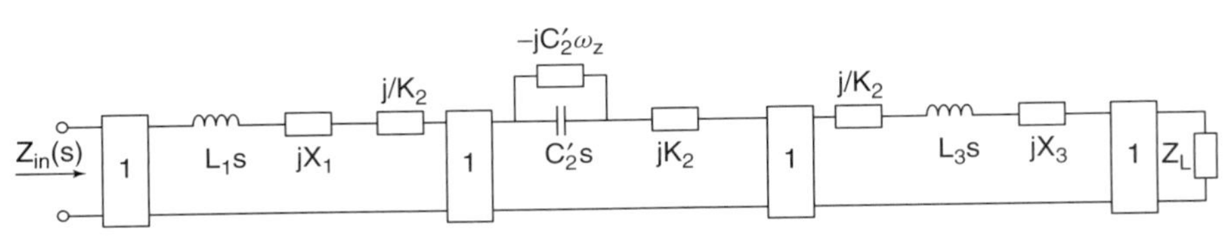

Figure 12.9 Intermediary circuit used for the extraction technique.

fundamental elements. For example, Figure 12.8 shows the structure that can realize a third-order generalized Chebyshev function with one zero of transmission. The extraction technique consists in replacing a fundamental element by the intermediary circuit as shown in Figure 12.9.

12.3.3 Synthesis by Simple Extraction Techniques

The input impedance $Z_{in}(s)$ of the network in Figure 12.9 is given by

$$Z_{in}(s) = \cfrac{1}{L_1 s + jX_1 + \cfrac{j}{K_2} + \cfrac{1}{jK_2 + \cfrac{1}{C_2'(s - j\omega_Z)} + \cfrac{1}{L_3 s + jX_3 + j/K_2 + 1/Z_L}}}$$

The input impedance for the generalized Chebychev function is defined in terms of $S_{11}(s)$ as

$$Z_{in}(s) = Z_L \frac{1 + S_{11}(s)}{1 - S_{11}(s)}$$

Note that for the procedure to start properly, the degree of numerator and denominator polynomials should be different by 1. Therefore, it might be necessary to perform some scaling of the numerator and denominator polynomials to match their highest-degree coefficients. In the case of complex-valued coefficients, one can also scale a polynomial by $\pm j$ without affecting the magnitude response.

The unity admittance inverter is first extracted, leaving the new input impedance as

$$Z(s) = L_1 s + jX_1 + \frac{j}{K_2} + \cfrac{1}{jK_2 + \cfrac{1}{C_2'(s - j\omega_Z)} + \cfrac{1}{L_3 s + jX_3 + j/K_2 + 1/Z_L}}$$

The inductance L_1 is extracted using

$$L_1 = \left.\frac{Z(s)}{s}\right|_{s\to\infty}$$

The input impedance becomes

$$Z_1(s) = Z(s) - L_1 s$$

The reactance X_1 is extracted using

$$jX_1 = Z_1(s)|_{s\to\infty}$$

The input impedance becomes

$$Z_2(s) = Z_1(s) - jX_1$$

The impedance inverter K_2 is extracted using

$$\frac{j}{K_2} = Z_2(s)|_{s\to j\omega_Z}$$

At this stage the input admittance $Y_2'(s)$ is computed as

$$Y_2'(s) = \frac{1}{Z_2(s) - j/K_2} - jK_2$$

The capacitance C_2' of the intermediary circuit is then extracted using

$$\frac{1}{C_2'} = (s - j\omega_Z)Y_2'(s)|_{s\to j\omega_z}$$

from which the inductance and reactance L_2 and X_2 are computed using

$$L_2 = C_2' K_2^2 \quad \text{and} \quad X_2 = -\left(\frac{1}{K_2} + \omega_Z C_2'\right) K_2^2$$

The input admittance becomes

$$Y_3(s) = Y_2'(s) - \frac{1}{C_2'(s - j\omega_Z)} \quad \text{and} \quad Z_3(s) = \frac{1}{Y_3(s)}$$

The inductance L_3 is then extracted using

$$L_3 = \left.\frac{Z_3(s)}{s}\right|_{s\to\infty}$$

The input impedance becomes

$$Z_4(s) = Z_3(s) - L_3 s - \frac{j}{K_2}$$

The reactance B_3 and final load can be found knowing that

$$Z_4(s) = jX_3 + \frac{1}{Z_L}$$

The procedure can be generalized using the flowchart shown in Figure 12.10.

12.3.4 Frequency Transformation

The low-pass prototype circuit can be modified to produce a bandpass response as shown in Figure 12.11. This circuit is obtained using the low pass-to-bandpass transformation technique:

$$s \to \alpha\left(\frac{s}{\omega_0} + \frac{\omega_0}{s}\right) \quad \text{with} \quad \begin{array}{l} \omega_0 = \sqrt{\omega_1\,\omega_2} \\ \alpha = \dfrac{\omega_0}{\omega_2 - \omega_1} \end{array}$$

The admittance inverters are independent of frequency, so that they remain the same. We then identify the impedances and their derivatives such that

$$L\left[\alpha\left(\frac{\omega}{\omega_0} - \frac{\omega_0}{\omega}\right)\right] + X = L_a\omega - \frac{1}{C_a\omega} \quad \text{and} \quad L\left[\alpha\left(\frac{1}{\omega_0} + \frac{\omega_0}{\omega^2}\right)\right] = L_a + \frac{1}{C_a\omega^2}$$

Solving the two equations at $\omega = \omega_0$, we find that

$$X = L_a\omega_0 - \frac{1}{C_a\omega_0} \quad \text{and} \quad \frac{2\alpha}{\omega_0}L = L_a + \frac{1}{C_a\omega_0^2}$$

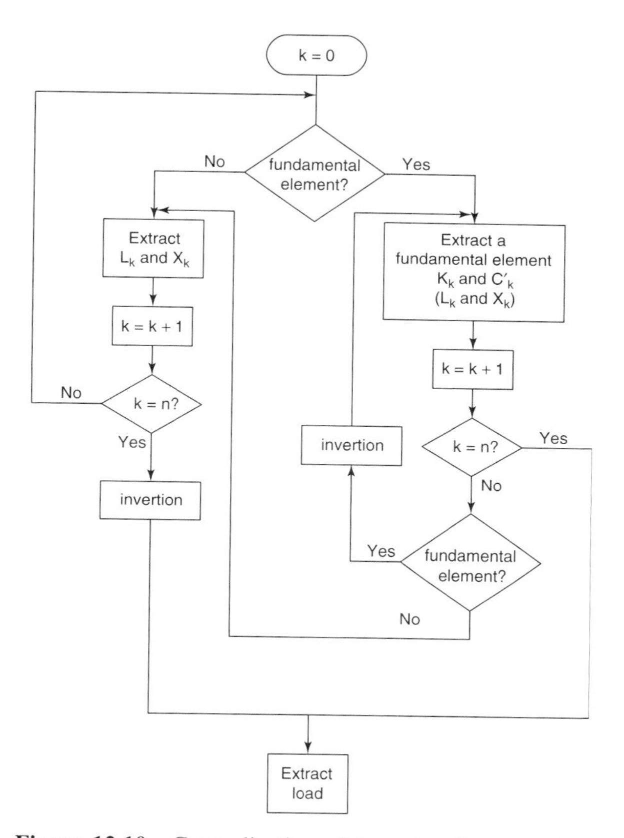

Figure 12.10 Generalization of the extraction procedure.

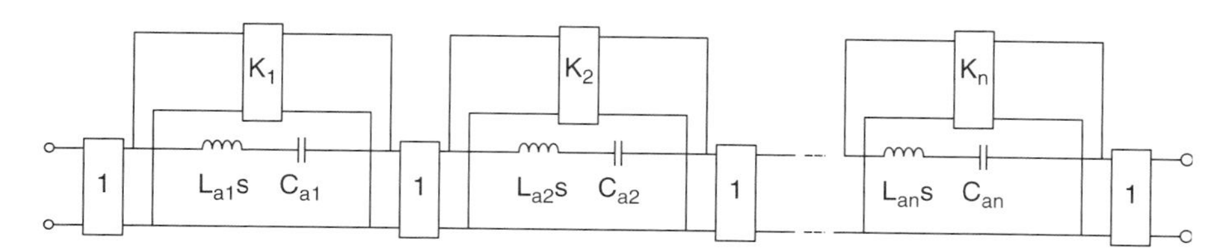

Figure 12.11 Bandpass in-line impedance inverter network.

so that to define the bandpass elements we will use the relations

$$L_{ak} = \frac{1}{2}\left(\frac{2\alpha L_k + X_k}{\omega_0}\right) \quad \text{and} \quad C_{ak} = \frac{2}{\omega_0(2\alpha L_k - X_k)}$$

The results are valid as long as $\omega/\omega_0 + \omega_0/\omega = 2$. This means that ω should be close to ω_0, and the design technique will work properly for narrowband applications.

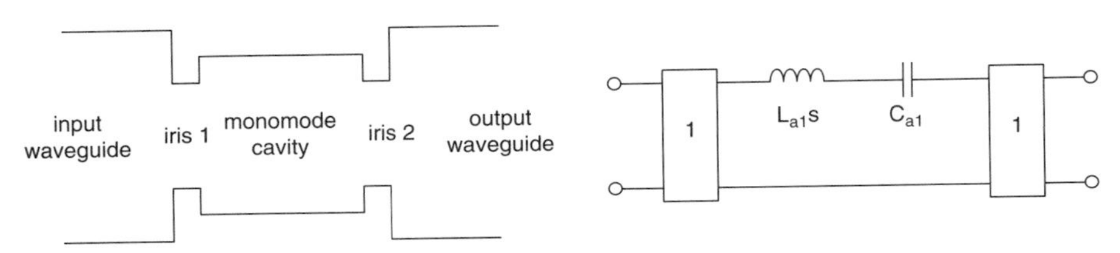

Figure 12.12 First-order filter without a transmission zero.

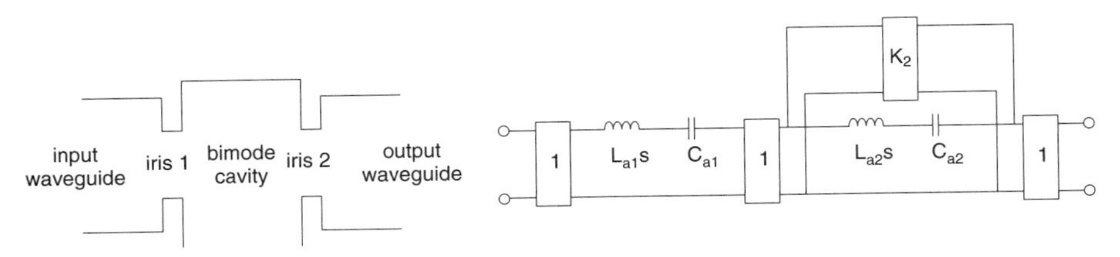

Figure 12.13 Second-order filter with one transmission zero.

12.4 DESIGN METHOD

12.4.1 Equivalent Circuit of Monomode and Bimode Cavities

In a monomode cavity, only the fundamental mode propagates. This cavity can generate one resonance (or pole). It is not possible to obtain a zero of transmission. The physical structure and its equivalent circuit for a first-order filter with no transmission zeros are shown in Figure 12.12. In a bimode cavity there will be two resonances (or poles), one for each mode [12.3]. In addition, by careful selection of the cavity, it will be possible to produce one transmission zero through coupling of the two modes. The physical structure and its equivalent circuit for a second-order filter with one transmission zero are shown in Figure 12.13.

By cascading monomode and bimode cavities, one can then obtain filters of various degrees and numbers of transmission zeros. For example, a third-order filter with one transmission zero is achieved using one bimode cavity (two poles, one zero) and one monomode cavity (one pole), as shown in Figure 12.14. By cascading two bimode cavities (two poles and one zero each), one can obtain a fourth-order filter with two transmission zeros, as shown in Figure 12.15.

It is clear that not all degrees and number of transmission zeros are possible, due to the equivalent circuit of the bimode cavity. In general, the maximum number of transmission zeros for a given order n is equal to $n/2$ for n even and $(n-1)/2$ for n odd.

12.4.2 Optimization Approach

At this stage one would traditionally define a set of equations relating the physical parameters of the structure with the equivalent circuit and define the values of

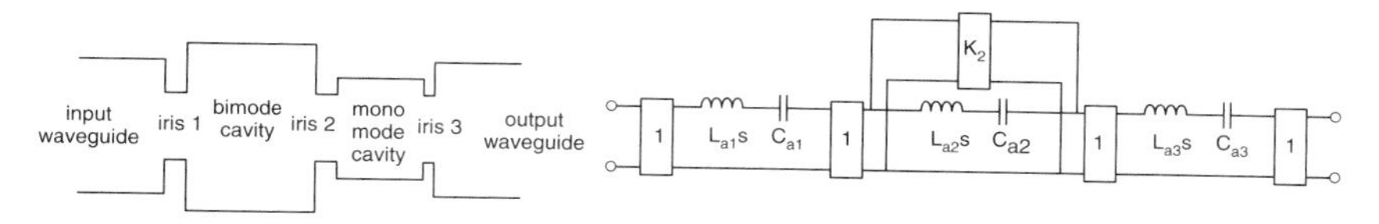

Figure 12.14 Third-order filter with one transmission zero.

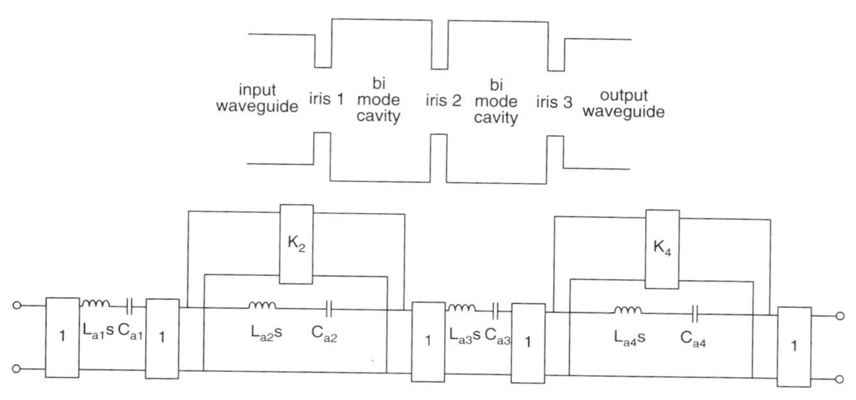

Figure 12.15 Forth-order filter with two transmission zeros.

the equivalent circuit based on the prototype circuit. The main problem with this analytical approach is that the equations involved should be kept simple by using a series of approximations and simplifications. This usually means that the measured response of the constructed filter will be different from that of the equivalent circuit, due to these simplifications. With the increased performance of computers, it is now possible to design microwave filters using numerical approaches. These are often referred to as optimizations techniques and are discussed in more detail in Chapter 14.

The optimization technique for the filters of this chapter makes use of an efficient electromagnetic simulator of the monomode and bimode rectangular cavities separated by thick irises in the filter. The simulator is based on a multimode impedance network representation of an inductive step for rectangular waveguides [12.4]. Briefly, the method starts by defining the limit conditions of the electromagnetic fields at the junctions between waveguides (in particular, those junctions created by the irises). It then computes a multimode impedance matrix. The elements are obtained by solving an integral equation independent of frequency [12.4]. Once all the impedance matrices of all the junctions between waveguides are computed, their cascading can provide the matrix of the overall structure (equivalent multimode network). The last step consists of computing the S parameters from the equivalent multimode network. A software package called WIND which implements this method was developed at the European Space Research and Technology Centre, Noordwijk, The Netherlands.

The design of a filter starts by defining the bandpass prototype circuit satisfying the filtering requirements using the techniques of Section 12.3. At this stage it is possible to define initial values for the physical structure. The dimensions of the input/output waveguides are selected from standard dimensions such as WR112 (28.5×12.62 mm) for 8 GHz operation, WR75 (19.05×9.53 mm) for 12 GHz operation, and so on. The input/output waveguide selection will define the height of all cavities since in this technique we take all the heights to be equal. The second step is to compute the length and width of the cavities. For a monomode cavity, the dimensions are computed such that the resonant frequency of the TE_{101} mode is equal to the center frequency of the filter:

$$f_{TE_{101}} = \frac{c}{2}\sqrt{\left(\frac{1}{a}\right)^2 + \left(\frac{1}{l}\right)^2} = f_0$$

where $a = 19.05$ mm for a WR75 guide, for example.

The dimensions of a bimode cavity are computed so that the resonant frequency of the TE_{102} and TE_{301} modes are both equal to the frequency of a transmission zero:

$$f_{TE_{102}} = \frac{c}{2}\sqrt{\left(\frac{1}{a}\right)^2 + \left(\frac{2}{l}\right)^2} = f_z$$

$$f_{TE_{301}} = \frac{c}{2}\sqrt{\left(\frac{3}{a}\right)^2 + \left(\frac{1}{l}\right)^2} = f_z$$

For both frequencies to be equal, a and l are related by the factor $\sqrt{8/3}$. The dimensions of the irises are, on the other hand, defined grossly. For instance, at 8 GHz, the initial iris dimensions were taken as 16×2 mm, whereas at 12 GHz they were taken as 8×2 mm.

The optimization technique then consists of optimizing the dimensions of the cavities and irises in a sequential manner based on the response of a section of the prototype circuit rather than optimizing all the dimensions at once using the entire response of the filter. This progressive approach is possible due to the equivalence between monomode and bimode cavities and sections of the prototype circuit. It has the advantage that only a few parameters are optimized at a given time.

For example, for a fourth-order filter with two transmission zeros, the procedure starts by comparing the elements that are introduced below. To compare the S parameters of the structure and equivalent circuit, it is important to use a termination impedance Z_{11} (Figure 12.16).

The physical parameters to be optimized, shown in Figure 12.17, are:

- a_1, l_1: width and length of the first bimode cavity
- t_1, w_1: depth and width of the first iris
- t_2, w_2: depth and width of the second iris

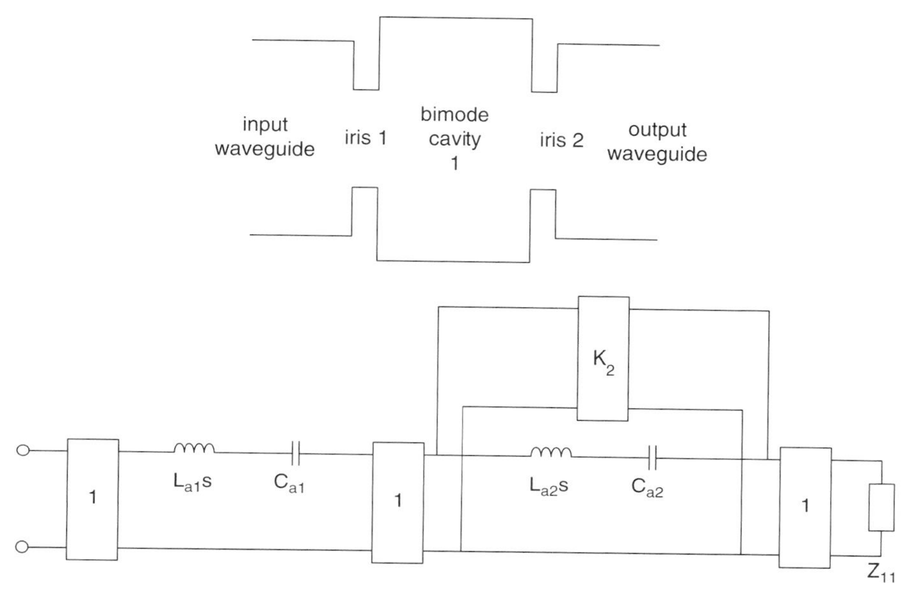

Figure 12.16 Starting point of the optimization.

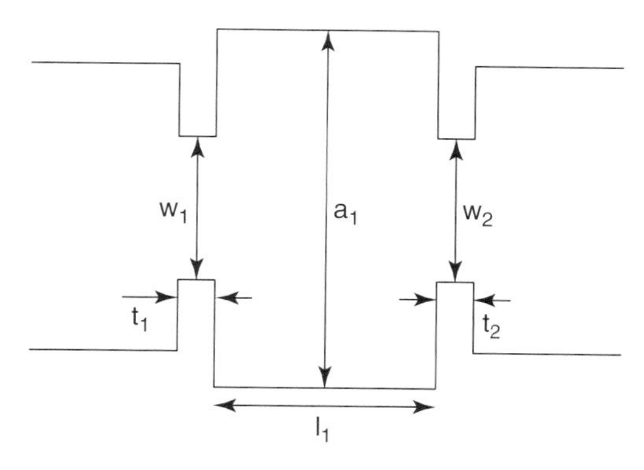

Figure 12.17 Physical parameters to optimize.

The next stage is to optimize the physical parameters of the second cavity by comparing the structure and prototype corresponding to the first and second cavities. In the case of a fourth-order filter, this results in optimizing the physical parameters of the second cavity using the optimized values of the first cavity and the entire prototype circuit, as shown in Figure 12.18.

The physical parameters being optimized are:

- a_2, l_2: width and length of the second bimode cavity
- t_3, w_3: depth and width of the third iris

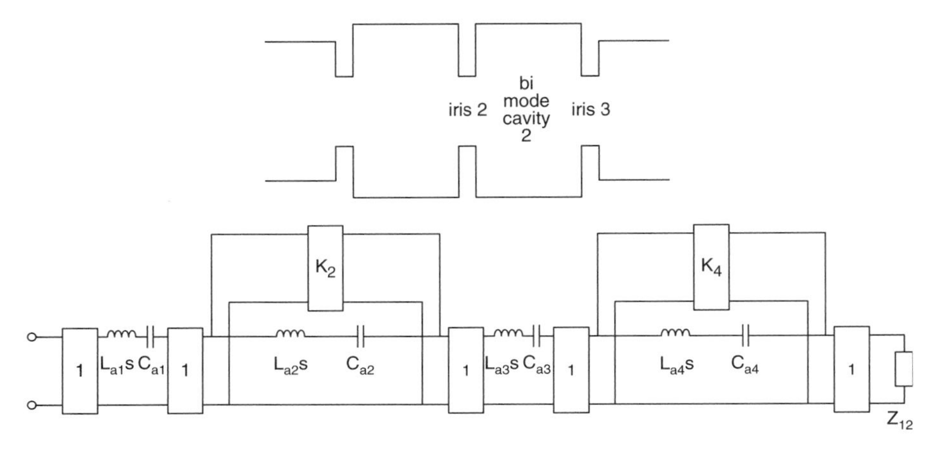

Figure 12.18 Next series of optimization.

Once every cavity and iris of the structure have been optimized using this progressive approach, the response is close to that of the target. It is then possible to have a final and general optimization of all parameters using the complete prototype circuit. The simplest optimization techniques are the Cauchy and Newton methods. Generally, these methods can apply to a multivariable function that returns a single value. For example,

$$f(x_1, x_2, x_3) = 7x_1x_2x_3 + 4x_1^2 + 3x_1x_2 + 5x_2^2 + x_3$$

The purpose of these algorithms is to find the values of x_1, x_2, and x_3 where the function $f(x_1, x_2, x_3)$ is at a minimum. There are, however, no guarantees that this could be a global minimum. They all use derivatives with respect to x_1, x_2, or x_3 and are referred to as *gradient methods*. The Newton method also requires the second derivatives.

The algorithms are based on defining the next set of parameters $x_1, x_2, \ldots, x_N$ following a general expression

$$\begin{aligned}
x_1^{(k+1)} &= x_1^{(k)} + \alpha^{(k)} s(x_1^{(k)}) \\
x_2^{(k+1)} &= x_2^{(k)} + \alpha^{(k)} s(x_2^{(k)}) \\
&\vdots \\
x_N^{(k+1)} &= x_N^{(k)} + \alpha^{(k)} s(x_N^{(k)})
\end{aligned}$$

where $x_i^{(k)}$ is the current estimate of the parameter x_i, $\alpha^{(k)}$ is a step length parameter, and $s(x_i^{(k)})$ is the search direction. The search directions for the Cauchy and Newton methods are

$$s(x_1^{(k)}) = -\left.\frac{\partial f(x_1, x_2, \ldots, x_N)}{\partial x_1}\right|_{x_1=x_1^{(k)}} \qquad s(x_1^{(k)}) = -\frac{\partial f(x_1, x_2, \ldots, x_N)/\partial x_1|_{x_1=x_1^{(k)}}}{\partial^2 f(x_1, x_2, \ldots, x_N)/\partial x_1^2|_{x_1=x_1^{(k)}}}$$

Cauchy Newton

The technique then consists of starting with initial values for all the parameters $x_1, x_2, \ldots, x_N$, and then modifying them at each iteration k using their current value and the derivatives at their current value. The Cauchy method works well far from the minimum but tends to converge slowly near the solution. On the other hand, the Newton method works well close to the solution. One could combine the two techniques in one optimization program.

There are several challenges in the case of microwave filter optimization. First, the EM simulator provides the frequency response of the structure at many different frequencies when the physical parameters $x_1, x_2, \ldots, x_N$ are given and not a single value. Therefore, one first needs to define an error function $e(x_1, x_2, \ldots, x_N)$ that gives a single value when the physical parameters $x_1, x_2, \ldots, x_N$ are given. This function can then be minimized using these techniques. The most common error functions are based on mean deviations or on maximum deviations between the simulated responses and the target or prototype circuit responses. For example, for optimizing the return loss, one could compute $|S_{11}(j\omega)|$ over 100 frequency points in the passband of the filter and express the mean and maximum error functions as

$$e(x_1, x_2, \ldots, x_N) = \frac{1}{N}\sum_{k=1}^{N} \left||S_{11}(j\omega_k)| - |\widehat{S_{11}}(j\omega_k)|\right|$$

$$e(x_1, x_2, \ldots, x_N) = \max \left||S_{11}(j\omega_k)| - |\widehat{S_{11}}(j\omega_k)|\right| \qquad k = 1, 2, \ldots, N$$

where $S_{11}(j\omega)$ is the EM simulated response that uses the physical parameters $x_1, x_2, \ldots, x_N$, and $\widehat{S_{11}}(j\omega)$ is the target or prototype circuit response. One could use the maximum criterion when far from the solution and the mean criterion when getting closer to the solution by using a combination of the error expressions with weight factors.

The second issue is that in microwave problems there will rarely be an analytical expression for the error function $e(x_1, x_2, \ldots, x_N)$. Therefore, defining analytical expressions for the derivatives is not possible and they must be evaluated numerically. For example the forward difference approximation

$$\left.\frac{\partial f(x_1, x_2, \ldots, x_N)}{\partial x_1}\right|_{x_1=x_1^{(k)}} \approx \frac{f(x_1 + \Delta, x_2, \ldots, x_N) - f(x_1, x_2, \ldots, x_N)}{\Delta}$$

where Δ will be a small increase in the parameter x_1, could be used. This means, however, that for 10 parameters to optimize, one needs to simulate 10 times the structure at each iteration to define the 10 search directions. If it takes 10 to 100 iterations to reach an acceptable solution, one would have simulated the structure at least 100 to 1000 times. If it takes the EM simulator 1 minute to simulate the structure, this means that it will take from 100 minutes (1 hour and 40 minutes) to 1000 minutes (16 hours and 40 minutes) for the optimization to complete. This is why it is often prohibitive to use a commercial general-purpose simulator, and one needs to develop a dedicated EM simulator.

For the Newton method, one can use a *quasi Newton method* such as the Davidon–Fletcher–Powell method. This technique uses previous computations of $x_i^{(k-1)}$ and previous derivatives to avoid direct computations of the second derivatives when defining the search direction.

In an optimization program there should also be a termination criterion. Usually, one would make the optimization program exit when the error function becomes very small. However, when developing an optimization program it is recommended that the user exit after a given number of iterations and plot the value of the error function versus the iteration number and plot the response achieved at that point. This can be used to adjust weighting functions and incremental change values Δ. It should be mentioned that, in general, the closer the initial filter is to the target, the faster the optimization will converge. Therefore, it is always important to have a design method that can provide a good starting point for optimization.

12.5 DESIGN EXAMPLE

We wish to design a third-order generalized Chebyshev filter centered at 12,000 MHz with a relative bandwidth of 300 MHz with one zero of transmission at 11,750 MHz. The return loss is specified at −20 dB [12.5]. The first step is to in define the location of the zero of transmission for the low-pass prototype. It is given by

$$\omega_Z = \alpha\left(\frac{\omega_{pz}}{\omega_0} - \frac{\omega_0}{\omega_{pz}}\right)$$

where ω_{pz} is the location of the bandpass zero (e.g., $2\pi \times 11{,}750 \times 10^6$ rad/s). In this case we find that $\omega_z = -1.71$ rad/s.

The technique for computing the low-pass transfer function is provided in Section 11.3. After scaling $Q(s)$ by 0.364061 so that $\max|S_{11}| = 1$ and $P(s)$ by j so that the extraction procedure can start properly, we have

```
S21(s):
D(s) = (0.0000+j-0.5848)s^1 + (1.0000+j0.0000)
Q(s) = (0.3641+j0.0000)s^3 + (0.8559+j0.1175)s^2
   + (1.2697+j0.4128)s^1 + (0.7123+j0.7043)
S11(s):
P(s) = (-0.3641+j0.0000)s^3 + (0.0000+j-0.1175)s^2
   + (-0.2636+j0.0000)s^1 + (0.0000+j-0.0588)
Q(s) = (0.3641+j0.0000)s^3 + (0.8559+j0.1175)s^2
   + (1.2697+j0.4128)s^1 + (0.7123+j0.7043)
```

```
Z(s):
Zn(s) = (0.7281+j0.0000)s^3 + (0.8559+j0.2351)s^2
   + (1.5333+j0.4128)s^1 + (0.7123+j0.7631)
Zd(s) = (0.8559+j-0.0000)s^2 + (1.0062+j0.4128)s^1
   + (0.7123+j0.6455)
```

The extraction procedure produces the following results:

```
L1 = 0.850676
X1 = -0.135554
K2 = 1.720567
L2 = 1.610383
X2 = 1.033187
L3 = 0.850676
X3 = -0.135554
load = 1.000000 + j0.000000
```

The capacitor and inductor values of the bandpass prototype are given by

$$L_{ak} = \frac{1}{2}\left(\frac{2\alpha L_k + X_k}{\omega_0}\right) \quad \text{and} \quad C_{ak} = \frac{2}{\omega_0(2\alpha L_k - X_k)}$$

Simulation of the bandpass prototype circuit using these values is shown in Figure 12.19.

For 12-GHz operation, the input/output waveguides are selected as WR75 (19.05×9.53 mm). Therefore, all the heights in the structure will be taken as 9.53 mm. The width and length of the bimode cavity are computed using

$$f_{\mathrm{TE}_{102}} = \frac{c}{2}\sqrt{\left(\frac{1}{a}\right)^2 + \left(\frac{2}{l}\right)^2} = 11,750 \text{ MHz}$$

$$f_{\mathrm{TE}_{301}} = \frac{c}{2}\sqrt{\left(\frac{3}{a}\right)^2 + \left(\frac{1}{l}\right)^2} = 11,750 \text{ MHz}$$

For both frequencies to be equal, a and l are related by the factor $\sqrt{8/3}$. This gives us $a_1 = 43.60$ mm and $l_1 = 26.70$ mm. The width and length of the monomode cavity are computed using

$$f_{\mathrm{TE}_{101}} = \frac{c}{2}\sqrt{\left(\frac{1}{a}\right)^2 + \left(\frac{1}{l}\right)^2} = 12,000 \text{ MHz}$$

Keeping the same width of the input/output guides, this gives us $a_2 = 19.05$ mm and $l_2 = 16.56$ mm.

The initial iris dimensions are chosen as $w = 8$ mm and $t = 2$ mm. The layout of the initial filter is shown in Figure 12.20. The bimode cavity is first optimized using the portion of the prototype circuit shown in Figure 12.21.

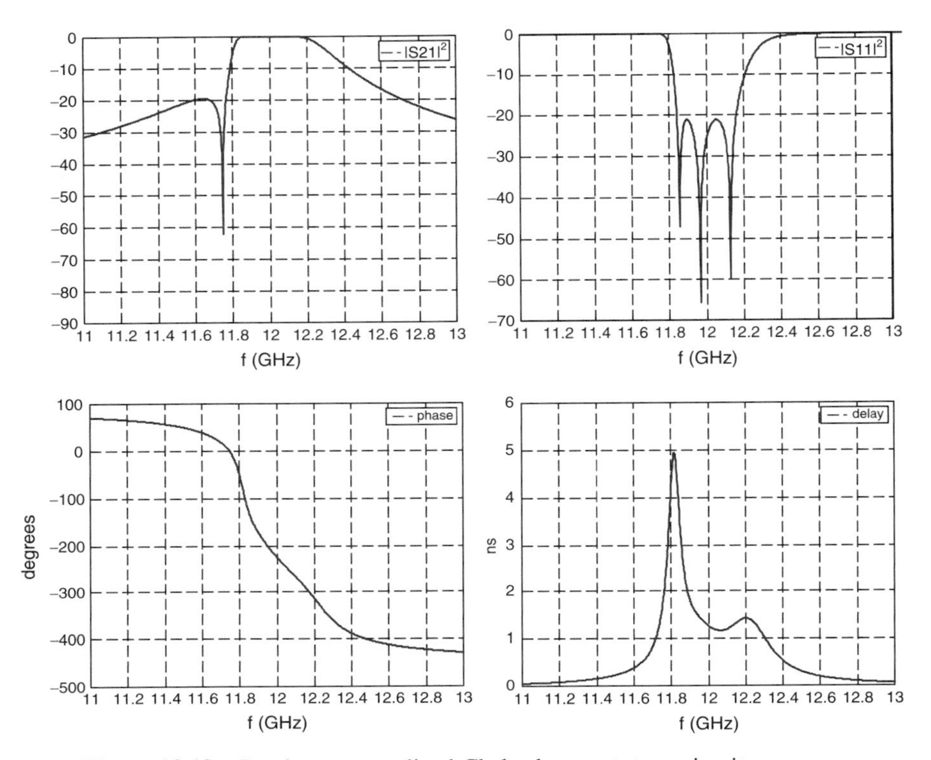

Figure 12.19 Bandpass generalized Chebyshev prototype circuit response.

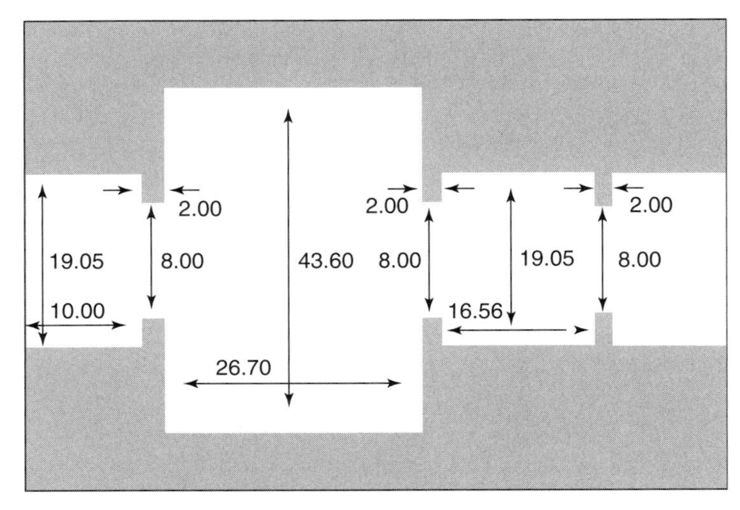

Figure 12.20 Initial physical parameter values.

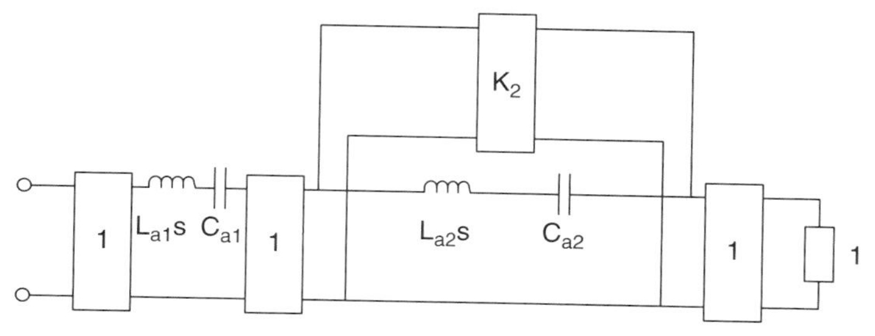

Figure 12.21 Prototype setup for optimizing the bimode cavity.

Figure 12.22 shows the EM simulation and the prototype simulation for the bimode cavity at the start of the optimization. After optimizing the bimode cavity, iris 1 and iris 2, the monomode cavity is added. Figure 12.23 shows the EM simulation and the prototype simulation for the bimode cavity and monomode cavity at the start of optimization.

The monomode cavity and iris 3 are then optimized. The values of the physical parameters after the optimization is complete are shown in Figure 12.24. The final simulated responses of the prototype and optimized structure are shown in Figure 12.25.

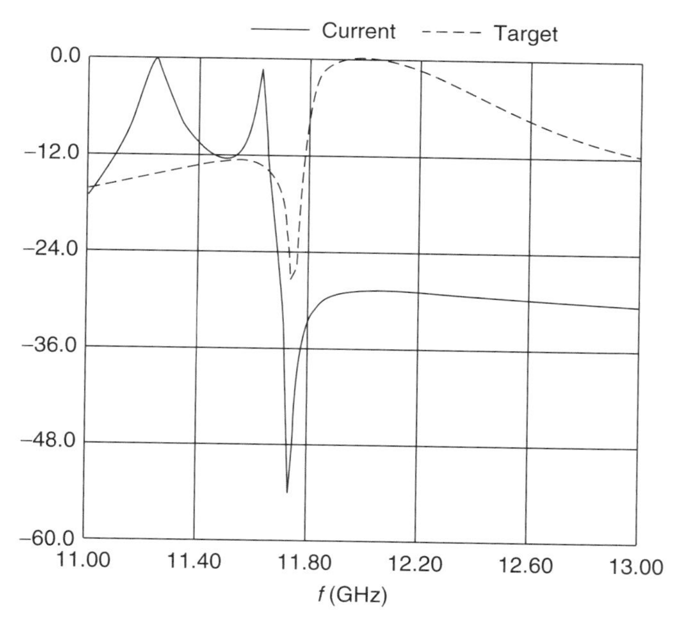

Figure 12.22 Simulation of the bimode cavity and its corresponding prototype portion.

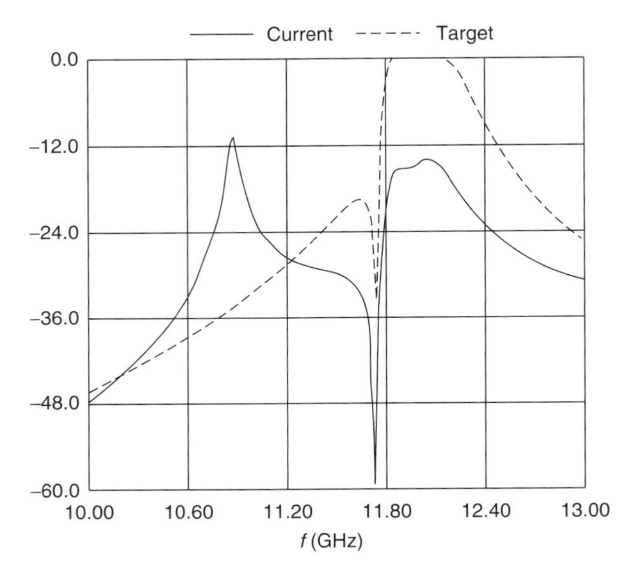

Figure 12.23 Adding the monomode cavity.

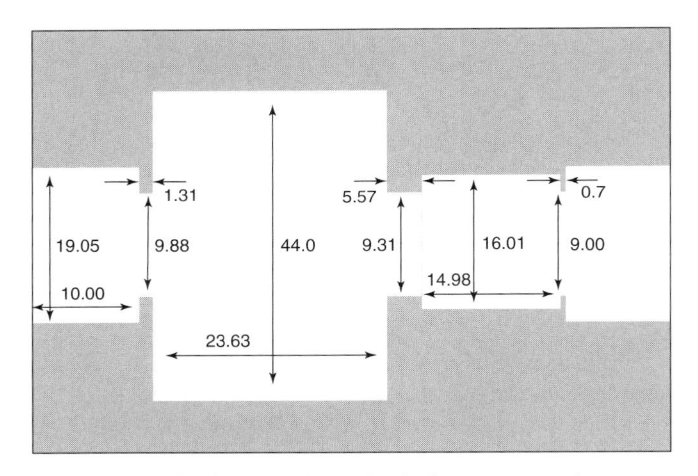

Figure 12.24 Optimized physical parameter values.

12.6 REALIZATIONS AND MEASURED PERFORMANCE

12.6.1 Third-Order Filter with One Transmission Zero

Several filters have been fabricated using the method described in this chapter [12.6,12.7]. The first filter was the third-order generalized Chebyshev with one transmission zero on the left of the passband described in Section 12.5. It uses one bimode (TE_{102}, TE_{301}) cavity and one monomode TE_{101} cavity. A picture of the filter constructed according to the physical parameters of Figure 12.24

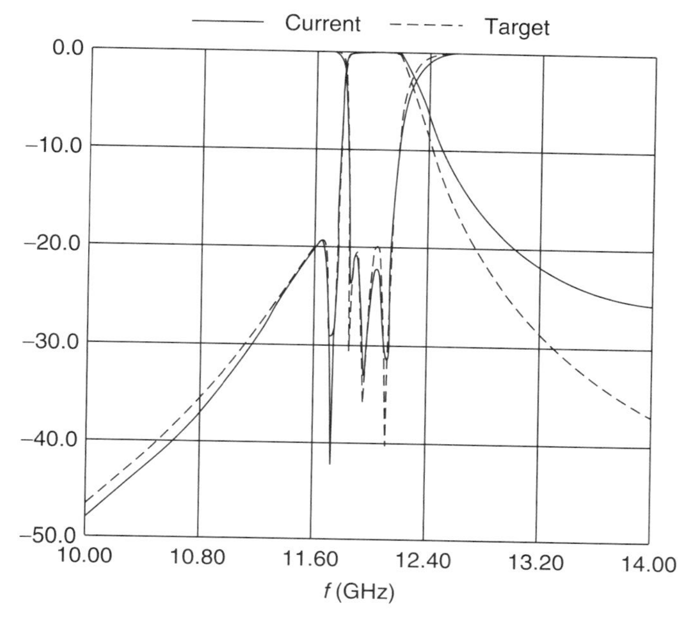

Figure 12.25 Target and optimized filter responses.

Figure 12.26 Third-order filter with one transmission zero.

is shown in Figure 12.26. The simulated and measured responses of this first moderate band filter are shown in Figure 12.27. The thin line shows the EM simulated response, and the thick line shows the measured response. The three resonances and the zero at 11.75 GHz are shown clearly. The measured response does not include any tuning screws.

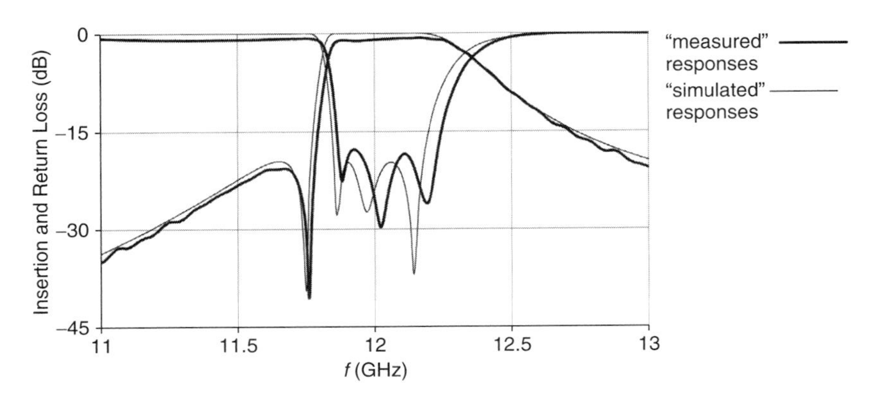

Figure 12.27 Simulated and measured responses of the third-order filter.

Figure 12.28 Fourth-order filter with two transmission zeros.

12.6.2 Fourth-Order Filter with Two Transmission Zeros

Using the method described in this chapter, a fourth-order filter with two transmission zeros was designed and built [12.6,12.7]. It uses two bimode (TE_{102}, TE_{301}) cavities. The zeros were specified at 11.76 and 12.26 GHz. The center frequency was specified at 12 GHz with a 300-MHz bandwidth and a −24-dB return loss. The filter is shown in Figure 12.28.

Figure 12.29 shows the simulated and measured responses of the filter in the vicinity of the passband. Due to construction tolerance problems, it was necessary to use a single tuning screw placed on the input waveguide to adjust the return loss without changing the transmission zeros that were correct. Figure 12.30 shows that the first spurious response is about 15.5 GHz.

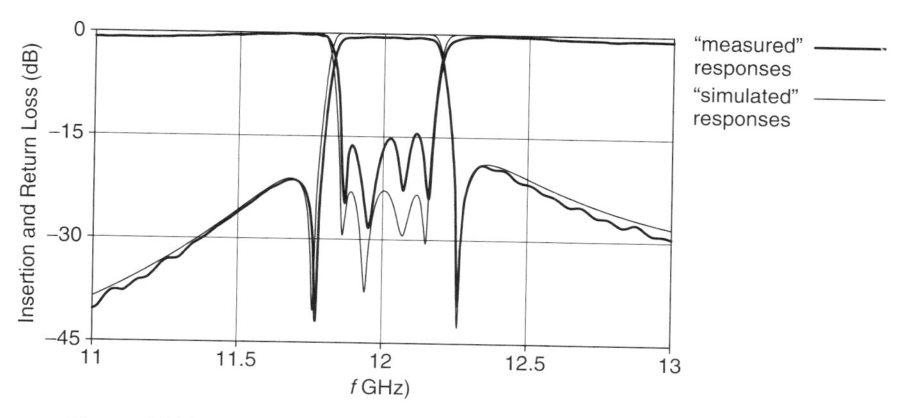

Figure 12.29 Simulated and measured responses of the fourth-order filter.

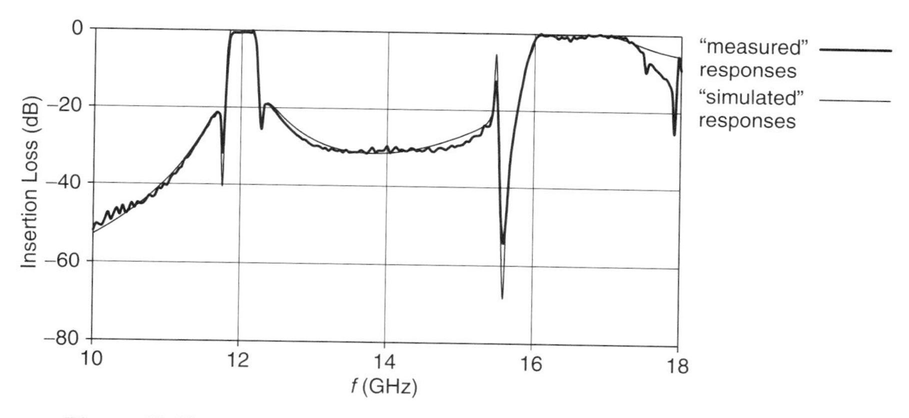

Figure 12.30 Evaluation of the spurious response of the fourth-order filter.

12.7 CONCLUSIONS

In this chapter we provided a design method for a new class of asymmetrical in-line dual-mode rectangular waveguide filters. These filters are based on the use of TE_{m0p} modes in rectangular cavities to provide transmission zeros. The cavities are coupled through thick irises and the structure can be simulated rapidly using the software package WIND, leading to a possible optimization-based design method. The realizations show that transmission zeros can be located on either the right and/or left side of the passband. At lower frequencies the construction tolerances have a minimal effect, resulting in filters that do not need additional tuning screws. At higher frequencies, however, the construction tolerances start having an effect.

In summary, these filters can handle high transmitting power, are easy to manufacture, have variable unloaded Q_S values, and meet the needs of satellite applications perfectly.

Acknowledgments

This work was carried out under contract with the European Space Agency and was patented in the United States [12.8].

REFERENCES

[12.1] M. Guglielmi, P. Jarry, E. Kerhervé, O. Roquebrun, and D. Schmitt, A new family of all-inductive dual-mode filters, *IEEE Trans. Microwave Theory Tech.*, pp. 1764–1769, Oct. 2001.

[12.2] P. Jarry, M. Guglielmi, J. M. Pham, E. Kerhervé, O. Roquebrun, and D. Schmitt, Synthesis of dual-modes in-line asymmetric microwave rectangular filters with higher modes, *Int. J. RF Microwave Comput. Aided Eng.*, pp. 241–248, Feb. 2005.

[12.3] P. Jarry, J. M. Pham, O. Roquebrun, E. Kerhervé, and M. Guglielmi, A new class of dual-mode asymmetric microwave rectangular filter, *Proc. IEEE International Symposium on Circuits and Systems*, ISCA'02, Phoenix, AZ, pp. 859–862, May 26–29, 2002.

[12.4] G. G. Conciauro, M. Guglielmi, and R. Sorrentino, Advanced Modal Analysis: CAD Techniques for Waveguide Components and Filters, Wiley, New York, 2000.

[12.5] P. Jarry, E. Kerhervé, O. Roquebrun, M. Guglielmi, and J. M. Pham, A new class of dual-mode microwave filter, *Proc. Asia Pacific Microwave Conference*, APMC′02, Kyoto, Japan, pp. 727–730, Nov. 19–22, 2002.

[12.6] M. Guglielmi, O. Roquebrun, P. Jarry, E. Kerhervé, M. Capurso, and M. Piloni, Low-cost dual-mode asymmetric filters in rectangular waveguide, *Proc. IEEE International Microwave Symposium*, IMS′01, Phoenix, AZ, pp. 1787–1790, vol. 3, May 20–25, 2001.

[12.7] P. Jarry, E. Kerhervé, O. Roquebrun, M. Guglielmi, D. Schmitt, and J. M. Pham, Rectangular realizations of a new class of dual-mode microwave filters, *Proc. IEEE International Microwave and Optoelectronics conference*, IMOC′03, Iguaçu, Brazil, pp. 9–12, Sept. 19–23 2003.

[12.8] M. Guglielmi, P. Jarry, and E. Kerhervé, All-inductive dual-mode filters, U.S. Patent, May 31, 2001.

Selected Bibliography

MacDonald, J., Electric and magnetic coupling through small apertures in shield walls of any thickness, *IEEE Trans. Microwave Theory Tech.*, vol. 20, no. 10, pp. 689–695, Oct. 1972.

Tang, W. C., and S. K. Chaudhuri, A true elliptic-function filter using triple-mode degenerate cavities, *IEEE Trans. Microwave Theory Tech.*, vol. 32, no. 11, pp. 1449–1454, 1984.

Rhodes, J. D., and I. H. Zabalawi, Synthesis of symmetrical dual mode in-line prototype networks, *Int. J. Circuit Theory APP.*, vol. 8, pp. 145–160, 1980.

Rosenberg, U., and D. Wolk, New possibilities of cavity-filter design by a novel TE-TM-mode iris-coupling, *MTT-S Int. Microwave Symp. Dig.*, vol. 89, no. 3, pp. 1155–1158, 1989.

Rosenberg, U., and D. Wolk, Filter design using in-line triple-mode cavities and novel iris couplings, *IEEE Trans. Microwave Theory Tech.*, vol. 37, no. 12, pp. 2011–2019, 1989.

13

Asymmetrical Cylindrical Dual-Mode Waveguide Filters with Transmission Zeros

13.1 INTRODUCTION

In this chapter we present a design method of iris-coupled dual-mode cylindrical cavity waveguide filters that have prescribed asymmetrical responses. The method is based on a cross-coupled prototype circuit and an extraction procedure based on chain matrices. The elements of the filter are defined by matching the coupling graphs and coupling matrices of the prototype circuit and filter. Prototype circuit

---

Advanced Design Techniques and Realizations of Microwave and RF Filters,
By Pierre Jarry and Jacques Beneat

layouts are provided that minimize the number of rotations needed for matching the coupling coefficients. Descriptions of cruciform irises and means for defining their physical parameters are also given. The realizations show that it is possible to achieve transmission zeros on either side of the passband.

13.2 DUAL-MODE CYLINDRICAL WAVEGUIDE FILTERS

The structure of the filter is shown in Figure 13.1 [13.1]. The resonators are cylindrical cavities that use the resonant modes TE_{11p}. The TE_{11p} modes provide for high-quality factors. These modes are polarized so that two orthogonal modes can be used independently inside each cavity. Coupling between the two modes is achieved using a coupling screw placed at a 45° angle that destroys the field symmetries inside the cavity. Coupling between cavities is assured by the cruciform iris. These can provide four couplings each. One should note, however, that all four couplings might not be required, depending on the number of transmission zeros to achieve. In this case some of the cruciform irises can be replaced by longitudinal irises. The input/output couplings can be realized using longitudinal slots when connecting to a waveguide system or when using coaxial RIM plugs.

The general form of the coupling graph of the bimode filter is shown in Figure 13.2 and the associated coupling matrix is

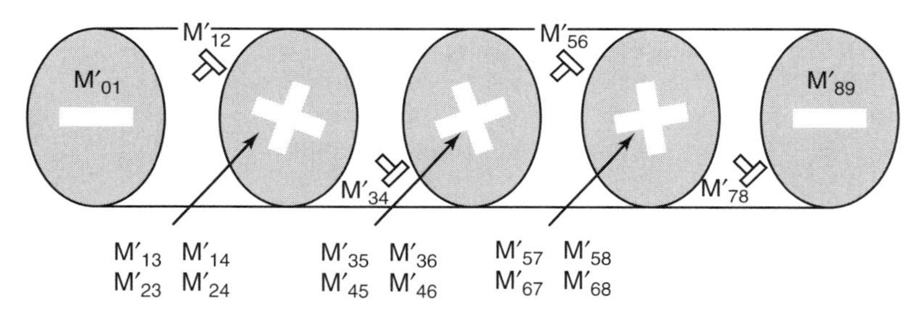

Figure 13.1 Structure of the bimode filter.

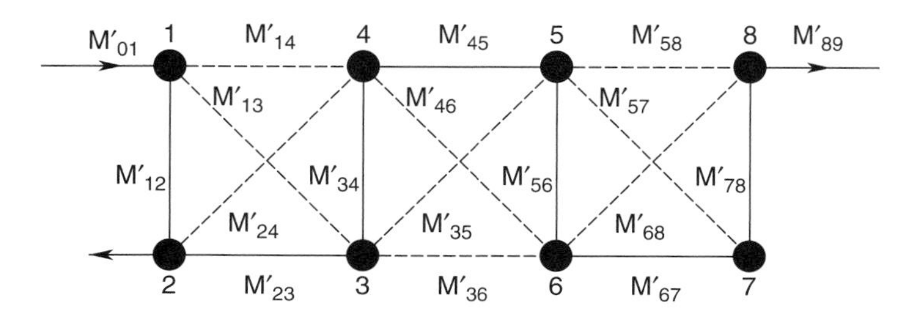

Figure 13.2 Coupling graph of the bimode filter.

$$\begin{bmatrix} 0 & M'_{12} & M'_{13} & M'_{14} & 0 & 0 & 0 & 0 \\ M'_{12} & 0 & M'_{23} & M'_{24} & 0 & 0 & 0 & 0 \\ M'_{13} & M'_{23} & 0 & M'_{34} & M'_{35} & M'_{36} & 0 & 0 \\ M'_{14} & M'_{24} & M'_{34} & 0 & M'_{45} & M'_{46} & 0 & 0 \\ 0 & 0 & M'_{35} & M'_{45} & 0 & M'_{56} & M'_{57} & M'_{58} \\ 0 & 0 & M'_{36} & M'_{46} & M'_{56} & 0 & M'_{67} & M'_{68} \\ 0 & 0 & 0 & 0 & M'_{57} & M'_{67} & 0 & M'_{78} \\ 0 & 0 & 0 & 0 & M'_{58} & M'_{68} & M'_{78} & 0 \end{bmatrix}$$

13.3 SYNTHESIS OF LOW-PASS ASYMMETRICAL GENERALIZED CHEBYSHEV FILTERS

The low-pass transmission function $|S_{21}(j\omega)|^2$ for the case of a symmetrical or asymmetrical generalized Chebyshev response is given by [13.2]

$$|S_{21}(j\omega)|^2 = \frac{1}{1 + \varepsilon^2 F_n^2(\omega)}$$

where ε provides the Chebyshev-type ripple in the passband and the function $F_n(\omega)$ is defined as

$$F_n(\omega) = \cosh\left(\sum_{k=1}^{n} \cosh^{-1} \frac{\omega - 1 \quad \omega_k}{1 - \omega\omega_k}\right)$$

where ω_k represents an arbitrary zero of transmission. Note that ω_k can be chosen finite or infinite. The transfer functions $S_{21}(s)$ and $S_{11}(s)$ can be obtained using the technique described in Chapter 11.

13.3.1 Synthesis from a Cross-Coupled Prototype

The first synthesis of dual-mode filters was given by Atia and Williams [13.3], but their method is concerned with symmetrical filters, while there exist several applications that requires asymmetrical responses (such as multiplexing and amplitude distortion). Designing asymmetrical response filters consists in using the cross-coupled network of Figure 13.3. It can be shown that the $S_{21}(s)$ and $S_{11}(s)$ parameters of this network can be written as [13.2]

$$S_{21}(s) = \frac{2}{\sqrt{R_1 R_2}} \frac{j\beta'_{N-2}s^{N-2} + \beta_{N-3}s^{N-3} + \cdots + \beta_0}{\gamma_N s^N + (\gamma_{N-1} + j\gamma'_{N-1})s^{N-1} + \cdots + (\gamma_0 + j\gamma'_0)}$$

$$S_{11}(s) = -1 + \frac{2}{R_1} \frac{\alpha_N s^N + (\alpha_{N-1} + j\alpha'_{N-1})s^{N-1} + \cdots + (\alpha_0 + j\alpha'_0)}{\gamma_N s^N + (\gamma_{N-1} + j\gamma'_{N-1})s^{N-1} + \cdots + (\gamma_0 + j\gamma'_0)}$$

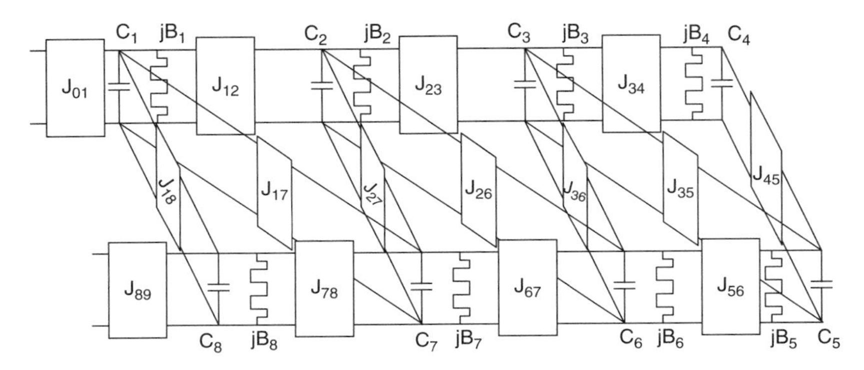

Figure 13.3 Example of a cross-coupled network of order 8.

where β_0 is imaginary if N is even and real if N is odd. For a symmetrical response,

$$\begin{aligned}
\alpha'_j &= 0 \qquad j : 0 \to N - 1 \\
\gamma'_j &= 0 \qquad j : 0 \to N - 1 \\
\beta'_{N-j} &= 0 \qquad \text{when } j \text{ is odd } (N \text{ even}) \text{ or} j \text{ even } (N\text{odd})
\end{aligned}$$

Compared to the synthesis technique of the in-line admittance or impedance networks of Chapters 11 and 12, the synthesis technique of the cross-coupled network is based on successive extraction of *ABCD* matrices from the general *ABCD* matrix of the structure. One starts by expressing the forward and return transfer functions found using the technique of Chapter 11 as

$$S_{21}(s) = \frac{P(s)}{Q(s)} \text{ and } S_{11}(s) = \frac{R(s)}{Q(s)}$$

It is important to scale these transfer functions properly so that $\max |S_{21}(j\omega)| = \max |S_{11}(j\omega)| = 1$. From [13.4] the scattering matrix can then be expressed as

$$[S(s)] = \begin{bmatrix} S_{11}(s) & S_{12}(s) \\ S_{21}(s) & S_{22}(s) \end{bmatrix} = \frac{1}{Q(s)} \begin{bmatrix} R(s) & P(s) \\ P(s) & \sigma R^*(s) \end{bmatrix}$$

where $\sigma = 1$ for $R(s)$ even and $\sigma = -1$ for $R(s)$ odd. The *ABCD* matrix is obtained using the conversion formulas

$$A = \sqrt{\frac{R_1}{R_2}} \frac{(1 + S_{11})(1 - S_{22}) + S_{12}S_{21}}{2S_{12}}$$

$$B = \sqrt{R_1 R_2} \frac{(1 + S_{11})(1 + S_{22}) - S_{12}S_{21}}{2S_{12}}$$

$$C = \frac{1}{\sqrt{R_1 R_2}} \frac{(1 - S_{11})(1 - S_{22}) - S_{12}S_{21}}{2S_{12}}$$

$$D = \sqrt{\frac{R_2}{R_1}} \frac{(1 - S_{11})(1 + S_{22}) + S_{12}S_{21}}{2S_{12}}$$

From conservation of energy, we have $S_{11}^* S_{11} + S_{12}^* S_{12} = 1$, and the *ABCD* polynomials can be expressed as

$$A(s) = \frac{1}{P(s)}\sqrt{\frac{R_1}{R_2}}\left\{\frac{Q(s) - \sigma Q^*(s)}{2} + \frac{R(s) - \sigma R^*(s)}{2}\right\}$$

$$B(s) = \frac{1}{P(s)}\sqrt{R_1 R_2}\left\{\frac{Q(s) + \sigma Q^*(s)}{2} + \frac{R(s) + \sigma R^*(s)}{2}\right\}$$

$$C(s) = \frac{1}{P(s)}\frac{1}{\sqrt{R_1 R_2}}\left\{\frac{Q(s) + \sigma Q^*(s)}{2} - \frac{R(s) + \sigma R^*(s)}{2}\right\}$$

$$D(s) = \frac{1}{P(s)}\sqrt{\frac{R_2}{R_1}}\left\{\frac{Q(s) - \sigma Q^*(s)}{2} - \frac{R(s) - \sigma R^*(s)}{2}\right\}$$

where

$$Q(s) = q_0 + q_1 s + q_2 s^2 + q_3 s^3 + \cdots$$
$$Q^*(s) = q_0^* - q_1^* s + q_2^* s^2 - q_3^* s^3 + \cdots$$

and

$$R(s) = r_0 + r_1 s + r_2 s^2 + r_3 s^3 + \cdots$$
$$R^*(s) = r_0^* - r_1^* s + r_2^* s^2 - r_3^* s^3 + \cdots$$

and for example,

$$Q(s) - Q^*(s) = 2j\mathrm{Im}(q_0) + 2Re(q_1)s + 2j\mathrm{Im}(q_2)s^2 + 2Re(q_3)s^3 \cdots$$
$$Q(s) + Q^*(s) = 2\mathrm{Re}(q_0) + 2jIm(q_1)s + 2\mathrm{Re}(q_2)s^2 + 2jIm(q_3)s^3 \cdots$$
$$R(s) - R^*(s) = 2j\mathrm{Im}(r_0) + 2Re(r_1)s + 2j\mathrm{Im}(r_2)s^2 + 2Re(r_3)s^3 \cdots$$
$$R(s) + R^*(s) = 2\mathrm{Re}(r_0) + 2jIm(r_1)s + 2\mathrm{Re}(r_2)s^2 + 2jIm(r_3)s^3 \cdots$$

13.3.2 Extracting the Elements from the Chain Matrix

The synthesis method consists of extracting elements of the network from the polynomials A,B,C,D,P of the chain matrix. The method can be summarized as

$$\frac{1}{P}\begin{bmatrix} A & B \\ C & D \end{bmatrix} = \begin{bmatrix} A_x & B_x \\ C_x & D_x \end{bmatrix} \frac{1}{P'}\begin{bmatrix} A' & B' \\ C' & D' \end{bmatrix}$$

where A,B,C,D, and P are the starting $ABCD$ matrix parameters, A_x, B_x, C_x, and D_x are the $ABCD$ parameters of a known element to extract, and A', B', C', D' and P' are $ABCD$ parameters of the new chain matrix after extraction of that element. The elements to extract are described below.

Parallel Capacitor

Circuit	*ABCD* Matrix	Extraction	New *ABCD* Matrix
C_k	$\begin{pmatrix} 1 & 0 \\ C_k s & 1 \end{pmatrix}$	$C_k = \left.\frac{D(s)}{sB(s)}\right\rvert_{s\to\infty}$	$\begin{cases} A' = A \\ B' = B \\ C' = C - AC_k s \\ D' = D - BC_k s \\ P' = P \end{cases}$

Matrix equation: $$\frac{1}{P}\begin{bmatrix} A & B \\ C & D \end{bmatrix} = \begin{bmatrix} 1 & 0 \\ C_k s & 1 \end{bmatrix} \frac{1}{P'}\begin{bmatrix} A' & B' \\ C' & D' \end{bmatrix}$$

Susceptance

Circuit	*ABCD* Matrix	Extraction	New *ABCD* Matrix
jB_k	$\begin{pmatrix} 1 & 0 \\ jB_k & 1 \end{pmatrix}$	$jB_k = \left.\frac{D(s)}{B(s)}\right\rvert_{s\to\infty}$	$\begin{cases} A' = A \\ B' = B \\ C' = C - AjB_k \\ D' = D - BjB_k \\ P' = P \end{cases}$

Matrix equation: $$\frac{1}{P}\begin{bmatrix} A & B \\ C & D \end{bmatrix} = \begin{bmatrix} 1 & 0 \\ jB_k & 1 \end{bmatrix} \frac{1}{P'}\begin{bmatrix} A' & B' \\ C' & D' \end{bmatrix}$$

Admittance Inverter

Circuit	*ABCD* Matrix	Extraction	New *ABCD* Matrix
J	$\begin{pmatrix} 0 & j/J \\ jJ & 0 \end{pmatrix}$	They will be taken as $J = 1$	$\begin{cases} A' = -jC/J \\ B' = -jD/J \\ C' = -jAJ \\ D' = -jBJ \\ P' = P \end{cases}$

Matrix equation: $$\frac{1}{P}\begin{bmatrix} A & B \\ C & D \end{bmatrix} = \begin{bmatrix} 1 & j/J \\ jJ & 1 \end{bmatrix} \frac{1}{P'}\begin{bmatrix} A' & B' \\ C' & D' \end{bmatrix}$$

Admittance Inverter in Parallel First, one needs to define the *ABCD* matrix of an admittance inverter J in parallel with an arbitrary $A_1B_1C_1D_1$ matrix, as shown in Figure 13.4. This matrix is defined by converting the *ABCD* matrices of the inverter and the arbitrary $A_1B_1C_1D_1$ element to their corresponding admittances matrices, adding the two admittance matrices, and converting this admittance matrix back to an *ABCD* matrix:

$$Y_1 = \begin{bmatrix} \dfrac{D_1}{B_1} & \dfrac{B_1C_1 - A_1D_1}{B_1} \\ -\dfrac{1}{B_1} & \dfrac{A_1}{B_1} \end{bmatrix} \qquad Y_J = \begin{bmatrix} 0 & \dfrac{-1}{j\ \ J} \\ \dfrac{-1}{j\ \ J} & 0 \end{bmatrix}$$

and

$$Y = Y_1 + Y_J = \begin{bmatrix} \dfrac{D_1}{B_1} & \dfrac{B_1C_1 - A_1D_1}{B_1} + jJ \\ -\dfrac{1}{B_1} + jJ & \dfrac{A_1}{B_1} \end{bmatrix}$$

The *ABCD* matrix is obtained converting the admittance matrix to the *ABCD* matrix:

$$\frac{1}{P}\begin{bmatrix} A & B \\ C & D \end{bmatrix} = \begin{bmatrix} \dfrac{-Y_{22}}{Y_{21}} & \dfrac{-1}{Y_{21}} \\ \dfrac{-(Y_{11}Y_{22} - Y_{12}Y_{21})}{Y_{21}} & \dfrac{-Y_{11}}{Y_{21}} \end{bmatrix}$$

For reciprocal networks, $A_1D_1 - B_1C_1 = 1$ and the *ABCD* matrix of this system reduces to

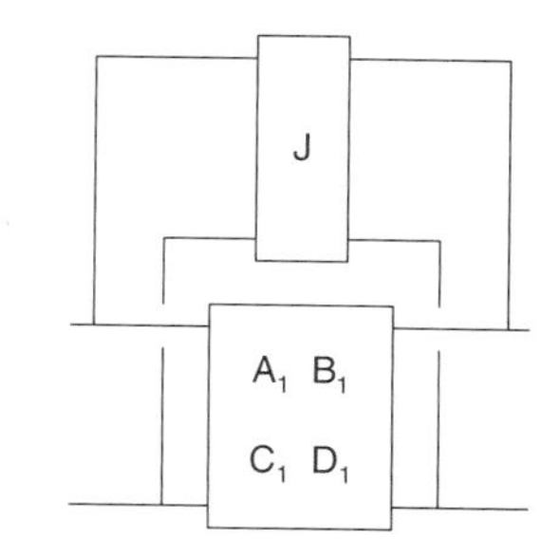

Figure 13.4

$$\frac{1}{P}\begin{bmatrix} A & B \\ C & D \end{bmatrix} = \frac{1}{1-jJB_1}\begin{pmatrix} A_1 & B_1 \\ C_1 + B_1 J^2 + j2J & D_1 \end{pmatrix}$$

The parallel admittance inverter is extracted using

$$jJ = -\left.\frac{P(s)}{B(s)}\right|_{s\to\infty}$$

After extraction of the parallel inverter the new matrix is given by reversing the expressions to find A_1, B_1, C_1, D_1, and P_1 in terms of A, B, C, D, and P. This is shown as

$$\frac{1}{P}\begin{bmatrix} A' & B' \\ C' & D' \end{bmatrix} = \frac{1}{P_1}\begin{bmatrix} A_1 & B_1 \\ C_1 & D_1 \end{bmatrix} = \frac{1}{P+jJB}\begin{bmatrix} A & B \\ C + BJ^2 - j2JP & D \end{bmatrix}$$

The general extraction procedure consists of extracting the elements described above following a rather specific order. For example, for a third-order generalized Chebyshev filter with one transmission zero, one needs to synthesize the structure shown in Figure 13.5.

First the admittance inverter J_{01} is taken as 1, so that one computes the new $ABCD$ matrix. Then the capacitance and susceptance C_1 and B_1 are extracted. At this stage one must continue the extraction from the output of the network. This is done by swapping parameters A and D in the $ABCD$ matrix. Then the admittance inverter J_{34} is taken as 1 and the new $ABCD$ matrix is computed. Then the capacitance and susceptance C_3 and B_3 are extracted. At this stage the admittance inverter in parallel J_{13} is extracted and the A_1, B_1, C_1, D_1, P_1 left represents the combination $J_{12} = 1$, C_2, B_2, and $J_{23} = 1$. One can extract these elements without additional swapping. For more general cases, one needs to extract the resonators C_k, and B_k on each side of a inverter in parallel before extracting the parallel inverter [13.5].

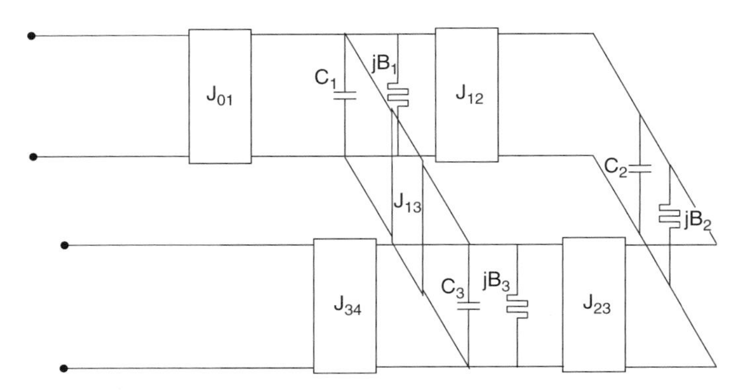

Figure 13.5 Prototype circuit for a third-order generalized Chebyshev function with one transmission zero.

For example, for a third-order low-pass generalized Chebyshev with a transmission zero at $\omega_z = 3.37$ rad/s and -20 dB return loss, the starting polynomials are given by

```
S21(s):
P(s) = (0.0000+j0.2970)s^1 + (1.0000+j0.0000)
Q(s) = (0.3929+j0.0000)s^3 + (0.9210+j-0.0597)s^2
   + (1.3717+j-0.2143)s^1 + (0.9302+j-0.3683)
S11(s):
R(s) = (0.3929+j0.0000)s^3 + (0.0000+j-0.0597)s^2
   + (0.2924+j0.0000)s^1 + (0.0000+j-0.0299)
Q(s) = (0.3929+j0.0000)s^3 + (0.9210+j-0.0597)s^2
   + (1.3717+j-0.2143)s^1 + (0.9302+j-0.3683)
```

The starting *ABCD* matrix polynomials are given by

```
A(s) = (0.0000+j0.9210)s^2 + (0.0000-j0.2143)s
   + (0.9302+j0.000)
B(s) = (0.7859+j0.0000)s^3 + (0.0000-j0.1194)s^2
   + (1.6641+j0.0000)s + (0.0000-j0.3981)
C(s) = (1.0793+j0.0000)s + (0.0000-j0.3384)
D(s) = (0.9210+j0.0000)s^2 + (0.0000-j0.2143)s
   + (0.9302+j0.0000)
P(s) = (0.0000+j0.2970)s^1 + (1.0000+j0.0000)
```

The extraction procedure produces the following results:

```
J01 = 1.000000
C1 = 0.853328
B1 = 0.068953
J34 = 1.000000
C3 = 0.853328
B3 = 0.068953
J13 = 0.275203
J12 = 1.000000
C2 = 1.190100
B2 = -0.373153
J23 = 1.000000
```

13.3.3 Coupling Graph and Frequency Transformation

The next stage consists of defining the coupling graph and the coupling matrix associated with the cross-coupled network. The coupling graph of the eighth-order cross-coupled network of Figure 13.3 is shown in Figure 13.6. The coupling

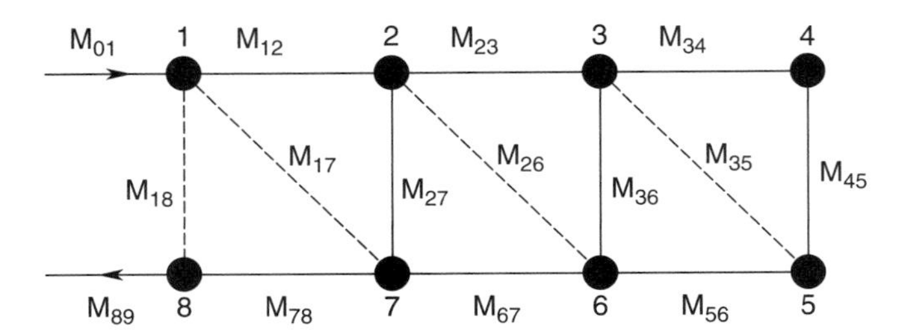

Figure 13.6 Coupling graph of the eighth-order cross-coupled network.

matrix of the eighth-order cross-coupled network, neglecting the input/output couplings, is given by

$$[M] = \begin{bmatrix} 0 & M_{12} & 0 & 0 & 0 & 0 & M_{17} & M_{18} \\ M_{12} & 0 & M_{23} & 0 & 0 & M_{26} & M_{27} & 0 \\ 0 & M_{23} & 0 & M_{34} & M_{35} & M_{36} & 0 & 0 \\ 0 & 0 & M_{34} & 0 & M_{45} & 0 & 0 & 0 \\ 0 & 0 & M_{35} & M_{45} & 0 & M_{56} & 0 & 0 \\ 0 & M_{26} & M_{36} & 0 & M_{56} & 0 & M_{67} & 0 \\ M_{17} & M_{27} & 0 & 0 & 0 & M_{67} & 0 & M_{78} \\ M_{18} & 0 & 0 & 0 & 0 & 0 & M_{78} & 0 \end{bmatrix}$$

The low-pass coupling coefficients of this matrix are given in terms of the elements from the synthesis of the low-pass generalized Chebsyhev function. These are

$$M_{01} = \frac{J_{01}}{\sqrt{C_1}} \qquad M_{j,j} = \frac{B_j}{C_j} \qquad j = 1, 2, \ldots, N$$

$$M_{N,N+1} = \frac{J_{N,N+1}}{\sqrt{C_N}} \qquad M_{j,j+1} = \frac{J_{j,j+1}}{\sqrt{C_j C_{j+1}}} \qquad j = 1, 2, \ldots, N-1$$

$$M_{i,2(k+1)-i+j} = \frac{J_{i,2(k+1)-i+j}}{\sqrt{C_i C_{2(k+1)-i+j}}} \qquad \begin{matrix} i = k \rightarrow 1 \\ j = 0 \rightarrow 1 \end{matrix}$$

where k is taken as the integer division of $(m+1)/2$.

The bandpass cross-coupled network is obtained using the low pass-to-bandpass transformation technique:

$$s \rightarrow \alpha\left(\frac{s}{\omega_0} + \frac{\omega_0}{s}\right) \qquad \text{with} \qquad \begin{aligned} \omega_0 &= \sqrt{\omega_1 \omega_2} \\ \alpha &= \frac{\omega_0}{\omega_2 - \omega_1} = \frac{\omega_0}{\Delta\omega} \end{aligned}$$

The admittance inverters are independent of frequency, so they remain the same. By matching the admittances and their derivatives, we obtain

$$C\left[\alpha\left(\frac{\omega}{\omega_0}-\frac{\omega_0}{\omega}\right)\right]+B=C_a\omega-\frac{1}{L_a\omega} \quad \text{and} \quad C\left[\alpha\left(\frac{1}{\omega_0}+\frac{\omega_0}{\omega^2}\right)\right]=C_a+\frac{1}{L_a\omega^2}$$

Solving the two equations at $\omega = \omega_0$, we find that

$$B=C_a\omega_0-\frac{1}{L_a\omega_0} \quad \text{and} \quad \frac{2\alpha}{\omega_0}C=C_a+\frac{1}{L_a\omega_0^2}$$

The bandpass elements are defined using the relations

$$C_{ak}=\frac{1}{2}\left(\frac{2\alpha C_k+B_k}{\omega_0}\right) \quad \text{and} \quad L_{ak}=\frac{2}{\omega_0(2\alpha C_k-B_k)}$$

The presence of the susceptances jB_k has the effect of moving the resonant frequency of the resonator away from the center frequency of the filter. This shifted resonant frequency is defined using

$$L_{ak}C_{ak}\omega_i^2=1$$

After some simplifications, it can be found that

$$\omega_i \approx \omega_0-\frac{B_k}{C_k}\frac{\Delta\omega}{2}$$

For a filter with N resonators (r, L, C) using magnetic coupling, we have the relations

$$L_iC_i\omega_{0i}^2=1 \qquad Q_{pri}=\frac{L_i\omega_{0i}}{r} \qquad Z_{0i}=\sqrt{\frac{L_i}{C_i}}$$

where ω_{0i}, Q_{pri}, and Z_{0i} are, respectively, the resonance frequency, the quality factor, and the characteristic impedance of the resonator. The terminations will be given by

$$R_1=\frac{\Delta\omega}{\omega_0}Z_0r_1 \quad \text{and} \quad R_2=\frac{\Delta\omega}{\omega_0}Z_0r_2 \quad \text{where} \quad r_1=M_{01}^{\prime 2} \quad \text{and} \quad r_2=M_{n,n+1}^{\prime 2}$$

and the coupling factor between two resonators is such that

$$K_{i,j}=\frac{M_{i,j}}{\sqrt{L_iL_j}} \quad \text{and} \quad K_{i,j}=M'_{i,j}\frac{\Delta\omega}{\omega_0}$$

and then $r = Z_0/Q_{pr}$.

13.4 DESIGN METHOD

The first step when designing bimode filters is to solve the comflict regarding the differences between the coupling graph and coupling matrix of the bimode filter of Figure 13.2 and the coupling graph and coupling matrix of the cross-coupled prototype network of Figure 13.6 [13.6]. This type of problem was first presented in Chapter 10 for the case of a symmetrical response.

13.4.1 Rotation Matrix

To find suitable values for all M' couplings of the bimode filter, one can use a method based on a rotation matrix R such that

$$[M'] = [R]^{\mathrm{T}}[M][R]$$

The matrix R should be such that it preserves the properties of the initial matrix M of the cross-coupled prototype. One such matrix is given in terms of a pivot $[x,y]$ and a rotation angle θ_r by

$$\begin{aligned} R_{kk} &= 1 && k \neq x, k \neq y \\ R_{xx} &= R_{yy} = C_r = \cos\theta_r && \\ R_{xy} &= -R_{yx} = S_r = \sin\theta_r && \\ R_{jk} &= 0 && j \neq x, k \neq y \end{aligned} \qquad [R] = \begin{bmatrix} 1 & 0 & 0 & 0 & 0 & 0 \\ 0 & C_r & 0 & 0 & S_r & 0 \\ 0 & 0 & 1 & 0 & 0 & 0 \\ 0 & 0 & 0 & 1 & 0 & 0 \\ 0 & -S_r & 0 & 0 & C_r & 0 \\ 0 & 0 & 0 & 0 & 0 & 1 \end{bmatrix}$$

In the example provided, the pivot was chosen as [2, 5] and the order was 6. The main problem consists of defining the number of rotations that are needed to go from the M couplings to the M' couplings. This means that one needs to define not only the correct pivots but also the correct rotation angles. For example, for a pivot $[n,m]$ and angle θ_r, the results of the matrix operation can be expressed as

$$\begin{aligned} M'_{i,j} &= M_{i,j} \qquad \text{for } i, j \neq (m, n), \text{ and } i, j = 1 \rightarrow N \\ M'_{i,m} &= M'_{m,i} = M_{m,i}C_r - M_{i,n}S_r \\ M'_{i,n} &= M'_{n,i} = M_{n,i}C_r + M_{i,m}S_r \\ M'_{m,m} &= M_{m,m}C_r^2 + M_{n,n}C_r^2 - 2M_{m,n}S_rC_r \\ M'_{n,n} &= M_{n,n}C_r^2 + M_{m,m}S_r^2 + 2M_{m,n}S_rC_r \\ M'_{m,n} &= M'_{n,m} = M_{m,n}(C_r^2 - S_r^2) + (M_{m,m} - M_{n,n})S_rC_r \end{aligned}$$

Using these expressions it is possible to define a value for the rotation angle. For example, if the coupling $M'_{i,m}$ does not exist (e.g., $M'_{i,m} = 0$) in the bimode filter, then using the second equation we should have

$$M'_{i,m} = M'_{m,i} = M_{m,i}C_r - M_{i,n}S_r = 0$$

This equation will be satisfied if we choose the rotation angle such that

$$\tan\theta_r = \frac{M_{m,i}}{M_{n,i}}$$

Clearly, in the general case of Figures 13.2 and 13.6, the problem of defining the sequence of rotations and their parameters can become very difficult. In order to simplify the problem, one can first reduce the number of couplings. For example, one does not need to have all the cross admittance inverters in the cross-coupled prototype of Figure 13.3. As for the bimode filter, one does not need all the couplings of Figure 13.2 that require complex cruciform irises. The next stage is to choose which layouts of the cross-coupled prototype can provide coupling graphs closer to that of the bimode filter. These layouts are provided in Figures 13.7, 13.8, and 13.9 and are sorted according to the number of transmission zeros.

For the case of a single zero corresponding to the coupling M_{13} this coupling is directly realizable by the bimode filter. For the case of two transmission zeros corresponding to couplings M_{13} and M_{14}, these couplings are also realizable by the bimode filter. There is also some flexibility in making $M_{13} = 0$ in favor of a coupling M_{24} for filters above order 4. In this case, a rotation is needed: for example, one with a pivot [2,3] and an angle $\tan\theta_r = -M_{12}/M_{23}$.

For the case of three transmission zeros, one would need two rotations. The first has a pivot [4,5] and angle $\tan\theta_r = -M_{15}/M_{14}$, and the second rotation has a pivot [2,3] and angle $\tan\theta_r = M_{25}/M_{35}$. Table 13.1 shows how the number of

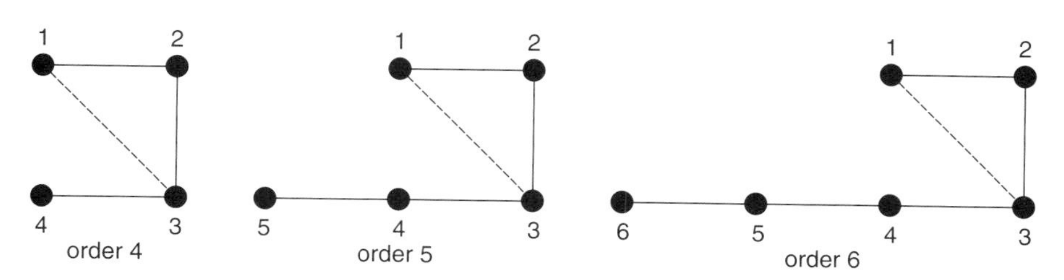

Figure 13.7 Cross-coupled layouts for one transmission zero.

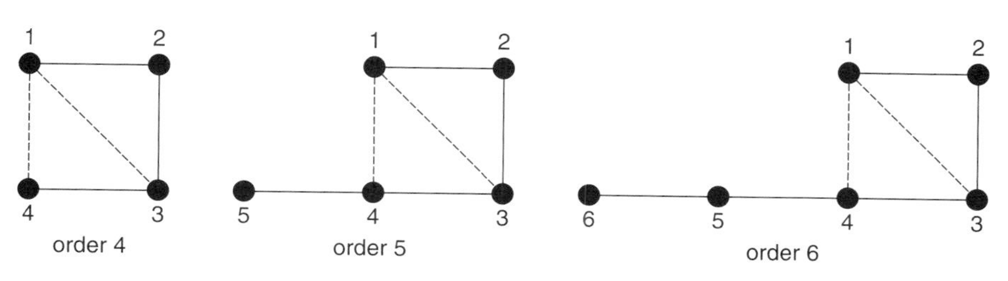

Figure 13.8 Cross-coupled layouts for two transmission zeros.

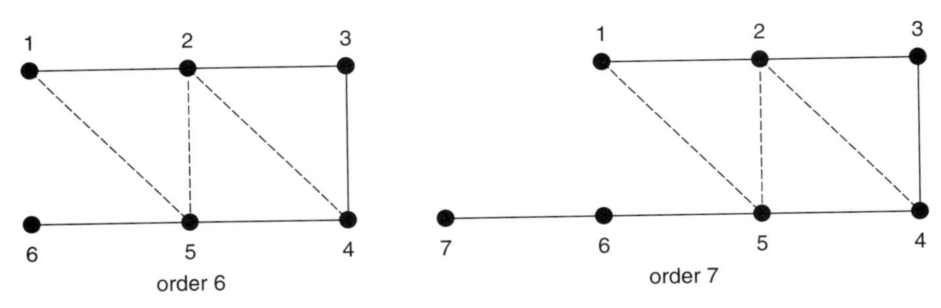

Figure 13.9 Cross-coupled layouts for three transmission zeros.

TABLE 13.1. Rotations and Transmission Zeros

Number of Transmission Zeros	Number of Rotations	Minimum Degree of the Filter
1	0	3
2	0	4
3	2	6
4	8	8
5	12	10

transmission zeros is related to the order of the filter and how the number of rotations is related to the number of transmission zeros.

13.4.2 Cruciform Iris

In the case of symmetrical responses the coupling irises are placed parallel to the magnetic fields of each mode. The electric polarizations are perpendicular to the magnetic polarizations. For asymmetric responses the coupling matrix has nondiagonal components such as M'_{13}, M'_{24}, and so on. These couplings can be achieved by rotating the iris and then the cavities [13.1]. The method for computing these two rotation angles θ and ψ and the couplings M_1 and M_2 characteristic of the cruciform iris to create the M' couplings of the bimode filter are given below. First, let's consider the case of a fourth-order filter made of two cavities and a single cruciform iris, as shown in Figure 13.10.

Figure 13.11 shows the cruciform iris and the associated M_1 and M_2 couplings. It also shows the orientation of the magnetic polarizations of all four modes (1 and 2 are in the first cavity and 3 and 4 are in the second cavity). Modes 1 and 2 are perpendicular to each other and modes 3 and 4 are perpendicular to each other. On the other hand, the orientation of the modes in the first cavity (1,2) and the modes in the second cavity (3,4) are neither parallel nor perpendicular

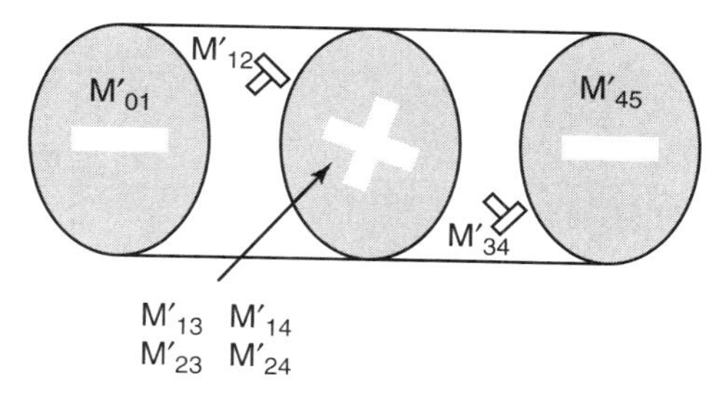

Figure 13.10 First coupling cruciform iris.

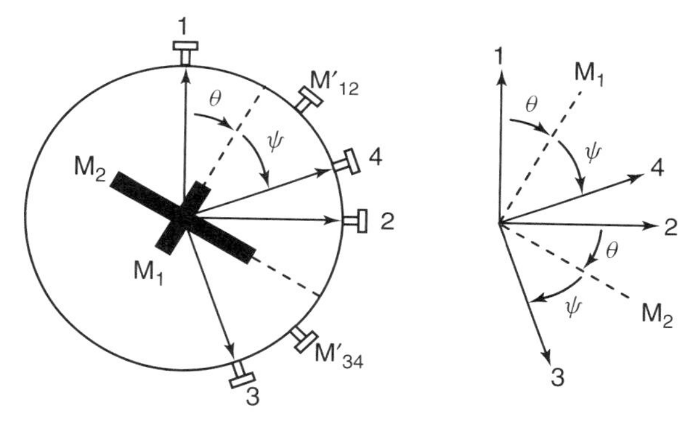

Figure 13.11 Description of the modes, angles, and couplings in the first cruciform iris.

but rotated by an angle. Modes 1 and 4 are rotated by $\theta + \psi$; this will be by how much the cavities are rotated. The angle θ shows by how much the iris is rotated with respect to mode 1 and coupling M_1.

The bimode filter couplings M'_{13}, M'_{14}, M'_{23}, M'_{24} are related to the cruciform iris couplings M_1 and M_2 by

$$M'_{13} = -M_1 \sin\psi \cos\theta - M_2 \cos\psi \sin\theta$$

$$M'_{23} = -M_1 \sin\psi \sin\theta + M_2 \cos\psi \cos\theta$$

$$M'_{14} = M_1 \cos\psi \cos\theta - M_2 \sin\psi \sin\theta$$

$$M'_{24} = M_1 \cos\psi \sin\theta + M_2 \sin\psi \cos\theta$$

To design the filter, we need to define values for the iris angles θ, and ψ and the iris couplings M_1 and M_2 that can produce the desired M'_{13}, M'_{14}, M'_{23}, M'_{24} couplings. It can be shown that

$$\tan 2\psi = -2\frac{M'_{23}M'_{24} + M'_{13}M'_{14}}{M'^2_{24} - M'^2_{23} + M'^2_{14} - M'^2_{13}}$$

$$\tan\theta = -\frac{M'_{13} + M'_{14}\tan\psi}{M'_{23} + M'_{24}\tan\psi}$$

$$M_1 = \frac{M'_{14}\cos\psi - M'_{13}\sin\psi}{\cos\theta}$$

$$M_1 = \frac{M'_{24}\cos\psi - M'_{13}\cos\psi}{\sin\theta}$$

$$M_2 = \frac{-M'_{14}\sin\psi - M'_{13}\cos\psi}{\sin\theta}$$

$$M_2 = \frac{M'_{24}\sin\psi + M'_{23}\cos\psi}{\cos\theta}$$

Note that their are two different ways to compute M_1 and M_2.

The technique is extended to other irises. For example the bimode filter in Figure 13.12 has two cruciform irises. For a study of the first iris, the polarization of mode 1 of the first cavity was taken as the reference. For the second iris, the polarization of mode 4 of the second cavity is taken as the reference and is shown in Figure 13.13. In this case we have

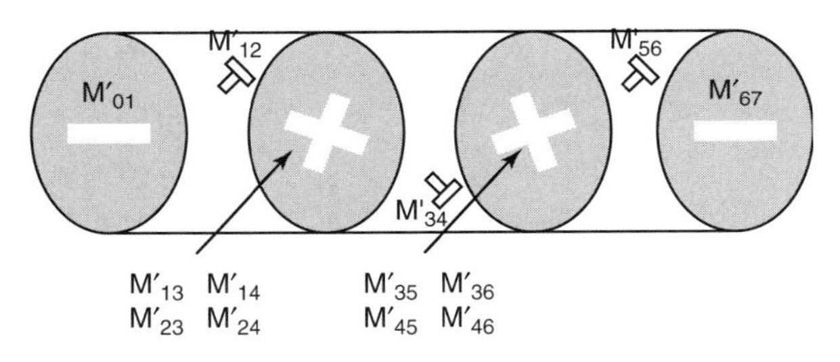

Figure 13.12 Second coupling cruciform iris.

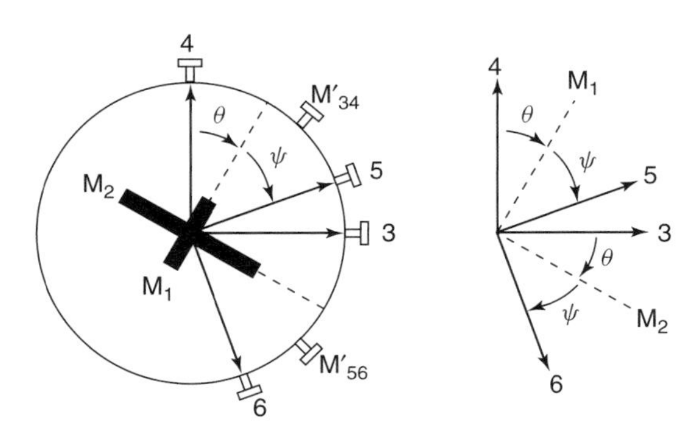

Figure 13.13 Description of the modes, angles, and couplings in the second cruciform iris.

$$M'_{46} = -M_1 \sin\psi \cos\theta - M_2 \cos\psi \sin\theta$$

$$M'_{36} = -M_1 \sin\psi \sin\theta + M_2 \cos\psi \cos\theta$$

$$M'_{45} = M_1 \cos\psi \cos\theta - M_2 \sin\psi \sin\theta$$

$$M'_{35} = M_1 \cos\psi \sin\theta + M_2 \sin\psi \cos\theta$$

The iris angles θ, and ψ and the iris couplings M_1 and M_2 for the couplings M'_{35}, M'_{36}, M'_{45}, and M'_{46} are computed from the formulas for the first iris using the following change of subscripts: $1 \to 4$, $2 \to 3$, $3 \to 6$, and $4 \to 5$.

The procedure can be extended to additional irises. The polarizations used as reference will be 1, 4, 5, 8, 9, and so on. The changes of subscripts to use are

$$1 \to 4 \to 5 \to 8 \to 9$$

$$2 \to 3 \to 6 \to 7 \to 10$$

$$3 \to 6 \to 7 \to 10 \to 11$$

$$4 \to 5 \to 8 \to 9 \to 12$$

It is possible in some cases to replace a cruciform iris by a simple iris. In this case the condition $M_2 = 0$ must be satisfied. This means that we must find suitable iris angles θ, and ψ and iris coupling M_1 to produce the desired M'_{13}, M'_{14}, M'_{23}, M'_{24} couplings. In other words, we need to solve

$$M'_{13} = -M_1 \sin\psi \cos\theta$$

$$M'_{23} = -M_1 \sin\psi \sin\theta$$

$$M'_{14} = M_1 \cos\psi \cos\theta$$

$$M'_{24} = M_1 \cos\psi \sin\theta$$

There exists a solution to this problem if a matrix rotation R is applied. The rotation with pivot 2,3 is selected so that it does not create any additional couplings either impossible to realize or to be realized by the other irises in the system. In this manner it will not be necessary to modify the other parts of the filter. The critical new couplings for the bimode filter from the matrix rotation are

$$M''_{13} = M'_{13} C_r + M'_{12} S_r$$

$$M''_{23} = M'_{23}(C_r^2 - S_r^2) - S_r C_r (B_3 - B_2)$$

$$M''_{14} = M'_{14}$$

$$M''_{24} = M'_{24} C_r - M'_{34} S_r$$

To solve the problem, the matrix rotation angle should be taken as

$$\tan\theta_r = \frac{-B \pm \sqrt{B^2 - 4AC}}{2A} \quad \text{where} \quad \begin{aligned} A &= M'_{12}M'_{34} - M'_{14}M'_{23} \\ B &= M'_{13}M'_{34} - M'_{12}M'_{24} - M'_{14}(B_3 - B_2) \\ C &= M'_{14}M'_{23} - M'_{24}M'_{13} \end{aligned}$$

There is, however, a reliability condition. One should have $B^2 - 4AC \geq 0$. If this condition is satisfied, the iris parameters are given by

$$\tan\theta = \frac{M''_{23}}{M''_{13}} \quad \text{or} \quad \tan\theta = \frac{M''_{24}}{M''_{14}} \quad \text{and} \quad \tan\psi = -\frac{M''_{23}}{M''_{24}} \quad \text{or} \quad \tan\psi = -\frac{M''_{13}}{M''_{14}}]$$

and

$$M_1 = -\frac{M''_{23}}{\sin\theta\sin\psi} = -\frac{M''_{14}}{\cos\theta\cos\psi} = -\frac{M''_{24}}{\sin\theta\cos\psi} = -\frac{M''_{13}}{\cos\theta\sin\psi}$$

Due to multiple cavity rotations, one would need to realign the input and output polarizations. The technique is to align the input/output irises rather than to use additional waveguides. If to each iris a rotation θ_k is applied and to each polarization a rotation ψ_k is applied, the first and last polarizations form an angle δ such that

$$\delta = \begin{cases} \dfrac{\pi}{2} + \displaystyle\sum_{k=1}^{m}\left[\theta_k + \psi_k + \frac{\pi}{2}(-1)^k\right] & N \text{ even} \\ \displaystyle\sum_{k=1}^{m}\left[\theta_k + \psi_k + \frac{\pi}{2}(-1)^{k+1}\right] & N \text{ odd} \end{cases}$$

where m is the integer division of $(N-1)/2$. The goal is to make $\delta = 0$. For even-order filters with at most two transmission zeros, one can, for example, use the additional matrix rotation with pivot [2,3] and angle

$$\tan\theta_r = \frac{M'_{24} - M'_{13}}{M'_{12} - M'_{34}}$$

13.4.3 Physical Parameters of the Irises

Once the coupling coefficients of the iris are computed, one must define the physical parameters of the iris [13.7,13.8]. This is done using theoretical expressions but often requires experimentation for more accurate results. In the case of the cruciform iris, one can think of two simple irises. One iris is perpendicular to the magnetic field as shown in Figure 13.14 (left), and the other iris is parallel to the magnetic field as shown in Figure 13.14 (right).

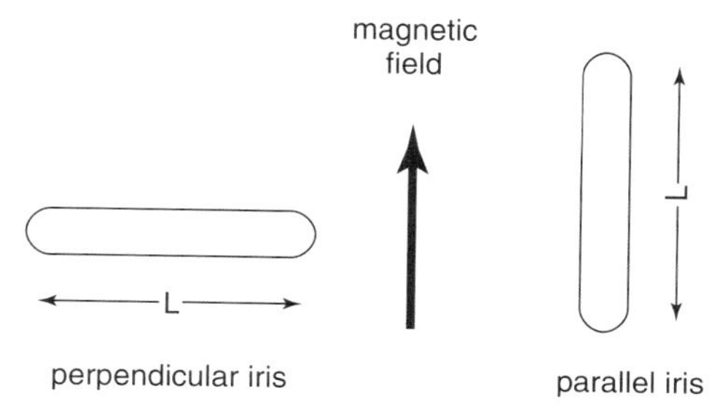

Figure 13.14 Perpendicular and parallel irises.

One method for computing the properties of the perpendicular iris is to model the rectangle as a sum of circular irises since the properties of circular openings are well known [13.9,13.10] . The results were compared to those of MacDonald [13.11] and showed good agreement. In addition, the results were compared to experimental data and are shown in Figure 13.15. In the case of the parallel iris, using the results of Bethe [13.9], the polarization of a circular opening can be expressed as V/π, where V represents the volume of a sphere created by the circle defining the opening. In the case of a rectangular iris, we made the assumption that the polarization can be expressed as V'/π, where V' represents the volume created by the contour of the rectangular iris rotating about the perpendicular axis to the iris and passing through its center. Correction factors were added to account for dimensions such as the thickness of the opening [13.12].

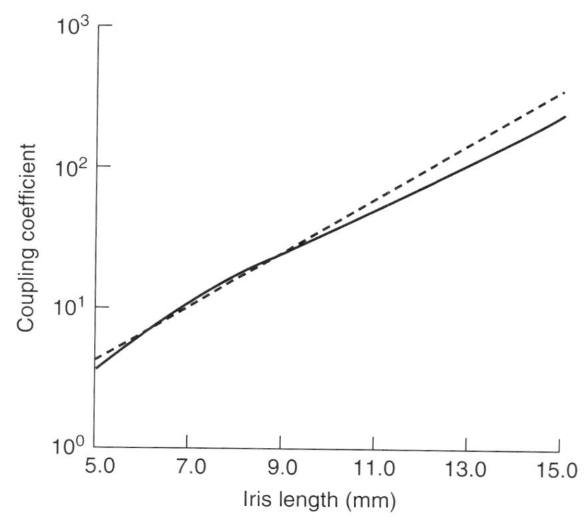

Figure 13.15 Coupling coefficients versus length for perpendicular iris (cavity length = 60 mm, radius = 19.5 mm, frequency = 8750 MHz, TE_{113} mode, iris width = 1.3 mm, thickness = 0.5 mm).

In cylindrical cavities, the resonating frequency of a $TE_{n,m,l}$ resonant mode is given by

$$f_{n,m,l} = \frac{c}{2\pi}\sqrt{\left(\frac{x'_{n,m}}{a}\right)^2 + \left(\frac{l\pi}{d}\right)^2}$$

where $x'_{n,m}$ is the mth root of the derivative of the nth Bessel function $[J'_n(x) = 0]$. The radius a and length d of the cavity are computed using the TE_{113} mode and the desired resonant frequency. In addition, it is recommended to choose a ratio a/d that provides the best frequency separation between the TE_{113} mode and the other modes as well as the highest quality factor, as mentioned in Chapter 10.

13.5 REALIZATIONS AND MEASURED PERFORMANCE

Several filters were built using the techniques described in this chapter [13.5, 13.7, 13.8, 13.13].

13.5.1 Fourth-Order Filter with One Transmission Zero on the Left

The first filter is a fourth-order generalized Chebychev with one transmission zero on the left side of the passband. The center frequency is 8670 MHz, with a relative bandwidth of 30 MHz, and the transmission zero is located at 8640 MHz. The filter is realized using two bimode cavities and one simple iris, as shown in Figure 13.16. The input and output to the filter were achieved using RIM connectors.

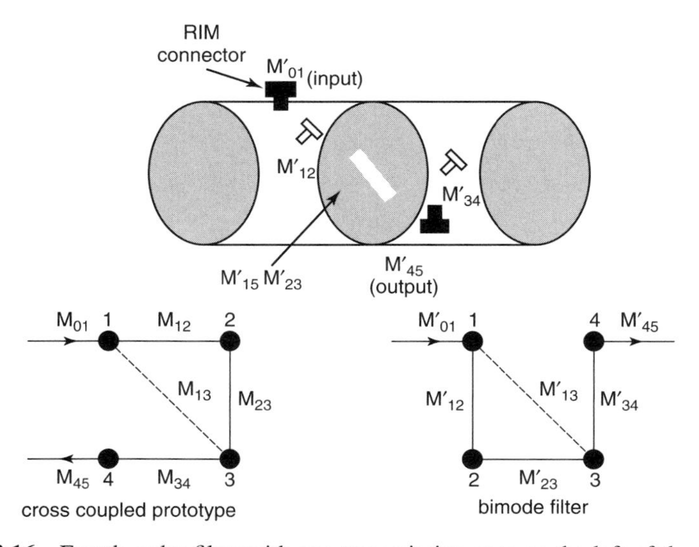

Figure 13.16 Fourth-order filter with one transmission zero on the left of the passband.

Synthesis of the filter produced one iris with $\theta = -52.89^\circ$, $\psi = 90^\circ$, $M_1 = -0.91716$, ($M_1 = -3.17 \times 10^{-3}$ denormalized), and $M_2 = 0$ with $\delta = 37.1^\circ$. The resonances were 8668.64, 8679.11, 8667.61, and 8668.64 MHz. The filter is shown in Figure 13.17. Note that several tuning screws were used in addition to the screws used for the M'_{12} and M'_{34} couplings. The expected and measured filter responses are shown in Figure 13.18. The losses are about −0.66 dB, which correspond to a quality factor of 7000.

13.5.2 Fourth-Order Filter with Two Transmission Zeros on the Right

The second filter is a fourth-order generalized Chebychev with two transmission zeros on the right side of the passband. The center frequency is 8660 MHz with

Figure 13.17 Fourth-order filter with one transmission zero.

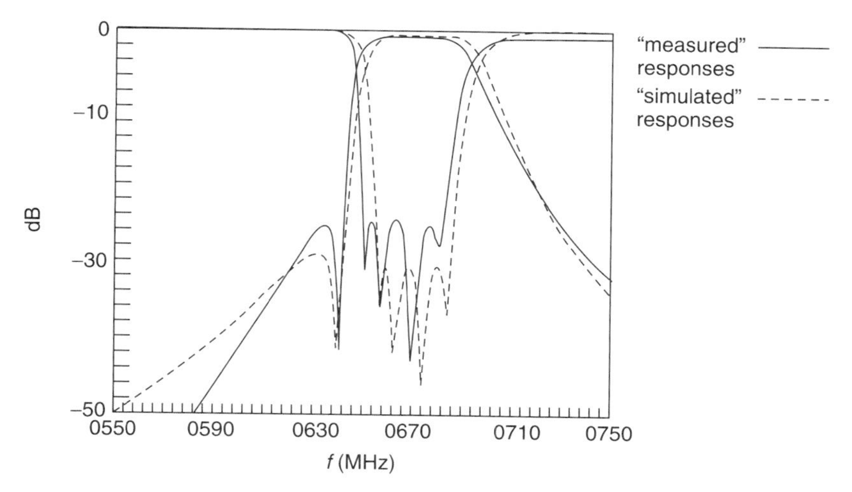

Figure 13.18 Fourth-order filter with one transmission zero on the left of the passband.

a relative bandwidth of 40 MHz, and the transmission zeros are located at 8695 and 8710 MHz. The filter is realized using two bimode cavities and one simple iris, as shown in Figure 13.19. The input and output to the filter were achieved using RIM connectors.

Synthesis of the filter produced one iris with $\theta = 52.17^\circ$, $\psi = -65.55^\circ$, $M_1 = -0.9576$, ($M_1 = 3.87 \times 10^{-3}$ denormalized), and $M_2 = 0$ with $\delta = -13.4^\circ$. The resonances were 8662.84, 8650.77, 8655.51, and 8662.84 MHz. Figure 13.20 shows the response of the filter. The losses are about −0.68 dB, which corresponds to a quality factor of 6000.

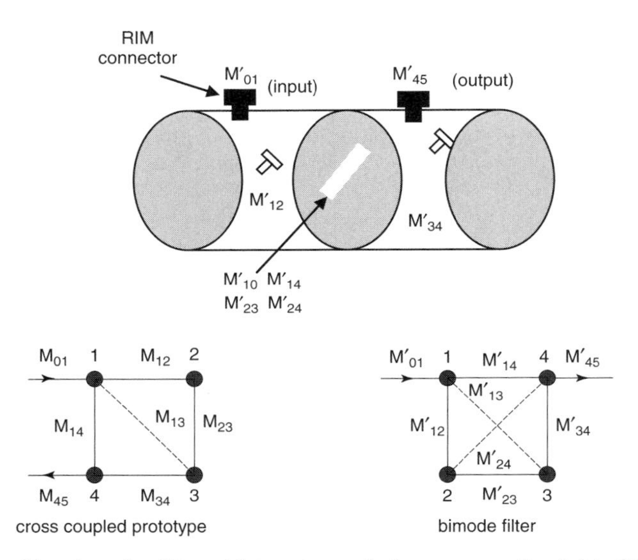

Figure 13.19 Fourth-order filter with two transmission zeros on the right of the passband.

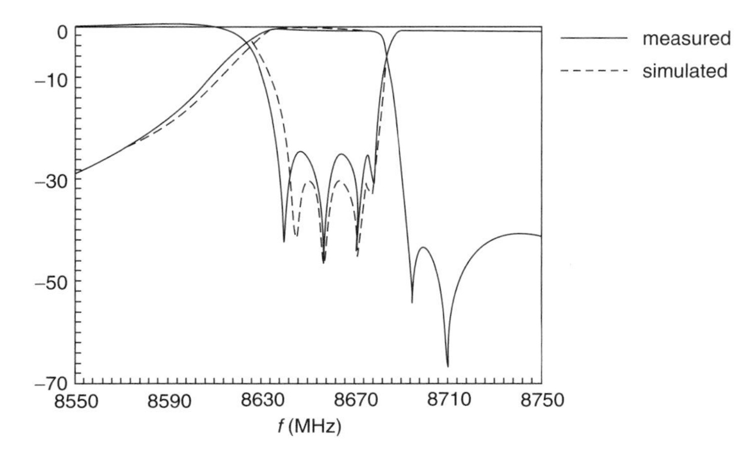

Figure 13.20 Fourth-order filter with two transmission zeros on the right of the passband.

13.5.3 Sixth-Order Filter with One Transmission Zero on the Right

A third filter is a sixth-order generalized Chebychev with one transmission zero on the right side of the passband. The center frequency is 8650 MHz with a relative bandwidth of 40 MHz, and the transmission zero is located at 8685 MHz. The filter is realized using three bimode cavities, one cruciform iris, and one simple iris, as shown in Figure 13.21.

Synthesis of the filter produced the first iris with $\theta = 0^\circ$, $\psi = -90^\circ$, $M_1 = -0.8619$ ($M_1 = -4.01 \times 10^{-3}$ denormalized), and $M_2 = -0.32028$ ($M_2 = -1.48 \times 10^{-3}$ denormalized), and the second iris with $\theta = 0^\circ$, $\psi = 0^\circ$, $M_1 = -0.68926$ ($M_1 = 3.19 \times 10^{-3}$ denormalized) and $M_2 = 0$, and with $\delta = 0^\circ$. The resonances were 8650.49, 8632.02, 8659.07, and 8651.02, 8650.62, and 8650.49 MHz. The input and output have the same polarization. Figure 13.22 shows the filter and an internal view. Figure 13.23 shows the responses of the sixth-order filter. The

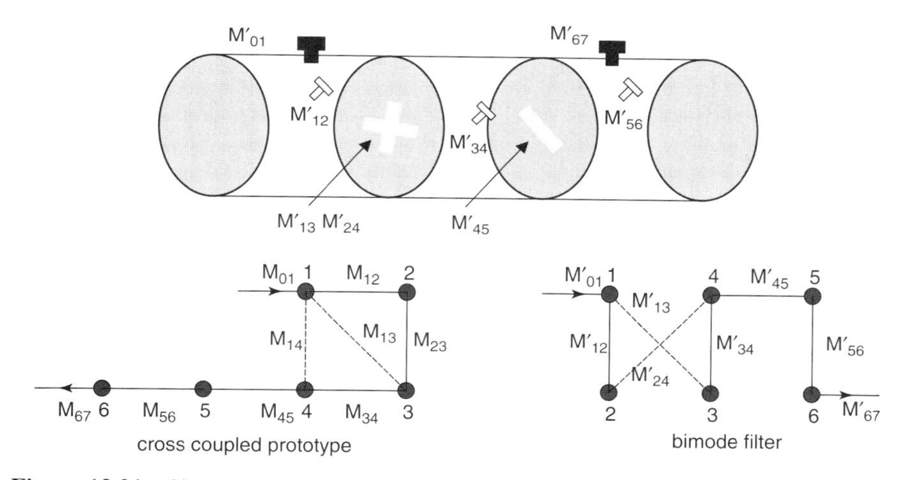

Figure 13.21 Sixth-order filter with one transmission zero on the right of the passband.

Figure 13.22 Sixth-order filter with the two irises and screws.

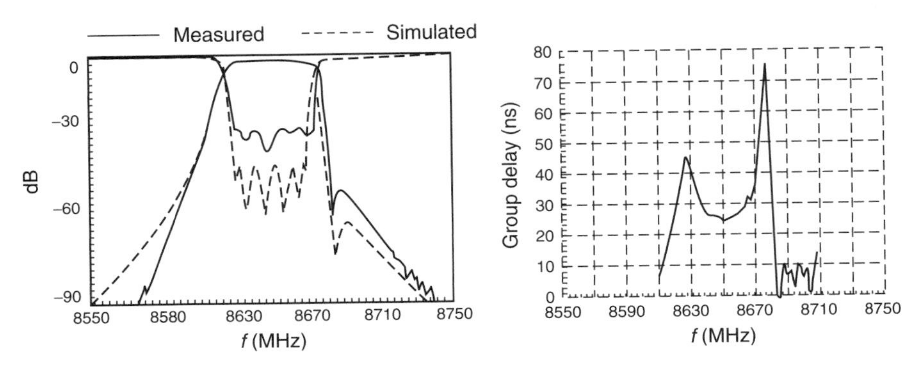

Figure 13.23 Sixth-order filter with one transmission zero on the right of the passband.

losses are about 1 dB in the passband, and the quality factor is about 7000. The group delay of this dual-mode filter is not constant in the passband and increases rapidly on the upper edge of the passband.

13.6 CONCLUSIONS

In this chapter we provide a practical method to design and realize microwave dual-mode filters with asymmetric responses. Prototype layouts are shown that minimize the number of rotations needed to match the coupling coefficients of the cross-coupled prototype network and the coupling coefficients of the dual-mode filter. The realizations around 8.6 GHz using TE_{113} modes show that the transmission zeros can be achieved on either side of the passband. It is possible to use other modes that provide good quality factors.

Acknowledgments

This work was carried out under contracts with the Centre National d'Études spatiales and Thomson.

REFERENCES

[13.1] R. J. Cameron and J. D. Rhodes, Asymmetric realisation for dual-mode bandpass filters, *IEEE Trans. Microwave Theory Tech.*, vol. 29, pp. 51–58, Jan. 1981.

[13.2] R. J. Cameron, General prototype network-synthesis methods for microwave filters, *ESA J.*, vol. 6, pp. 192–206, 1982.

[13.3] A. E. Atia, and A. E. Williams, Narrow band-pass waveguide filters, *IEEE Trans. Microwave Theory Tech.*, vol. 20, pp. 258–265, Apr. 1972.

[13.4] H. Baher, *Synthesis of Electrical Networks*, Wiley, New York, 1984.

[13.5] C. Guichaoua, P. Jarry, J. Sombrin, and C. Boschet Synthesis and realizations of bi-mode cavities filters with an asymmetric amplitude, *Ann. Telecommun.*, vol. 45 no. 5–6, pp. 351–360, 1990.

[13.6] J. D. Rhodes and I. H. Zabalawi, Synthesis of symmetric dual-mode in-line prototype network, *Circuit Theory Appl.*, no. 8, pp. 145–160, 1980.

[13.7] C. Guichaoua and P. Jarry, Asymmetric frequency dual-mode satellite filters: simple realizations using slot irises and input/output realignment, *Proc. 8th B. Franklin Symposium on Advances in Antenna and Microwave Technology*, IEEE AP/MTT-S, University of Pennsylvania, Philadelphia, PA, pp. 32–36, Mar. 1990.

[13.8] C. Guichaoua, P. Jarry, and C. Boschet, Approximation, in *Synthèse et réalisation de filtres microondes à réponses pseudo elliptiques dissymétrique,* Tech. Rep. 124, CNES, DCAF 02583, pp. 1–192, 1990.

[13.9] H. A. Bethe, Theory of diffraction by small holes, *Phys. Rev.*, vol. 66, pp. 163–182, Oct. 1944.

[13.10] R. E. Collin, *Field Theory of Guided Waves*, McGraw-Hill, New York, 1950.

[13.11] J. MacDonald, Simple approximations for the longitudinal magnetic polarisabilities of some small apertures, *IEEE Trans. Microwave Theory Tech.*, vol. 36, no. 7, pp. 1141–1144, 1988.

[13.12] S. B. Cohn, Microwave coupling apertures, *Proc. IRE*, vol. 40, pp. 696–699, June 1952.

[13.13] C. Guichaoua, G. Forterre, P. Jarry, and C. Boschet, Conception of microwave satellite diplexers using asymmetrical responses filters, *Proc. ESA/ESTEC Workshop on Microwave Filters for Space Applications*, Noordwijk, The Netherlands, pp. 287–293, June, 7–8, 1990.

Selected Bibliography

Accatino, L., G. Bertin, and M. Mongiardo, A four-pole dual mode elliptic filter realized in circular cavity without screws, *IEEE Trans. Microwave Theory Tech.*, vol. 44, no. 12, pp. 2680–2687, 1996.

Atia, A. E. and A. E. Williams, New types of waveguides bandpass filters for satellites transponders, *Comsat Tech. Rev.*, vol. 1, no. 1, pp. 21–43, 1971.

Atia, A. E., A. E. Williams, and R. W. Newcomb, Synthesis of dual-mode filters, *IEEE Trans. Circuits Syst.*, vol. 21, pp. 649–655, Sept. 1974.

Cameron, R. J. Fast generation of Chebyshev filter prototypes with asymmetrically prescribed transmission zeros, *ESA J.* vol. 6 pp. 83–95, 1982.

Chang, H. C., and K. A. Zaki, Evanescent-mode coupling of dual-mode rectangular waveguide filters *IEEE Trans. Microwave Theory Tech.*, vol. 39, no. 8, pp. 1307–1312, 1991.

Couffignal, P., H. Baudrand, and B. Theron, A new rigorous method for the determination of iris dimensions in dual-mode cavity filters, *IEEE Trans. Microwave Theory Tech.*, vol. 42, no. 7, pp. 1314–1320, 1994.

Franti, L., and G. Paganuzzi, Odd-degree pseudo elliptic phase-equalised filter with asymmetric band-pass behaviour, *Proc. European Microwave Conference*, EMC′ 81, Amsterdam, The Netherlands, pp. 111–116, Sep 1981

Liang, J. F., X. P. Liang, K. A. Zaki, and A. E. Atia, Dual-mode dielectric or air-filled rectangular waveguide filters, *IEEE Trans. Microwave Theory Tech.*, vol. 42, no. 7, pp. 1330–1336, July 1994.

Liang, X. P., K. A. Zaki, and A. E. Atia, Dual mode coupling by square corner cut in resonators and filters, *IEEE Trans. Microwave Theory Tech.*, vol. 40, no. 12, pp. 2294–2302, 1992.

Orta, R., P. Savi, R. Tascone, and D. Trinchero, Rectangular waveguide dual-mode filters without discontinuities inside the resonators, *Microwave Guided Wave Lett.*, vol. 5, no. 9, pp. 302–304, 1995.

Williams, A. E. and A. E. Atia Dual-mode canonical waveguide filters. *IEEE Trans. Microwave Theory Tech.*, vol. 25, no. 12, pp. 1021–1026, 1977.

Williams, A. E. and R. R. Bonetti, A mixed dual-quadruple mode 10-pole filter, *Proc. European Microwave Conference*, EMC'88, Stockholm, Sweden, pp. 966–968, Sept. 1988.

14

ASYMMETRICAL MULTIMODE RECTANGULAR BUILDING BLOCK FILTERS USING GENETIC OPTIMIZATION

14.1 INTRODUCTION

Classical filter design often requires calculating the coupling coefficients between the various electromagnetic modes used [14.1]. In this chapter we describe a new approach to designing microwave filter structures based on rectangular waveguide building blocks. The process does not require a priori knowledge of the couplings involved. It also relies on a new coupling scheme where all resonant modes are used near their cutoff frequencies. The filter is composed of a cascade of basic nonloaded rectangular cavities where single and dual modes are used alternately. The cavities are coupled directly without an iris.

Advanced Design Techniques and Realizations of Microwave and RF Filters,
By Pierre Jarry and Jacques Beneat

The technique is based on defining the multimode scattering matrices of the basic building blocks representing rectangular waveguides and two types of discontinuities: a change in dielectric material and a change in cross-sectional dimensions in the waveguides. The change in cross-sectional dimensions between a large waveguide and multiple smaller waveguide sections is characterized using the moment method. The physical dimensions and locations of the waveguide sections for a given filter response are computed using an optimization method based on the genetic algorithm [14.2]. Filters based on these basic building blocks present three main advantages: the length reduction due to resonant modes operating near cutoff frequency, the ease of manufacture due to the simplicity of the building blocks, and the absence of screws, iris, or plugs for the cavities.

A fourth-order filter with two transmission zeros was realized and the response measured agrees well with the expected response even if no tuning screws were needed. Another realization is given for a seventh-order filter with four transmission zeros. The method was extended to a tenth-order filter with six transmission zeros.

14.2 MULTIMODE RECTANGULAR WAVEGUIDE FILTERS

The filters described in this chapter are made of sections of rectangular waveguide sections. The waveguide sections can be either empty or filled with a homogeneous isotropic dielectric material. The walls of the various waveguide pieces are assumed perfectly conducting when they do not intersect with each other. Figure 14.1 illustrates some structures based on these rectangular waveguide building blocks.

The main building blocks considered are:

- Straight rectangular waveguide sections characterized by their length
- Discontinuities created by a change in dielectric within a section of rectangular waveguide
- Discontinuities created by a change in the cross-sectional dimensions of the waveguide sections
- Shorts

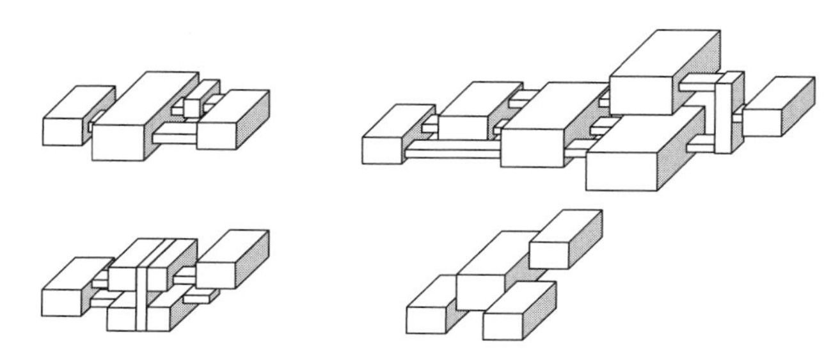

Figure 14.1 Possible filter structures.

The restricted composition of these structures allows for simplified analysis and reduced simulation times [14.3]. Figure 14.2 shows how a potential filter structure is decomposed into basic building blocks. The filter of Figure 14.3 has a waveguide section followed by a change in dimensions discontinuity, a waveguide section, a change in dimensions discontinuity, a waveguide section, a bisection change in dimensions, a bisection, a bisection change in dimensions discontinuity, and a waveguide section. The bisection refers to two waveguide sections considered together.

The scattering matrices for several of the building blocks in these structures were described in Chapter 6. However, the bisection change in dimensions corresponds to a discontinuity in dimensions from one waveguide section to multiple waveguide sections and was not considered in Chapter 6. The analysis of this discontinuity is based on a modal decomposition at the junction in the various waveguide sections, as shown in Figure 14.3. The tangential electric and magnetic fields are decomposed according to the transverse electric (TE) and transverse magnetic (TM) eigenmodes.

By using an appropriate scaling of the reference modes, the tangential electric field will not change sign, while the tangential magnetic field will change sign when the wave propagation direction is reversed. This homogenizes the expressions for the TE and TM modes and is accomplished by choosing reference magnitudes that are independent of frequency. This also improves the computation speeds. The field continuity equations at the junction are then solved using

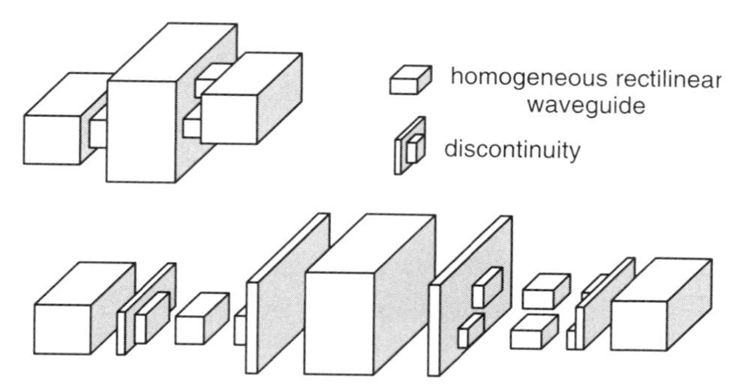

Figure 14.2 Structure decomposition into basic building blocks.

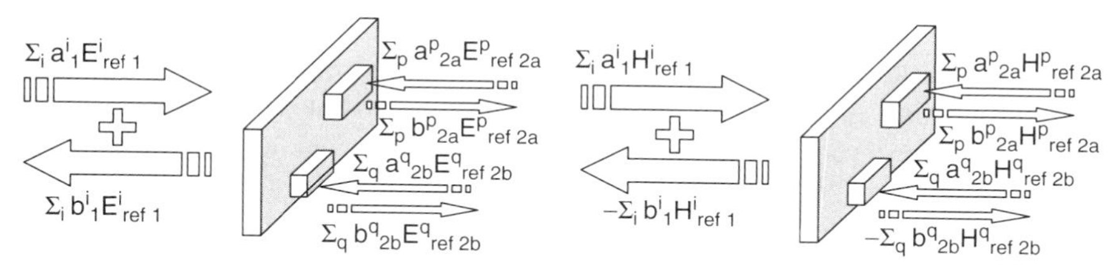

Figure 14.3 Modal decomposition of the tangential electric and magnetic fields.

the moment method. The scattering matrices for multiple waveguide sections, for a change in dielectric permittivity, for a change in dimensions, and the reference modes are given in Appendix 5.

The only approximation in the results is due to the number of modes considered. In the case of multimode realizations, one must keep track of a few accessible modes throughout the entire structure. However, at a discontinuity such as a change in dimensions, distant modes are created by the discontinuity and interact with the other modes. The distant modes that were not incident on the discontinuity will affect the scattering parameters of the ones that were. However, the distant modes that were created locally are assumed to be sufficiently attenuated and will not be considered as incident at neighboring discontinuities. In practice, we considered 10-fold more distant modes than accessible modes. The scattering matrices representing every building block in the structure are then combined together to form the scattering matrix of the filter with respect to the input and output access guides [14.4].

14.3 OPTIMIZATION-BASED DESIGN

A general view of optimization is seen in Figure 14.4. To improve the speed of the process, one can start with a rough analysis of the filter based on simplified equations or equivalent circuits. When the filter has satisfied the requirements using this rough simulator, it is simulated using an accurate electromagnetic simulator. If the filter shows the same level of performance when using the accurate simulator, the process is finished. On the other side, if the level of performance is degraded, the rough simulator must be improved by using more complex and time-consuming calculations.

14.3.1 Genetic Algorithm

Several optimization algorithms can be used, such as the gradient-based methods described briefly in Chapter 12. In this chapter the genetic algorithm is presented in some detail. It is a randomly guided process that mimics Darwin's natural law of evolution [14.5,14.6]. Figure 14.5 provides an overview of the optimization process using the genetic algorithm.

The objective of the reproduction cycle is to produce a new population that hopefully has members with better chromosomes (better filter responses) than those of the older population. The reproduction cycle is based on the concepts of selection, crossover, and mutation, as shown in Figure 14.6. The first step in the genetic algorithm is coding. This step consists of associating the parameters to optimize with a chromosome, as shown in Figure 14.7. A chromosome is made of several genes. For example, one could make a one-to-one correspondence between one parameter to optimize and one gene in the chromosome. Therefore, the chromosome is nothing but the ordered set of parameters to optimize. On the other side, it might be advantageous to restrict a gene to have only integer

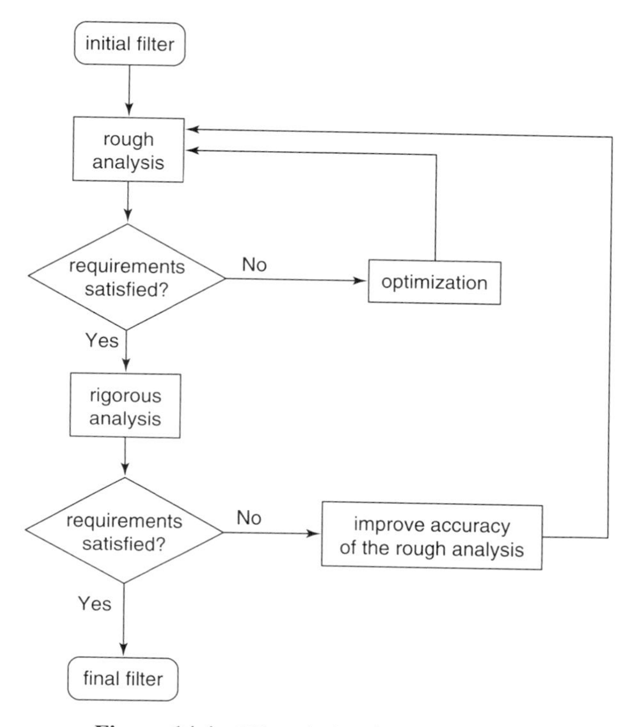

Figure 14.4 Filter design by optimization.

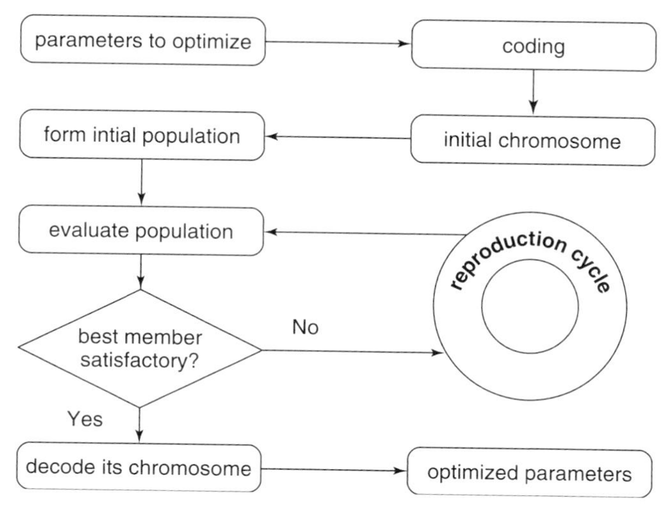

Figure 14.5 Overview of optimization by the genetic algorithm.

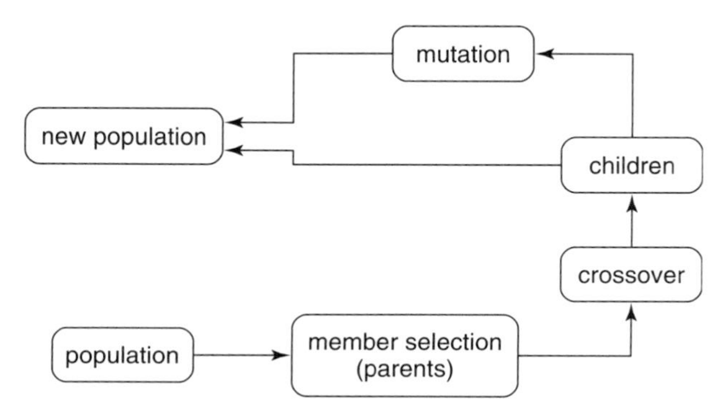

Figure 14.6 Overview of the reproduction cycle.

values. In this case, the coding would consist of defining a multiplicative factor to be applied to a physical parameter. For instance, if the physical parameter represents a waveguide length, then due to construction tolerances, one might like to express this parameter as an integer value in terms of the minimum possible value that can be machined. For example, a particular length should never be smaller than 1 mm. Then the gene will represent the physical parameter value in integer values of 1 mm. If other physical parameters have a 0.1-mm tolerance, the gene would represent the physical parameter value in integer values of 0.1 mm. This might not appear to be significant, but it would make a difference when gene values are modified during the reproduction cycle, mainly through mutation. The incremental change in a gene value can be 1 across the board

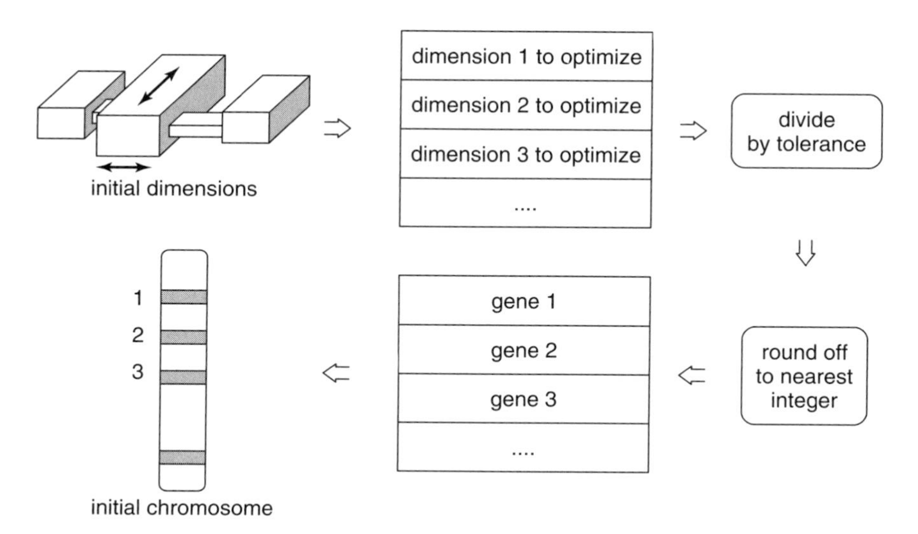

Figure 14.7 Coding operation.

using this technique, regardless of whether some physical parameters could be more precise than others.

The initial chromosome is obtained after deciding the coding technique and the initial values of the parameters to optimize. It is clear that the closer the initial parameters are to produce the desired frequency response, the faster will be the optimization. It is therefore useful to at least define the order of magnitude of these parameters (e.g., in the range of a few millimeters, in the range of a few tens of millimeters, etc.). Rough filter design techniques can be used for defining suitable initial parameters.

The initial population is based entirely on the initial chromosome as shown in Figure 14.8. It is possible to use crossover and mutation as in the reproduction cycle. However, the simplest is to take the initial chromosome and to create new chromosomes simply by increasing or decreasing the value of the genes by a given increment in a random fashion. This is a form of mutation.

At this stage, the algorithm goes into a loop. The population is evaluated by computing a quality factor for each member (chromosome), as shown in Figure 14.9. For each member, the chromosome provides the physical parameters of the microwave filter to be simulated. The simulation of the microwave filter is then compared to a target response such as a Butterworth, Chebyshev, or generalized Chebyshev response. One needs to quantify in a single number how close the optimized filter response is to the target filter response.

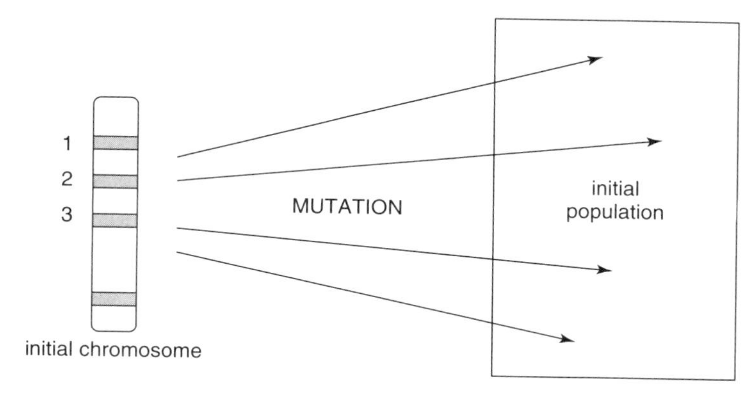

Figure 14.8 Generation of the initial population.

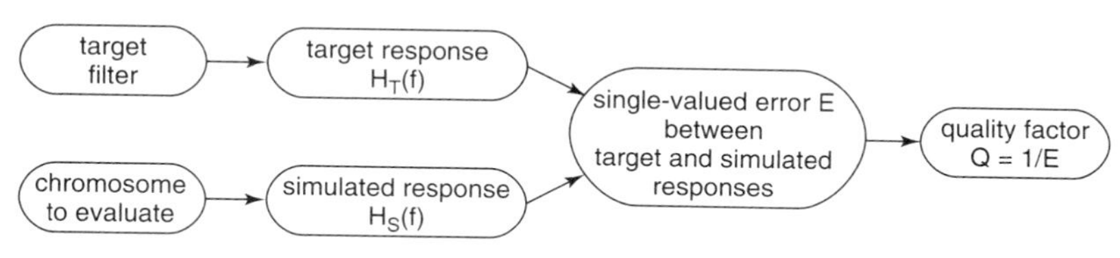

Figure 14.9 Computation of the error or quality factor of a chromosome.

In the case of filters, one needs to check how good the match is over a range of frequencies, and it will be necessary to compute the simulated and target filter responses at several frequencies. The single error number needed for the optimization program can be taken as the mean or maximum error functions:

$$E = \begin{cases} \dfrac{1}{N}\displaystyle\sum_{k=1}^{N} \left| |S_{21}(j\omega_k)|^2 - |\widehat{S_{21}}(j\omega_k)|^2 \right| \\ \max \left| |S_{21}(j\omega_k)|^2 - |\widehat{S_{21}}(j\omega_k)|^2 \right| \end{cases} \qquad k = 1, 2, \ldots, N$$

where $S_{21}(j\omega)$ is the EM simulated response that uses the physical parameters, and $\widehat{S_{21}}(j\omega)$ is the target or prototype circuit response. The quality factor can then be taken as $1/E$.

In the case of microwave filters, the main problem is that an average error of 1 dB in the stopband would be acceptable, but not in the passband (e.g., for a −20-dB return loss, the ripple in the passband is about 0.04 dB). Using decibel scales rather than linear scales can improve the problem. The response in the passband is rather complex (ripples), so more points should be used. The response in the stopband is less complex (monotonic with a few possible transmission zeros), so fewer points can be used. One could have a combined error that sums the errors in the lower and upper stopbands and in the passband. However, for the 0.04-dB issue, the error in the passband should be multiplied by a weight factor (on the order of 10, for example) so that stopband errors do not drown the passband error. Using S_{11} rather than S_{21} in the passband is another possibility. One can either try to make sure that the poles are being reproduced accurately or make sure that the return loss does not exceed −20 dB. This leads to different error criterions to be used (e.g., only provide the error between the simulated S_{11} and −20 dB and count the errors only where it is above −20 dB). The error criterion can make big differences in the results of optimization and should be investigated when the optimization is not satisfactory.

After computing the quality factors for each member of the population, one can check the quality factor of the best member. If it is not satisfactory, the loop consists of creating a new population with the hopes that the new population can produce members with higher quality factors.

The reproduction cycle starts by selecting the members of the current population that will be used to generate the next population. There are several possibilities for the selection. Depending on the crossover and mutation used, one can select many or few members of the current population. If the crossover produces two children for two parents, it is good to retain many parents, if the crossover produces more than two children for two parents, the number of parents can be less. The easiest selection process consists in taking the best members in the current population. Another technique consists of selecting the best members and then choosing some of them randomly. Another technique consists in random choice of some members and then selecting the best among them. Another technique consists in choosing the members in a random fashion, but where the

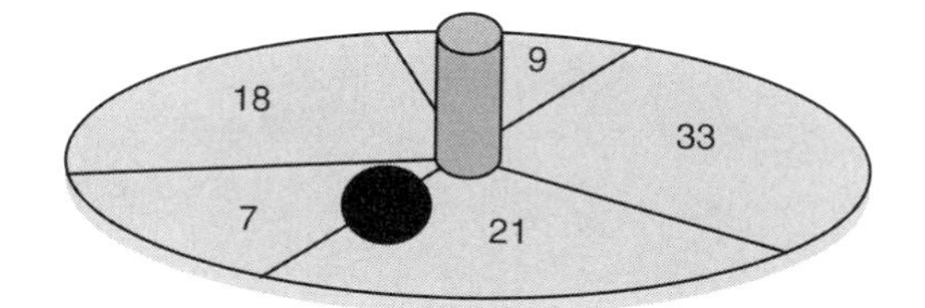

Figure 14.10 Selection process based on the concept of a spinning wheel.

probability of being chosen is proportional to the quality factor. This is done using the concept of a spinning wheel, as shown in Figure 14.10.

The ball has a uniform probability of falling anywhere on the surface of the wheel. However, the surface of one member is proportional to its quality factor. In other words, a member with a high quality factor has a larger surface than a member with a small quality factor and therefore there is a higher probability that the ball will fall on its surface and be selected.

Compared to gradient optimization methods, the genetic algorithm depends greatly on randomness. It will be very important to master random number generators such as `rand.m` in MATLAB. Also, one should make sure that all random generators are reset [e.g., `rand('state', 0)`] at the start of the program; otherwise, one could run the program one day and obtain the perfect filter, and run it another day and have rather different results.

Once the parents have been selected, the new population can be created. There are two principal techniques: crossover and mutation. The crossover operation consists of taking some genes from parent 1 and some genes from parent 2 to form child 1 and child 2, as shown in Figure 14.11. This is done by taking k genes from the first parent and $N - k$ genes from the second parent, where N is the total number of genes in a chromosome. For example, parents P_a and P_b below have $N = 6$ genes, and $k = 3$ is chosen to form the two children C_a and C_b.

$$P_a = (g_{a1}, g_{a2}, g_{a3}, g_{a4}, g_{a5}, g_{a6}) \qquad C_a = (g_{a1}, g_{a2}, g_{a3}, g_{b4}, g_{b5}, g_{b6})$$

$$P_b = (g_{b1}, g_{b2}, g_{b3}, g_{b4}, g_{b5}, g_{b6}) \qquad C_b = (g_{b1}, g_{b2}, g_{b3}, g_{a4}, g_{a5}, g_{a6})$$

Recombination of the genes can also be chosen differently (e.g., $C_a = g_{a1}, g_{b2}, g_{a3}, g_{b4}, g_{a5}, g_{b6}$, $C_b = g_{b1}, g_{a2}, g_{b3}, g_{a4}, g_{b5}, g_{a6}$). The value of k can be chosen randomly at each group of two parents.

If only crossover were applied, only the genes (parameter values) of the initial population are available for the final filter. It is unclear how this could lead to a great filter. This is why it is important to be able to create new genes (or new parameter values). The mutation consists of randomly selecting a gene to mutate in a child chromosome. The gene is then increased or decreased randomly by a specified amount, as shown in Figure 14.12. Therefore, there are two acts of randomness: the choice of the gene to mutate, and that to either increase or decrease the gene value. One can see that there is a lot of flexibility in writing

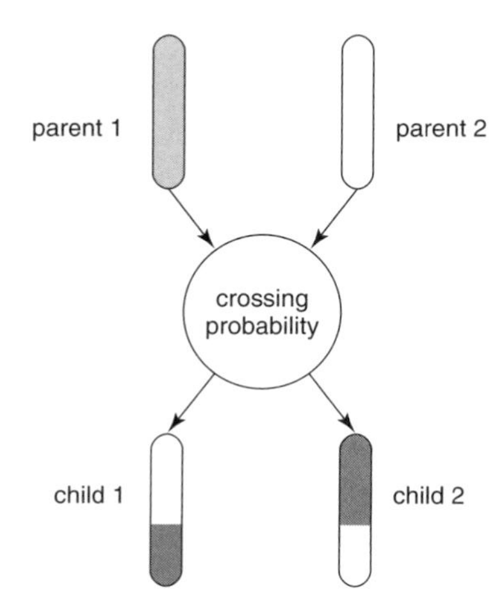

Figure 14.11 Cross-over operation.

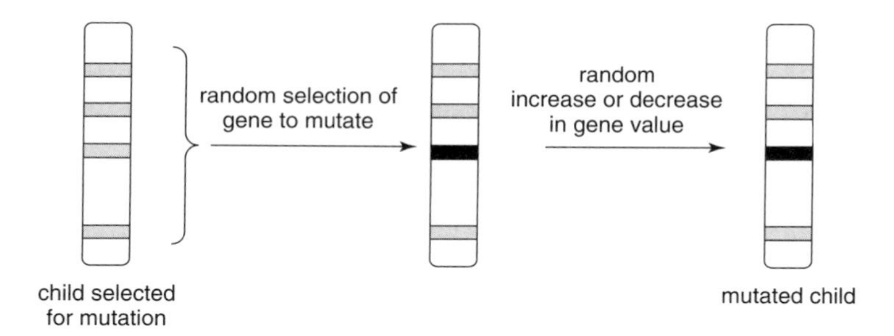

Figure 14.12 Mutation operation.

a genetic optimization program. More details on the genetic algorithm are given in the book by Rahmat-Samii and Michielssen [14.2].

14.3.2 Example

A simple optimization program to illustrate all the components of the genetic algorithm is given in this section. The program tries to find the parameters of a polynomial to approximate a Chebyshev low-pass response of order 5 with a −20-dB return loss. The purpose is not to compete with the function `polyfit.m` of MATLAB but mainly to show how to implement a genetic algorithm on an example that is pertinent to microwave filter design.

The program makes use of several functions. The first function populates the initial population from the initial chromosome. There was no coding, as the six polynomial coefficients correspond to the six first values in a chromosome. The seventh value in the chromosome was initialized at -1 and corresponds to the error value rather than the quality factor.

```
function [pop] = fct_initpop(chromosome, N)
dd = 5;

pop(1,:) = chromosome;
for k=2:N
  for i=1:length(chromosome)-1
      if rand(1)<0.5;
         pop(k,i) = abs(chromosome(i)-rand(1)*dd);
      else
         pop(k,i) = abs(chromosome(i)+rand(1)*dd);
      end;
      pop(k, length(chromosome)) = -1;
   end;
end;
```

The second function returns the error rather than the quality factor for each member in the population. Note that this function will be critical for the speed of the optimization and could be constructed with a system call to a C-based program representing the EM simulator. It is also critical since it defines the error criterion. In this case, three bands were used, with the passband error being weighted by a factor of 10. Note that the error value can be appreciated since it represents an average decibel error value between optimized and target responses.

```
function [pop_full] = fct_qfactor(pop)

[row, col] = size(pop);
loss_db = -20;
X = 10^(loss_db/10);
eps = sqrt( X/(1-X) );
[numt, dent] = fct_cheby(5, eps);
wlow    = linspace(-2, -1.4, 11);
wcenter = linspace(-1.4, 1.4, 101);
whigh   = linspace(1.4, 2, 11);
S21mtlow    = 20*log10( abs(freqs(numt, dent, wlow)) );
S21mtcenter = 20*log10( abs(freqs(numt, dent, wcenter)) );
S21mthigh   = 20*log10( abs(freqs(numt, dent, whigh)) );
warning off
for k=1:row
   chromosome = pop(k,:);
   for j=1:col-1
     den(j) = chromosome(j);
   end;
```

```
    S21mlow    = 20*log10( abs(freqs([1], den, wlow)) );
    S21mcenter = 20*log10( abs(freqs([1], den, wcenter)) );
    S21mhigh   = 20*log10( abs(freqs([1], den, whigh)) );
    Eklow    = (1/11) *sum(abs(S21mlow-S21mtlow));
    Ekcenter = (1/101)*sum(abs(S21mcenter-S21mtcenter));
    Ekhigh   = (1/11) *sum(abs(S21mhigh-S21mthigh));
    chromosome(col) = (Eklow + 10*Ekcenter + Ekhigh)/(12);
    pop_full(k,:) = chromosome;
end;
```

The third function selects the members to participate in the reproduction cycle. This is simply selecting the first best members, but other techniques have been mentioned.

```
function [pop_sel] = fct_select(pop)

[row, col] = size(pop);
sort_pop   = sortrows(pop, col);

for k=1:20
    pop_sel(k,:) = sort_pop(k,:);
end;
```

The fourth function creates the new population from the members selected. There is flexibility to increase or decrease the number of mutated chromosomes.

```
function [pop] = fct_reprod(pop_sel)

[row, col] = size(pop_sel);

k = 1;
while (k<=row-1)
    cra = pop_sel(k,:);
    crb = pop_sel(k+1,:);
    [child1, child2] = fct_cross(cra, crb);
    pop(k,:)   = child1;
    pop(k+1,:) = child2;
    k = k+2;
end;

pop(k,:)   = fct_mutate(pop_sel(1,:));
pop(k+1,:) = fct_mutate(pop_sel(2,:));
pop(k+2,:) = fct_mutate(pop_sel(3,:));
pop(k+3,:) = fct_mutate(pop_sel(4,:));
pop(k+4,:) = fct_mutate(pop_sel(5,:));
```

This function depends on two functions: one returns the two children when two parents are provided, and the other returns a mutated chromosome when a starting chromosome is provided.

```
function [child1, child2] = fct_cross (cra, crb)

N = length(cra)-1;
isel = 1+round((N-2)*rand(1));


for i=1:isel
   child1(i) = cra(i);
   child2(i) = crb(i);
end;
for i=(isel+1):N
   child1(i) = crb(i);
   child2(i) = cra(i);
end;
child1(N+1) = -1;
child2(N+1) = -1;

function [mutant] = fct_mutate(cra)
dd = 0.05*cra(7);

gene_i = 1+ round(5*rand(1));
if rand(1)<0.5
    sig = -1;
else
   sig = 1;
end;
mutant = cra;
mutant(gene_i) = abs (mutant(gene_i) +sig*dd);
```

The main program sets the initial chromosome and starts the iterative process. The process is stopped after a predefined number of iterations.

```
seed = 0;
rand('state', seed);
col = 7;
chromosome_initial = [1 1 1 1 1 1  -1];
pop_initial  = fct_initpop(chromosome_initial, 40);
pop_complete = fct_qfactor(pop_initial);
pop_sel      = fct_select(pop_complete);
err(1)      = pop_sel(1,7);
```

```
Niter = 500;
for ni=2:Niter
   pop_new       = fct_reprod(pop_sel);
   pop_complete = fct_qfactor(pop_new);
   pop_sel       = fct_select(pop_complete);
   err(ni)        = pop_sel(1,7);
end;
```

The optimized filter responses along with the theoretical Chebyshev response of order 5 are given in Figure 14.13. It is often found that the forward coefficient (part a) often looks great, while the main difficulty is in matching the poles in the return coefficient (part b). Figure 14.14 shows the convergence of the error versus the iteration number. The initial parameters were totally arbitrary (e.g., all taken as 1) and one can see that after 500 iterations, a good result is obtained. This is a case of blind optimization where the initial guess was far from the target.

One should note that this program is not very sophisticated and that there is a lot of flexibility in implementing the genetic algorithm. A very rudimentary mechanism for decreasing the amount of change on a gene during mutation was used and is related to the value of the error (the smaller the current error, the smaller the change). When running the program, it is found that the algorithm is sensitive to the random generator seed, the initial population values, the error criteria, the amount of mutation, the amount of crossover, and the size of the population. Also, no attempts were made to speed the computations, such as avoiding recomputing the target function values.

It is worthwhile to note that one could also use the program with large incremental changes in a first run and then use the results as the initial parameters for a second run where the incremental changes are smaller. It is also possible to use the genetic algorithm for defining a suitable initial guess and then use the gradient methods to fine-tune the parameters.

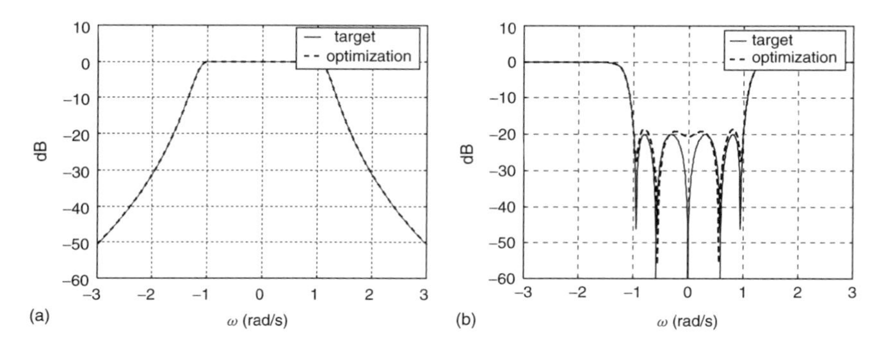

Figure 14.13 Example of optimization using the genetic algorithm.

14.4 REALIZATIONS

Several filters have been designed and constructed using the method described in this chapter [14.5,14.7,14.8].

14.4.1 Fourth-Order Filter with Two Transmission Zeros

The structure of the first filter is shown in Figure 14.15. It consists of two terminating waveguides [(A,B) and (I,J)], a central building block [(E,F)], and two small connecting waveguides [(C,D) and (G,H)]. The structure is defined entirely by the coordinates of the various points. The cross-sectional dimensions of the terminating waveguides were taken as the WR75 standard cross-section dimensions. WR75 was chosen for applications around 14 GHz so that the TE_{10} mode will be above cutoff while higher modes would vanish before reaching the first step discontinuity.

The modes and couplings considered for this filter are shown in Figure 14.16. Note that the the TE_{10} mode does not resonate in the central building block. This is accomplished through the TM_{12} and TM_{21} modes. In addition, the TE_{10} mode can be used with each of the TM modes to create a transmission zero. Therefore, the structure leads to a fourth-order filter with two transmission zeros. This type of filter is small, as the resonant modes are used close to their cutoff frequencies. In this case the height and width are thus smaller and the length of the waveguides does not affect the resonance frequency as much as in the classical resonance condition.

The optimization program based on the genetic algorithm was used, and the physical parameters of the structure are given below. It was found that using 10 modes throughout the entire structure and about 100 modes in the vicinity

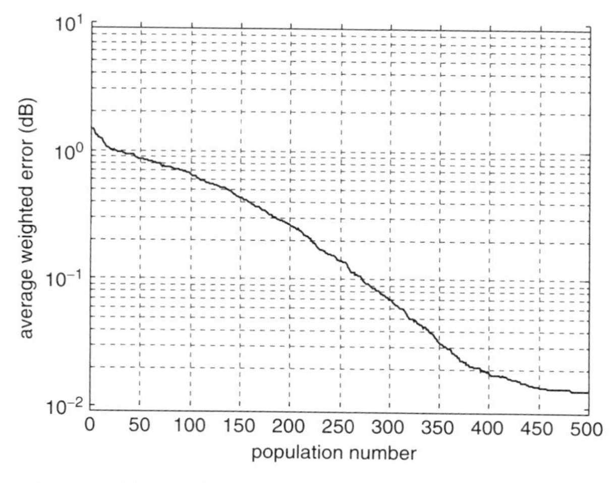

Figure 14.14 Convergence of the genetic algorithm.

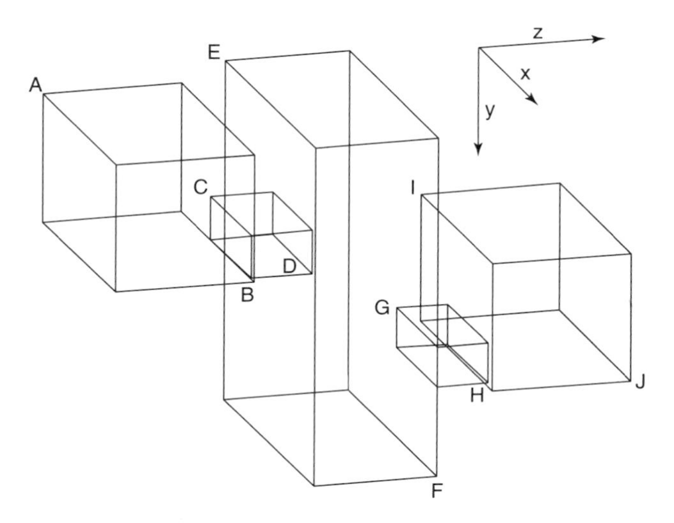

Figure 14.15 Structure of the fourth-order filter with two transmission zeros.

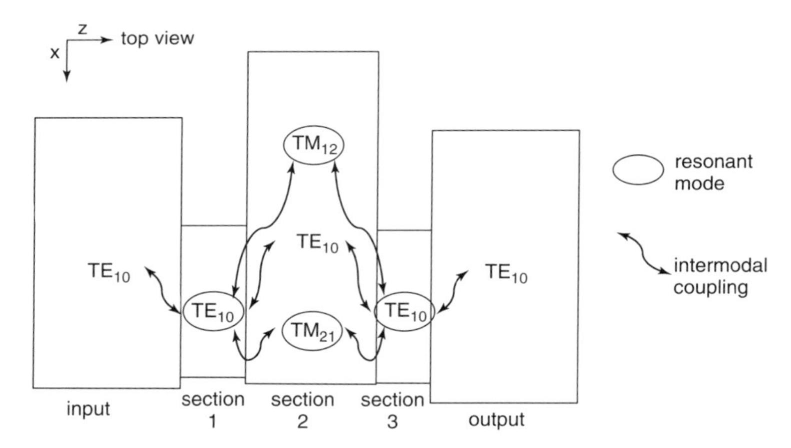

Figure 14.16 Top view of the filter with various modes and couplings.

of the step discontinuities provided a good compromise between the accuracy achieved and the computation times. All coordinates are given in millimeters.

```
A(0.00, 0.00, 0.00)          B(19.05, 9.525, 10.00)
C(7.59, 6.28, 10.00)         D(18.21, 9.52, 14.48)
E(-4.64, -0.06, 14.48)       F(18.63, 25.25, 23.40)
G(7.95, 15.63, 23.40)        H(18.56, 18.61, 27.09)
I(0.88, 9.32, 27.09)         J(19.93, 18.845, 37.09)
```

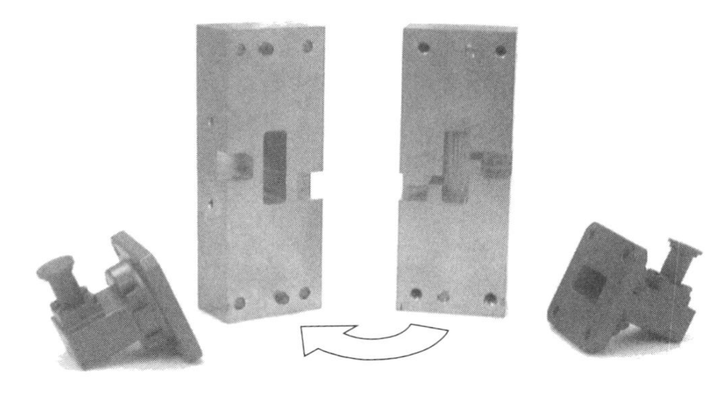

Figure 14.17 Fourth-order filter with two transmission zeros.

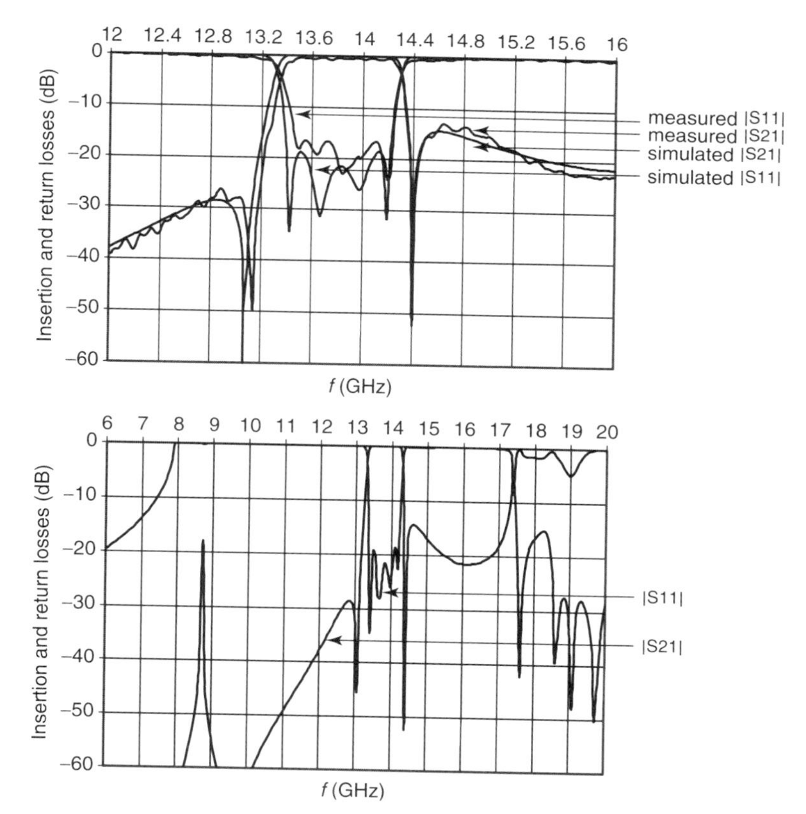

Figure 14.18 Simulated and measured responses (top) and wide-frequency-range measurements (bottom).

The filter realized is shown in Figure 14.17. The filter is made of brass, and the responses of the filter are shown in Figure 14.18.

14.4.2 Seventh-Order Filter with Four Transmission Zeros

A second filter [14.7,14.8] was designed around 20 GHz using two central building blocks, as shown in Figure 14.19. The design method was applied to the case of two multimodal cavities. In this case we obtained a seventh-order filter with four transmission zeros [14.8]. In fact, these were two double-order transmission zeros. The physical parameters are provided in millimeters below.

```
A(0.00, 0.00, 0.00)       F(13.11, 16.29, 18.14)  K(5.62, 3.14, 29.13)
B(12.95, 6.48, 10.00)     G(5.90, 10.53, 18.14)   L(12.82, 5.58, 31.09)
C(5.55, 2.96, 10.00)      H(13.05, 11.87, 22.60)  M(0.02, 0.21, 31.09)
D(12.70, 5.69, 11.67)     I(-2.50, -0.69, 22.60)  N(12.97, 6.69, 41.09)
E(-2.50, -0.73, 11.67)    J(13.11, 16.31, 29.13)
```

A top view of the filter with the resonances and coupling between modes is shown in Figure 14.20. The filter constructed is shown in Figure 14.21. The filter was made of brass.

The simulated response using our own simulator with 30 modes throughout the entire structure and 300 modes in the vicinity of the step discontinuities is shown in Figure 14.22 along with the simulated response using the commercial electromagnetic simulator HFSS (high-frequency structure simulator) and the measured response. The HFSS simulations required 48 minutes using 13,673 tetrahedrons. The mismatches are due primarily to construction problems in making perfectly square blocks. The corners, in particular, suffer from this fillet effect, which leads to smaller effective lengths [14.9]. Another problem is the horizontal split of the filter (e.g., at a constant value of y) and the smoothing effects on the TE_{10} modes. In addition, the two halves are not forced into each other, therefore, the contacts are not perfect, leading to some perturbation of the electromagnetic fields.

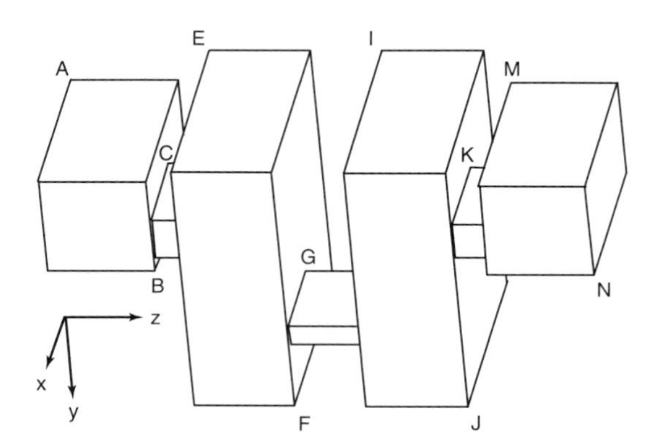

Figure 14.19 Seventh-order filter with four transmission zeros.

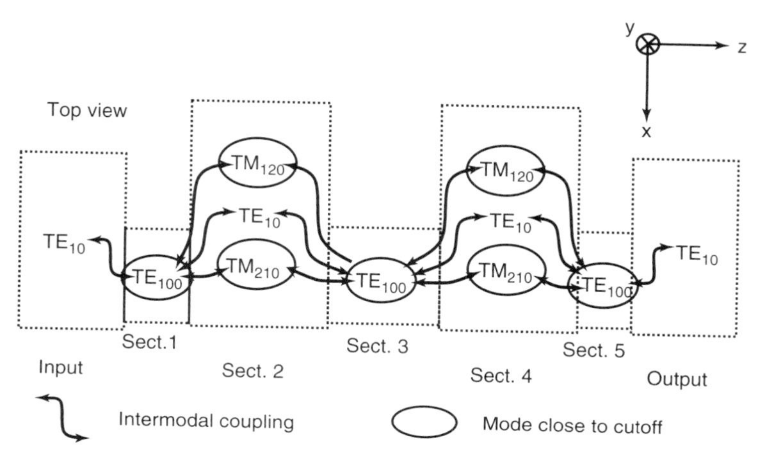

Figure 14.20 Top view of the filter with the various modes and couplings.

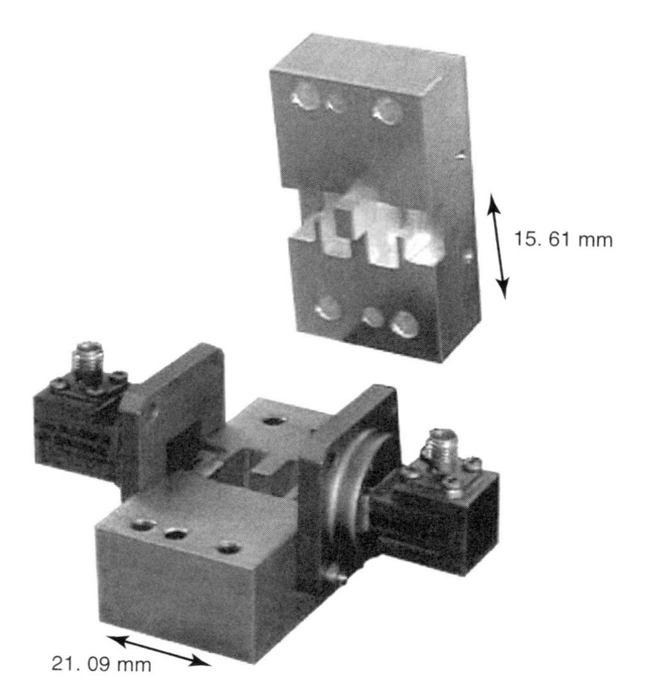

Figure 14.21 Seventh-order filter with two double transmission zeros.

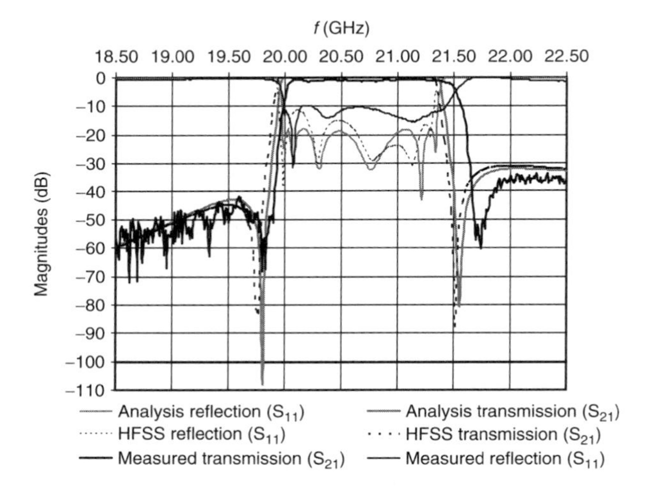

Figure 14.22 Simulated and measured responses.

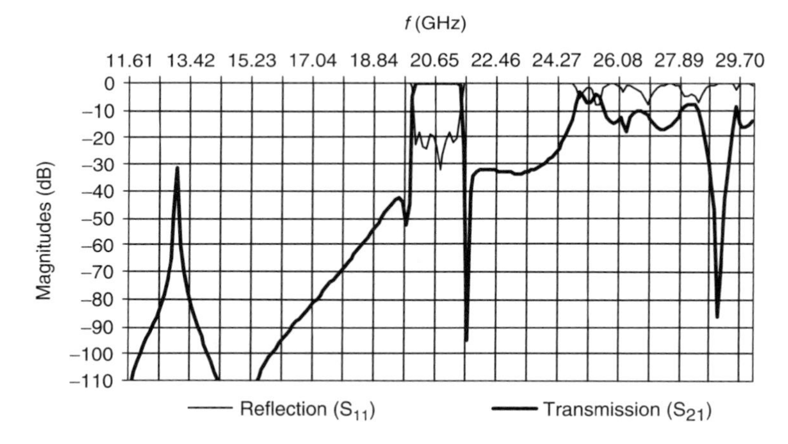

Figure 14.23 Wide-frequency-range measurements.

The response on a wider frequency range is shown in Figure 14.23. The first spurious response appears around 25 GHz and corresponds to the first resonance of the TE_{10} mode in the two central building blocks. The spurious response around 13 GHz corresponds to the cutoff of the TE_{11} mode in these building blocks.

14.4.3 Extension to a Tenth-Order Filter with Six Transmission Zeros

The coupling principle has been extended to a third example. The filter is composed of three dual-TM and four TE_{100} cavities. We therefore expected

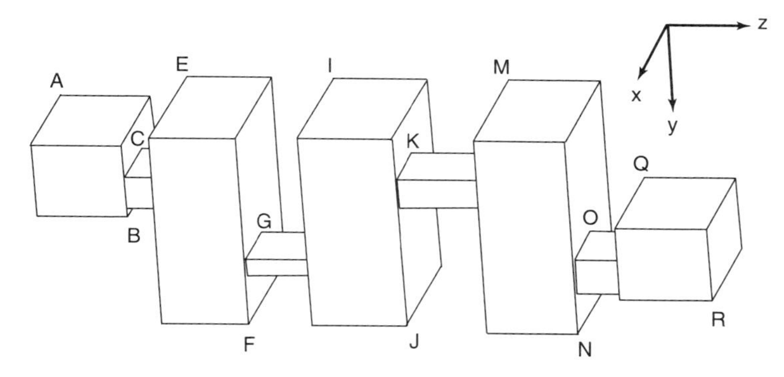

Figure 14.24 Tenth-order filter with six transmission zeros.

10 poles and six transmission zeros. The structure of this third filter is shown in Figure 14.24. The physical parameters found using the optimization program are given below in millimeters.

```
A(0.026, 0.00, 0.00)     G(5.90, 10.37, 15.18)    M (-2.50, 0.68, 32.72)
B(12.98, 6.477, 6.80)    H(13.01, 11.76, 19.74)   N(13.11, 16.75, 39.49)
C(5.52, 2.91, 6.80)      I (-2.50, 0.73, 19.74)   O(5.62, 10.17, 39.49)
D(12.80, 5.81, 8.79)     J(13.11, 16.28, 26.78)   P(12.85, 13.30, 42.65)
E(-2.50, -0.75, 8.79)    K(5.91, 2.99, 26.78)     Q(0.02, 7.21, 42.65)
F(13.11, 16.29, 15.18)   L(13.03, 5.53, 32.72)    R(12.974, 13.69, 49.45)
```

The access guides are taken as WR51. The entire length of the filter is less than 36 mm. The response obtained with our analysis software is shown in Figure 14.25

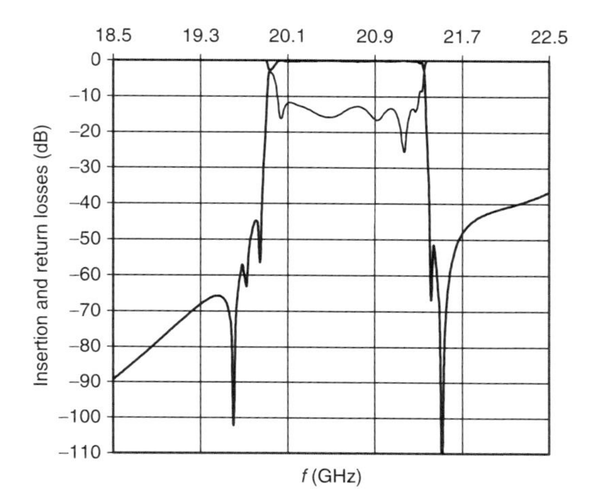

Figure 14.25 Simulated responses of the tenth-order filter with six transmission zeros.

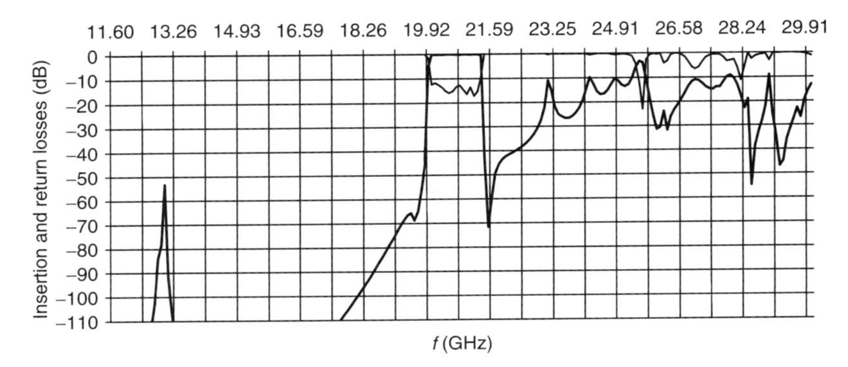

Figure 14.26 Simulated responses on a wide frequency range.

and was confirmed by HFSS. One can see about five transmission zeros. The zero on the right side of the passband is probably a double-order zero. The analysis on a wider frequency range is shown in Figure 14.26.

14.5 CONCLUSIONS

A new coupling scheme was introduced that uses all resonant modes near their cutoff frequencies with direct coupling without any iris. Screws and other artifacts are not needed for the cavities. The synthesis method employs a direct optimization technique based on the genetic algorithm. Coupled with a fast analysis process, it leads to a perfectly suitable technique for investigating new coupling concepts for rectangular waveguide microwave filters. The method is well suited for filters of relative bandwidth around 7%. The selectivity of the filters is high since numerous transmission zeros can be achieved. The sensitivity of these filters is also low, they are easily manufactured, and they are compact.

Acknowledgments

This work (and the corresponding patents) was carried out with the collaboration of Alcatel Space [14.10,14.12].

REFERENCES

[14.1] R. R. Bonetti and A. E. Williams, Application of dual TM modes to triple and quadruple-mode filters, *IEEE Trans. Microwave Theory Tech.*, vol. 35, no. 12, pp. 1143–1149, 1987.

[14.2] Y. Rahmat-Samii and E. Michielssen, *Electromagnetic Optimization by Genetic Algorithms*, Wiley, New York, 1999.

[14.3] M. Lecouve, P. Jarry, E. Kerhervé, N. Boutheiller, and F. Marc, Genetic algorihm optimisation for evanescent mode waveguide filter design, *(Proc. IEEE International Symposium on Circuits and Systems*, ISCAS'00, GNEVA, Switzerland, pp. 411–414, May 28–31, 2000.

[14.4] M. Lecouve, N. Boutheiller, P. Jarry, and E. Kerhervé, Genetic algorithm based CAD of microwave waveguide filters with resonating bend, *Eur. Microwave Conf. Dig.*, vol. 3, pp. 300–303, Oct. 2000.

[14.5] N. Boutheiller, P. Jarry, E. Kerhervé, J. M. Pham, and S. Vigneron, A two zeros fourth order microwave waveguide filter using a simple rectangular quadruple-mode cavity, *Proc. IEEE International Microwave Symposium*, IMS'02, Seattle, WA, pp. 1777–1780, June 3–7, 2002.

[14.6] N. Boutheiller, M. Lecouve, P. Jarry, J. M. Pham, Synthesis and optimization using the genetic algorithm (GA) of straight and curved evanescent microwave filters, invited paper, *Transworld Research Network*, Special Issue on Microwave Theory and Techniques, ed. P. Jarry, pp. 1–16, June 2004.

[14.7] N. Boutheiller, P. Jarry, E. Kerhervé, J. M. Pham, and S. Vigneron, A two zero fourth pole basic building block using all modes at their cutoff frequency resonance, presented at the European Workshop on Microwave Filters, CNES/ESA, Toulouse, France, June 24–26 2002.

[14.8] N. Boutheiller, P. Jarry, E. Kerhervé, J. M. Pham, and S. Vigneron, Two zero four poles building block for microwave filters using all resonant modes at their cut-off frequency, *Int. J. RF Microwave Comput. Aided Eng.*, vol. no. 13, Issue 6, pp. 429–437, 2003

[14.9] P. Jarry, E. Kerhervé, J. M. Pham, N. Boutheiller, and M. Lecouve, *Synthesis and optimisation of straight and curved evanescent microwave filters using the genetic algorithm (GA), Proc. IEEE Mediterranean Microwave Symposium*, Marseille, France, pp. 36–39 June 1–3, 2004.

[14.10] N. Boutheiller, P. Jarry, E. Kerhervé, Y. Latouche, J. M. Pham, and S. Vigneron, Filtre hyperfréquence quadri-mode en guide d'ondes sans réglage et possédant des zéros de transmission, French patent, 01 15 252, Alcatel Space, Nov. 2001.

[14.11] S. Vigneron, P. Jarry, Y. Latouche, E. Kerhervé, J. M. Pham, and N. Boutheiller, Filtre hyperfréquence quadrimodes en guide d'ondes et possédant des zeros de transmission, European patent, 02803825.5–220-FR 0203988, July 12, 2004.

[14.12] S. Vigneron, Y. Latouche, P. Jarry, E. Kerhervé, J. M. Pham, and N. Boutheiller, New microwave quadri-modes filters without tuning but with transmission zeros patents U.S. and Canadian Sugrue, Mion, Dec. 2002.

Selected Bibliography

Bila, S., D. Baillargeat, S. Verdeyme, M. Aubourg, P. Guillon, F. Seyfert, J. Grimm, L. Baratchart, C. Zanchi, and J. Sombrin, Direct electromagnetic optimization of microwave filters, *IEEE Microwave Mag.*, vol. 2, 2001.

Bonetti, R. R., and A. E. Williams, A 6-pole, self-equalized triple-mode filter with a single cross-coupling, *Proc. 19th European Microwave Conference*, EMC'89, London, pp. 881–885, Sept. 1989.

Budimir, D., Design of general filters for modern digital communication systems, *Microwave Eng. Eur.*, Mar. 1998.

Kurzrok, R.M., General three-resonator filters in waveguide, *IEEE Trans. Microwave Theory Tech.*, vol. 14, no. 1, pp. 46–47, 1966.

Kurzrok, R. M., General four-resonator filters at microwave frequencies, *IEEE Trans. Microwave Theory Tech.*, vol. 14, no. 1, pp. 295–296, 1966.

Lastoria, G., and G. Gerini, M. Guglielmi, and F. Emma, CAD of triple-mode cavities in rectangular waveguide, *Microwave Guided Wave Lett.*, vol. 8, no. 10, pp. 339–341, 1998.

Lin, W., Microwave filters employing a single cavity excited in more than one mode, *J. App. Phys.*, vol. 22, no. 8, pp. 989–1011, 1951.

Sheng-Li, L., and L. Wei-Gan, A five mode single spherical cavity microwave filter, *MTT-S Int. Microwave Symp. Dig.*, vol. 92, no. 2, pp. 909–912, 1992.

Swanson, D., and G. Macchiarella, Microwave filter design by synthesis and optimization, *IEEE Microwave Mag.*, vol. 8, no. 2, pp. 55–69, Apr. 2007.

Swanson, D., and R. J. Wenzel, Fast analysis and optimization of combline filters using FEM, presented at the IEEE MTT-S international Microwave Symposium, Phoenix, AZ, May 20–25, 2001.

Tang, W. C., and S. K. Chaudhuri, Triple-mode true elliptic-function filter realization for satellite transponders, *Proc. IEEE-MTT Symposium*, pp. 83–85, May 1983.

Tang, W. C., and S.K. Chaudhuri, A true elliptic function filter using triple mode degenerate cavities, *IEEE Trans. Microwave Theory Tech.*, vol. 32 no. 11, pp. 1449–1454, 1984.

Williams, A. E., and R. R. Bonetti, A mixed dual-quadruple mode 10-pole filter, *Proc. European Microwave Conference*, Stockholm, Sweden, pp. 966–968, Sept. 1988.

APPENDIX 1

LOSSLESS SYSTEMS

In a lossless system (Figure A1.1) the input power P_{in} is equal to the power P_L dissipated on the load R_L. In addition the input power is given by

$$P_{\text{in}} = \text{Re}[Z_{\text{in}}(j\omega)]I_1(j\omega)I_1(-j\omega) = \frac{1}{2}[Z_{\text{in}}(j\omega) + Z_{\text{in}}(-j\omega)]I_1(j\omega)I_1(-j\omega)$$

and the source voltage V_G is given by

$$V_G = [R_G + Z_{\text{in}}(j\omega)]I_1(j\omega)$$

Therefore, the input power can be expressed as

$$P_{\text{in}} = \frac{1}{2}\left\{\frac{Z_{\text{in}}(j\omega) + Z_{\text{in}}(-j\omega)}{[R_G + Z_{\text{in}}(j\omega)][R_G + Z_{\text{in}}(-j\omega)]}\right\} V_G\, V_G^*$$

It is seen that the input power will be at its maximum when there is impedance matching:

$$P_{\text{in}} = P_{\text{max}} \qquad \text{when} \quad Z_{\text{in}}(j\omega) = R_G$$

$$P_{\text{max}} = \frac{V_G\, V_G^*}{4R_G}$$

and

$$|S_{12}(j\omega)|^2 = \frac{P_{\text{in}}}{P_{\text{max}}} = 2R_G\frac{Z_{\text{in}}(j\omega) + Z_{\text{in}}(-j\omega)}{[R_G + Z_{\text{in}}(j\omega)][R_G + Z_{\text{in}}(-j\omega)]}$$

Advanced Design Techniques and Realizations of Microwave and RF Filters,
By Pierre Jarry and Jacques Beneat

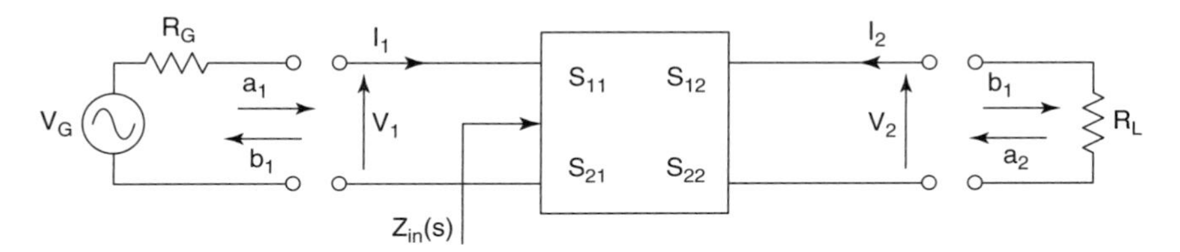

Figure A1.1 Lossless two-port system.

for a lossless system:

$$|S_{11}(j\omega)|^2 = 1 - |S_{12}(j\omega)|^2 = \left|\frac{R_G - Z_{\text{in}}(j\omega)}{R_G + Z_{\text{in}}(j\omega)}\right|^2$$

There are therefore two solutions for defining the input impedance from the reflection coefficient:

$$\frac{Z_{\text{in}}(j\omega)}{R_G} = \frac{1 \pm S_{11}(j\omega)}{1 \mp S_{11}(j\omega)}$$

The synthesis steps are then:

1. $|S_{12}(j\omega)|^2$ is given and provides the desired filter response.
2. Compute $|S_{11}(j\omega)|^2 = 1 - |S_{12}(j\omega)|^2$.
3. Form $S_{11}(s)S_{11}(-s)$ by letting $s = j\omega$.
4. Find the zeros and poles of $S_{11}(s)S_{11}(-s)$ and select the poles in the left plane to form a stable $S_{11}(s)$.
5. Form $Z_{\text{in}}/(s)R_G = [1 \pm S_{11}(s)]/[1 \mp S_{11}(s)]$ [either $Z_{\text{in}}(s)/R_G$ or $R_G/Z_{\text{in}}(s)$]
6. Synthesize $Z_{\text{in}}(s)$ using a method such as the Darlington.

REFERENCE

Jarry, P., *Microwave Filters and Couplers*, University of Bordeaux, Bordeaux, France 2003.

APPENDIX 2

REDUNDANT ELEMENTS

If we consider a system described by its $ABCD$ matrix and terminated on R_G and R_L, the scattering parameters can be defined as in Chapter 1. For example, the transmission coefficient S_{12} is given by

$$S_{12} = \frac{2\sqrt{R_G R_L}}{AR_L + B + CR_G R_L + DR_G}$$

Two impedance inverters are now placed on each side of this $ABCD$ system as shown in Figure A2.1. The $A'B'C'D'$ matrix representing the new system made of the cascade of the K_A inverter, the initial $ABCD$ system, and the K_B inverter is given by

$$\begin{aligned}
\begin{pmatrix} A' & B' \\ C' & D' \end{pmatrix} &= \begin{pmatrix} 0 & jK_A \\ j/K_A & 0 \end{pmatrix} \begin{pmatrix} A & B \\ C & D \end{pmatrix} \begin{pmatrix} 0 & jK_B \\ j/K_B & 0 \end{pmatrix} \\
&= -\begin{pmatrix} D\dfrac{K_A}{K_B} & CK_A K_B \\ \dfrac{B}{K_A K_B} & A\dfrac{K_B}{K_A} \end{pmatrix}
\end{aligned}$$

The new transmission coefficient is given by

$$\begin{aligned}
S'_{12} &= \frac{2\sqrt{R_G R_L}}{A'R_L + B' + C'R_G R_L + D'R_G} \\
&= \frac{-2\sqrt{R_G R_L}}{A(K_B/K_A)R_G + B(R_G R_L/K_A K_B) + CK_A K_B + D(K_A/K_B)R_L}
\end{aligned}$$

Advanced Design Techniques and Realizations of Microwave and RF Filters,
By Pierre Jarry and Jacques Beneat

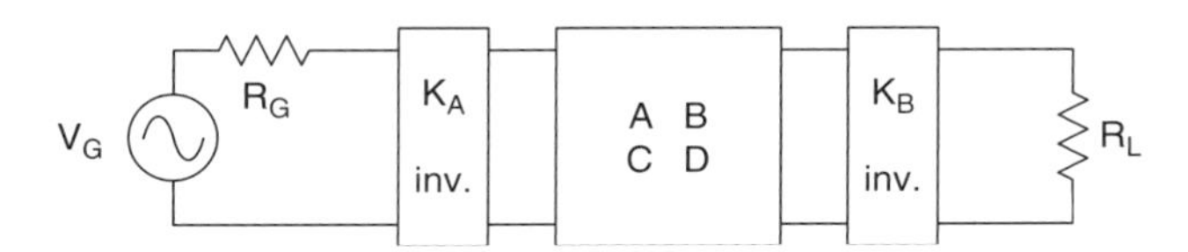

Figure A2.1 Placing impedance inverters at the input and output of a system.

If we take $K_A = R_G$ and $K_B = R_L$, then

$$S'_{12} = -S_{12}$$

The minus sign is not important since it does not affect the magnitude but only introduces an additional constant phase shift of $\pm\pi$, which does not affect the delay, which is the derivative of the phase.

Using the same approach, we find that when taking $K_A = R_G$ and $K_B = R_L$, the scattering matrix of the new system is given by

$$\begin{pmatrix} S'_{11} & S'_{12} \\ S'_{21} & S'_{22} \end{pmatrix} = \begin{pmatrix} S_{11} & -S_{12} \\ -S_{21} & S_{22} \end{pmatrix}$$

In conclusion, it is possible to add an inverter at the input of the system that has the value of the input termination and an inverter at the output of the system that has the value of the output termination without changing the response of the system.

The demonstration can be applied to the case of unit elements. For example, consider a system made of the initial system described by its *ABCD* matrix but where a unit element of length θ and impedance Z_A is placed at the input and a unit element of length θ and impedance Z_B is placed at the output, as shown in Figure A2.2.

Using an approach similar to that taken earlier, the scattering matrix of the new system when taking $Z_A = R_G$ and $Z_B = R_L$ is given by

$$\begin{pmatrix} S'_{11} & S'_{12} \\ S'_{21} & S'_{22} \end{pmatrix} = \begin{pmatrix} S_{11}e^{-j4\theta} & S_{12}e^{-j2\theta} \\ S_{21}e^{-j2\theta} & S_{22} \end{pmatrix}$$

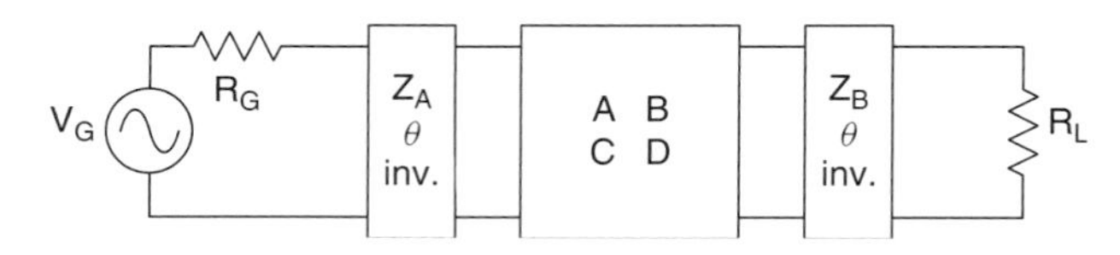

Figure A2.2 Placing unit elements at the input and output of a system.

When $\theta = \pi/2$ we find that

$$\begin{pmatrix} S'_{11} & S'_{12} \\ S'_{21} & S'_{22} \end{pmatrix} = \begin{pmatrix} S_{11} & -S_{12} \\ -S_{21} & S_{22} \end{pmatrix}$$

In conclusion, it is possible to add a unit element at the input of the system with impedance that of the input termination and length of $\theta = \pi/2$, and a unit element at the output of the system with impedance that of the output termination and length of $\theta = \pi/2$ without changing the response of the system.

APPENDIX 3

MODAL ANALYSIS OF WAVEGUIDE STEP DISCONTINUITIES

A3.1 DOUBLE STEP DISCONTINUITY

Figure A3.1 shows the configuration of a double step discontinuity in a rectangular waveguide. There are two step discontinuities, one to account for a change in width (x direction) and the other to account for a change in height (y direction). The waveguide in region I has width a_1 and height b_1, while the waveguide in region II has width a_2 and height b_2. The widths and heights are related such that $a_1 = a_2 - 2c$ and $b_1 = b_2 - 2d$. In the case under study, the waveguides are also assumed to be centered.

Figure A3.2 shows the scattering waves at the discontinuity or the junction of the two waveguides. The main goal is to define the scattering parameters of this junction related to the fundamental mode. The main problem is that this type of discontinuity tends to generate higher-order modes. This means that the scattering parameters related to the fundamental depend on what is happening to all the modes, including the fundamental.

An infinite number of incident and reflected waves are assumed to be present in the vicinity of each region surrounding the discontinuity, as shown in Figure A3.2. Each wave is carried by a corresponding field configuration or mode solution to the waveguide characteristic equations. The wave can therefore be characterized by a mode and a wave amplitude coefficient a or b. Furthermore, the power carried by a given wave should be proportional to the square of the wave amplitude coefficient so that functions pertinent to the corresponding mode are suitably normalized. Under these conditions, wave amplitude coefficients, a and b, are related to each other by the scattering matrix of the discontinuity. These relations can be found by matching the electric and magnetic tangential fields that support the waves at the discontinuity.

Advanced Design Techniques and Realizations of Microwave and RF Filters,
By Pierre Jarry and Jacques Beneat

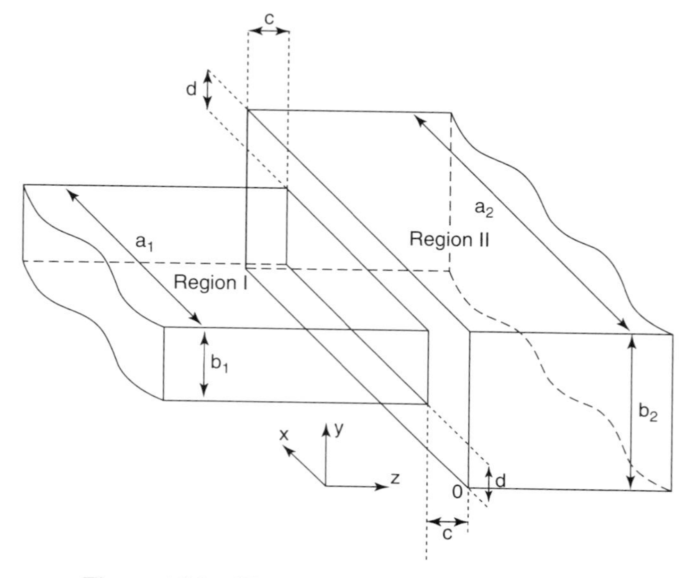

Figure A3.1 Waveguide double step discontinuity.

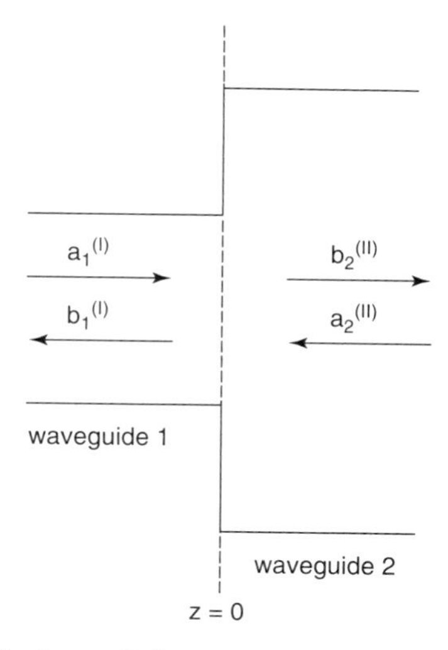

Figure A3.2 Wave amplitude coefficients at the junction between the two waveguides.

A3.2 MODAL ANALYSIS EQUATIONS

With reference to Figures A3.1 and A3.2, the continuity equations for the tangential fields on the surface of the junction plane can be expressed as

$$\vec{E}_{t1} = \vec{E}_{t2} \tag{A3.1}$$

$$\vec{H}_{t1} = \vec{H}_{t2} \tag{A3.2}$$

The complete electric and magnetic fields are made of a vectorial addition of the electric and magnetic fields of all the TE and TM modes with their associated incident or reflected wave amplitudes (a or b). In addition, one should take into account a common direction, so that the wave amplitudes are at times counted positively or negatively. This results in the following equations for the complete tangential fields:

$$\vec{E}_{t1} = \sum_{n=0}^{\infty}\sum_{m=0}^{\infty}(a_{1h\ nm} + b_{1h\ nm})\vec{e}_{1th\ nm} + \sum_{n=0}^{\infty}\sum_{m=0}^{\infty}(a_{1e\ nm} - b_{1e\ nm})\vec{e}_{1te\ nm}$$

$$\vec{E}_{t2} = \sum_{n=0}^{\infty}\sum_{m=0}^{\infty}(a_{2h\ nm} + b_{2h\ nm})\vec{e}_{2th\ nm} + \sum_{n=0}^{\infty}\sum_{m=0}^{\infty}(-a_{2e\ nm} + b_{2e\ nm})\vec{e}_{2te\ nm}$$

$$\vec{H}_{t1} = \sum_{n=0}^{\infty}\sum_{m=0}^{\infty}(a_{1h\ nm} - b_{1h\ nm})\vec{h}_{1th\ nm} + \sum_{n=0}^{\infty}\sum_{m=0}^{\infty}(a_{1e\ nm} + b_{1e\ nm})\vec{h}_{1te\ nm}$$

$$\vec{H}_{t2} = \sum_{n=0}^{\infty}\sum_{m=0}^{\infty}(-a_{2h\ nm} + b_{2h\ nm})\vec{h}_{2th\ nm} + \sum_{n=0}^{\infty}\sum_{m=0}^{\infty}(a_{2e\ nm} + b_{2e\ nm})\vec{h}_{2te\ nm}$$

- The coefficients $a_{ih\ nm}$ and $a_{ie\ nm}$ are the wave amplitude coefficients of an incident wave carried, respectively, by a TE_{nm} and TM_{nm} mode solution of the characteristic equations of waveguide i.
- The coefficients $b_{ih\ nm}$ and $b_{ie\ nm}$ are the wave amplitude coefficients of a reflected wave carried, respectively, by a TE_{nm} and TM_{nm} mode solution of the characteristic equations of waveguide i.
- The vectors $\vec{e}_{ith\ nm}$ and $\vec{e}_{ite\ nm}$ are, respectively, the tangential electric fields of a TE_{nm} and TM_{nm} mode solution of the characteristic equations of waveguide i.
- The vectors $\vec{h}_{ith\ nm}$ and $\vec{h}_{ite\ nm}$ are, respectively, the tangential electric fields of a TE_{nm} and TM_{nm} mode solution of the characteristic equations of waveguide i.

To satisfy the normalization criterion, the tangential fields must be such that

$$\frac{1}{2}\left|\iint_{S_i} (\vec{e}_{ith\ nm} \times \vec{h}_{ith\ nm}) \cdot \vec{e}_z\ dS\right| = 1$$

$$\frac{1}{2}\left|\iint_{S_i} (\vec{e}_{ite\ nm} \times \vec{h}_{ite\ nm}) \cdot \vec{e}_z\ dS\right| = 1$$

where S_i is the surface of the waveguide i cross section and $\vec{e}_z$ is a unit vector on the z-axis. It is clear that for practical cases, the number of modes used in describing the complete tangential fields must be restricted to finite values. Thus, in each region the first modes that have the lowest cutoff frequencies are selected. By substituting the subscripts n and m by a single subscript u for region I or subscript v for region II representing the rank by increasing order of cutoff frequency for a given TE_{nm} or TM_{nm} mode, equations A3.1 and A3.2 can be written as:

$$\sum_{u=1}^{\infty} (a_{1h\ u} + b_{1h\ u})\vec{e}_{1th\ u} + \sum_{u=1}^{\infty} (a_{1e\ u} - b_{1e\ u})\vec{e}_{1te\ u}$$
$$= \sum_{v=1}^{\infty} (a_{2h\ v} + b_{2h\ v})\vec{e}_{2th\ v} + \sum_{u=1}^{\infty} (-a_{2e\ v} + b_{2e\ v})\vec{e}_{2te\ v} \qquad \text{(A3.1)}$$

$$\sum_{u=1}^{\infty} (a_{1h\ u} - b_{1h\ u})\vec{h}_{1th\ u} + \sum_{u=1}^{\infty} (a_{1e\ u} + b_{1e\ u})\vec{h}_{1te\ u}$$
$$= \sum_{v=1}^{\infty} (-a_{2h\ v} + b_{2h\ v})\vec{h}_{2th\ v} + \sum_{v=1}^{\infty} (a_{2e\ v} + b_{2e\ v})\vec{h}_{2te\ v} \qquad \text{(A3.2)}$$

At this stage it is possible to reduce the complexity of these equations by applying the following integrals that make use of the orthogonality properties of the modes:

$$\iint_{S_2} (\text{Eq. A3.1}) \cdot \vec{e}^{\,*}_{2th\ v}\ dS \qquad \iint_{S_2} (\text{Eq. A3.1}) \cdot \vec{e}^{\,*}_{2te\ v}\ dS$$

$$\iint_{S_1} (\text{Eq. A3.2}) \cdot \vec{h}^{*}_{1th\ u}\ dS \qquad \iint_{S_1} (\text{Eq. A3.2}) \cdot \vec{h}^{*}_{1te\ u}\ dS$$

This provides the following equations:

$$a_{2h\ v} + b_{2h\ v} = \sum_{u=1}^{\infty} (a_{1h\ u} + b_{1h\ u})V_{hh\ vu} + \sum_{u=1}^{\infty} (a_{1e\ u} - b_{1e\ u})V_{he\ vu}$$

$$-a_{2e\,v} + b_{2e\,v} = \sum_{u=1}^{\infty} (a_{1h\,u} + b_{1h\,u}) V_{eh\,vu} + \sum_{u=1}^{\infty} (a_{1e\,u} - b_{1e\,u}) V_{ee\,vu}$$

$$a_{1h\,u} - b_{1h\,u} = \sum_{v=1}^{\infty} (-a_{2h\,v} + b_{2h\,v}) V_{hh\,uv} + \sum_{v=1}^{\infty} (a_{2e\,v} + b_{2e\,v}) V_{he\,uv}$$

$$a_{1e\,u} + b_{1e\,u} = \sum_{v=1}^{\infty} (-a_{2h\,v} + b_{2h\,v}) V_{eh\,uv} + \sum_{v=1}^{\infty} (a_{2e\,v} + b_{2e\,v}) V_{ee\,uv}$$

These relations can then be expressed in matrix form,

$$[M_1][A] = [M_2][B]$$

where

$$(A) = \begin{pmatrix} [A_{1h}] \\ [A_{1e}] \\ [A_{2h}] \\ [A_{2e}] \end{pmatrix} \quad \text{and} \quad (B) = \begin{pmatrix} [B_{1h}] \\ [B_{1e}] \\ [B_{2h}] \\ [B_{2e}] \end{pmatrix}$$

and $[M_1]$ and $[M_2]$ are given by

$$(M_1) = \begin{pmatrix} [I] & [0] & [V_{hhuv}] & -[V_{heuv}] \\ [0] & [I] & [V_{ehuv}] & -[V_{eeuv}] \\ [V_{hhvu}] & [V_{hevu}] & -[I] & [0] \\ -[V_{ehvu}] & -[V_{eevu}] & [0] & -[I] \end{pmatrix}$$

$$(M_2) = \begin{pmatrix} [I] & [0] & [V_{hhuv}] & [V_{heuv}] \\ [0] & -[I] & [V_{ehuv}] & [V_{eeuv}] \\ -[V_{hhvu}] & [V_{hevu}] & [I] & [0] \\ [V_{ehvu}] & -[V_{eevu}] & [0] & -[I] \end{pmatrix}$$

Then by definition of the scattering parameters, the S matrix of the discontinuity, referred to the characteristic impedances of each waveguide, is given by

$$[S] = [M_2]^{-1}[M_1]$$

However, there is no need to evaluate these matrices in their entirety. Rather, only the submatrices in the two columns corresponding to the fundamental mode are required. These two columns can be found while avoiding the inversion of the entire matrix.

A3.3 APPLICATION TO RECTANGULAR WAVEGUIDES

In the case of rectangular waveguides, the field equations at $z = 0$ are given below.

TE modes:

$$h_{iz\,nm} = C_{ih\,nm} \cos\left(\frac{n\pi}{a_i}x_i\right)\cos\left(\frac{m\pi}{b_i}y_i\right)$$

$$\vec{h}_{ith\,nm} = C_{ih\,nm}\frac{-\gamma_{ih\,nm}}{k^2_{ith\,nm}}\left[-\frac{n\pi}{a_i}\sin\left(\frac{n\pi}{a_i}x_i\right)\cos\left(\frac{m\pi}{b_i}y_i\right)\vec{x} - \frac{m\pi}{b_i}\cos\left(\frac{n\pi}{a_i}x_i\right)\sin\left(\frac{m\pi}{b_i}y_i\right)\vec{y}\right]$$

$$\vec{e}_{ith\,nm} = Z_{ih\,nm}C_{ih\,nm}\frac{-\gamma_{ih\,nm}}{k^2_{ith\,nm}}\left[\frac{m\pi}{b_i}\cos\left(\frac{n\pi}{a_i}x_i\right)\sin\left(\frac{m\pi}{b_i}y_i\right)\vec{x} - \frac{n\pi}{a_i}\sin\left(\frac{n\pi}{a_i}x_i\right)\cos\left(\frac{m\pi}{b_i}y_i\right)\vec{y}\right]$$

with

$$\gamma_{ih\,nm} = \sqrt{k^2_{ith\,nm} - \left(\frac{\omega}{c}\right)^2\varepsilon_{ri}} \qquad k_{ith\,nm} = \sqrt{\left(\frac{n\pi}{a_i}\right)^2 + \left(\frac{m\pi}{b_i}\right)^2}$$

$$Z_{ih\,nm} = \frac{j\omega\mu_0}{\gamma_{ih\,nm}} \qquad C_{ih\,nm} = \frac{2k^2_{ith\,nm}c_{ih\,nm}}{\sqrt{\omega\mu_0|\gamma_{ih\,nm}|}}$$

$$c_{ih\,nm} = \begin{cases} \dfrac{1}{\sqrt{a_i b_i}\;k^2_{ith\,nm}} & \text{if n = 0 or m = 0} \\ \dfrac{\sqrt{2}}{\sqrt{a_i b_i}\;k^2_{ith\,nm}} & \text{otherwise} \end{cases}$$

TM modes:

$$e_{iz\,nm} = C_{ie\,nm} \sin\left(\frac{n\pi}{a_i}x_i\right)\sin\left(\frac{m\pi}{b_i}y_i\right)$$

$$\vec{e}_{ite\,nm} = C_{ie\,nm}\frac{-\gamma_{ie\,nm}}{k^2_{ite\,nm}}\left[\frac{n\pi}{a_i}\cos\left(\frac{n\pi}{a_i}x_i\right)\sin\left(\frac{m\pi}{b_i}y_i\right)\vec{x} + \frac{m\pi}{b_i}\sin\left(\frac{n\pi}{a_i}x_i\right)\cos\left(\frac{m\pi}{b_i}y_i\right)\vec{y}\right]$$

$$\vec{h}_{ite\,nm} = Y_{ie\,nm}C_{ie\,nm}\frac{-\gamma_{ie\,nm}}{k^2_{ite\,nm}}\left[-\frac{m\pi}{b_i}\sin\left(\frac{n\pi}{a_i}x_i\right)\cos\left(\frac{m\pi}{b_i}y_i\right)\vec{x} + \frac{n\pi}{a_i}\cos\left(\frac{n\pi}{a_i}x_i\right)\sin\left(\frac{m\pi}{b_i}y_i\right)\vec{y}\right]$$

with

$$\gamma_{ih\,nm} = \sqrt{k^2_{ite\,nm} - \left(\frac{\omega}{c}\right)^2\varepsilon_{ri}} \qquad k_{ite\,nm} = \sqrt{\left(\frac{n\pi}{a_i}\right)^2 + \left(\frac{m\pi}{b_i}\right)^2}$$

$$Y_{ie\,nm} = \frac{j\omega\varepsilon_0}{\gamma_{ie\,nm}}$$

$$C_{ie\,nm} = \frac{2k_{ite\,nm}^2\, c_{ie\,nm}}{\sqrt{\omega\varepsilon_0|\gamma_{ie\,nm}|}} \qquad c_{ie\,nm} = \frac{\sqrt{2}}{\sqrt{a_i b_i}k_{ite\,nm}^2}$$

For the case described in Figure A3.1, we have

$$x_1 = x - c \qquad x_2 = x \qquad y_1 = x - d \qquad y_2 = y$$

Under these conditions, the coefficients of the matrices are

$$V_{hh\,uv} = 2c_{1h\,u}c_{2h\,v}\frac{\gamma_{2h\,v}}{\gamma_{1h\,u}}\sqrt{\frac{|\gamma_{1h\,u}|}{|\gamma_{2h\,v}|}} \qquad \text{integral 1}$$

$$V_{he\,uv} = -2j\omega\sqrt{\mu_0\varepsilon_0}c_{1h\,u}c_{2e\,v}\frac{1}{\gamma_{1h\,u}}\sqrt{\frac{|\gamma_{1h\,u}|}{|\gamma_{2e\,v}|}} \qquad \text{integral 2}$$

$$V_{eh\,vu} = \frac{-2}{j\omega\sqrt{\mu_0\varepsilon_0}}c_{1e\,u}c_{2h\,v}\ \gamma_{2h\,v}\sqrt{\frac{|\gamma_{1e\,v}|}{|\gamma_{2h\,u}|}} \qquad \text{integral 3}$$

$$V_{ee\,uv} = 2c_{1e\,u}c_{2e\,v}\sqrt{\frac{|\gamma_{1e\,u}|}{|\gamma_{2e\,v}|}} \qquad \text{integral 4}$$

$$V_{hh\,vu} = 2c_{1h\,u}c_{2h\,v}\sqrt{\frac{|\gamma_{2h\,v}|}{|\gamma_{1h\,u}|}} \qquad \text{integral 1}$$

$$V_{he\,vu} = \frac{-2}{j\omega\sqrt{\mu_0\varepsilon_0}}c_{1e\,u}c_{2h\,v}\ \gamma_{1e\,u}\sqrt{\frac{|\gamma_{2h\,v}|}{|\gamma_{1e\,u}|}} \qquad \text{integral 3}$$

$$V_{eh\,vu} = -2j\omega\sqrt{\mu_0\varepsilon_0}c_{1h\,u}c_{2e\,v}\frac{1}{\gamma_{2e\,v}}\sqrt{\frac{|\gamma_{2e\,v}|}{|\gamma_{1h\,u}|}} \qquad \text{integral 2}$$

$$V_{ee\,vu} = 2c_{1e\,u}c_{2e\,v}\frac{\gamma_{1e\,u}}{\gamma_{2e\,v}}\sqrt{\frac{|\gamma_{2e\,v}|}{|\gamma_{1e\,u}|}} \qquad \text{integral 4}$$

where

$$\text{Integral 1} = k_{1x}\,k_{2x}\ f(k_1,k_2) + k_{1y}k_{2y}\ g(k_1,k_2)$$
$$\text{with} \quad k_1 = k_{1th\,u} \text{ and } k_2 = k_{2th\,v}$$

$$\text{Integral 2} = k_{1x}\,k_{2y}\ f(k_1,k_2) - k_{1y}k_{2x}\ g(k_1,k_2)$$
$$\text{with} \quad k_1 = k_{1th\,u} \text{ and } k_2 = k_{2te\,v}$$

$$\text{Integral } 3 = k_{1y}\, k_{2x}\ f(k_1, k_2) - k_{1x} k_{2y}\ g(k_1, k_2)$$
$$\text{with} \quad k_1 = k_{1te\ u} \text{ and } k_2 = k_{2th\ v}$$
$$\text{Integral } 4 = k_{1y}\, k_{2y}\ f(k_1, k_2) + k_{1x} k_{2x}\ g(k_1, k_2)$$
$$\text{with} \quad k_1 = k_{1te\ u} \text{ and } k_2 = k_{2te\ v}$$

and

$$f(k_1, k_2) = \int_c^{a_2 - c} \sin[k_{1x}(x - c)]\ \sin(k_{2x} x)\ dx \int_d^{b_2 - d} \cos[k_{1y}(y - d)]\ \cos(k_{2y} y)\ dy$$
$$g(k_1, k_2) = \int_c^{a_2 - c} \cos[k_{1x}(x - c)]\ \cos(k_{2x} x)\ dx \int_d^{b_2 - d} \sin[k_{1y}(y - d)]\ \sin(k_{2y} y)\ dy$$

where k_{ix} and k_{iy} are related to the wavenumber of a mode for rectangular waveguides. In other words

$$k_{ix} = \frac{n\pi}{a_i} \quad \text{and} \quad k_{iy} = \frac{m\pi}{b_i} \quad \text{and} \quad k_i = \sqrt{k_{ix}^2 + k_{iy}^2}$$

A3.4 APPLICATION TO CIRCULAR WAVEGUIDES

For circular waveguides, the field equations at $z = 0$ are given below.

TE modes:

$$h_{iz\ nm} = C_{ih\ nm}\ J_n(k_{ith\ nm}\ r) \cos n\phi$$
$$\vec{h}_{ith\ nm} = C_{ih\ nm} \frac{-\gamma_{ih\ nm}}{k_{ith\ nm}^2} \Big[k_{ith\ nm}\ J_n'(k_{ith\ nm}\ r) \cos(n\phi) \vec{r} - \frac{n}{r} J_n(k_{ith\ nm}\ r) \sin(n\phi) \vec{\phi} \Big]$$
$$\vec{e}_{ith\ nm} = -Z_{ih\ nm} C_{ih\ nm} \frac{-\gamma_{ih\ nm}}{k_{ith\ nm}^2} \Big[\frac{n}{r} J_n(k_{ith\ nm}\ r) \sin(n\phi) \vec{r} + k_{ith\ nm}\ J_n'(k_{ith\ nm}\ r) \cos(n\phi) \vec{\phi} \Big]$$

with

$$\gamma_{ih\ nm} = \sqrt{k_{ith\ nm}^2 - \left(\frac{\omega}{c}\right)^2 \varepsilon_{ri}} \qquad k_{ith\ nm} = \frac{x_{nm}'}{a_i} \qquad Z_{ih\ nm} = \frac{j\omega\mu_0}{\gamma_{ih\ nm}}$$

$$C_{ih\,nm} = \frac{2k_{ith\,nm}^2\, c_{ih\,nm}}{\sqrt{\omega\mu_0|\gamma_{ih\,nm}|}}$$

$$c_{ih\,nm} = \begin{cases} \dfrac{1/\sqrt{2}}{\sqrt{\pi(a_i^2 k_{ith\,nm}^2 - n^2) J_n^2(a_i k_{ith\,nm})}} & \text{if } n = 0 \\ \dfrac{1}{\sqrt{\pi(a_i^2 k_{ith\,nm}^2 - n^2) J_n^2(a_i k_{ith\,nm})}} & \text{otherwise} \end{cases}$$

TM modes:

$$\begin{aligned} e_{iz\,nm} &= C_{ie\,nm}\ J_n(k_{ite\,nm}\ r)\sin n\phi \\ \vec{e}_{ite\,nm} &= C_{ie\,nm}\frac{-\gamma_{ie\,nm}}{k_{ite\,nm}^2}\Big[k_{ite\,nm}\ J_n'(k_{ite\,nm} r)\ \sin(n\phi)\vec{r} \\ &\quad + \frac{n}{r} J_n(k_{ite\,nm} r)\ \cos(n\phi)\vec{\phi}\Big] \\ \vec{h}_{ite\,nm} &= Y_{ie\,nm} C_{ie\,nm}\frac{-\gamma_{ie\,nm}}{k_{ite\,nm}^2}\Big[-\frac{n}{r} J_n(k_{ite\,nm} r)\ \cos(n\phi)\vec{r} \\ &\quad + k_{ite\,nm}\ J_n'(k_{ite\,nm} r)\ \sin(n\phi)\vec{\phi}\Big] \end{aligned}$$

with

$$\gamma_{ie\,nm} = \sqrt{k_{ite\,nm}^2 - \left(\frac{\omega}{c}\right)^2 \varepsilon_{ri}} \qquad k_{ite\,nm} = \frac{x_{nm}}{a_i} \qquad Y_{ie\,nm} = \frac{j\omega\varepsilon_0}{\gamma_{ie\,nm}}$$

$$C_{ie\,nm} = \frac{2k_{ite\,nm}^2\, c_{ie\,nm}}{\sqrt{\omega\varepsilon_0|\gamma_{ie\,nm}|}} \qquad c_{ie\,nm} = \frac{1}{\sqrt{\pi(a_i^2 k_{ite\,nm}^2 - n^2) J_n'^2(a_i k_{ite\,nm})}}$$

Under these conditions, the coefficients of the matrices are

$$V_{hh\,uv} = 2c_{1h\,u} c_{2h\,v}\frac{\gamma_{2h\,v}}{\gamma_{1h\,u}}\sqrt{\frac{|\gamma_{1h\,u}|}{|\gamma_{2h\,v}|}} \qquad \text{integral 1}$$

$$V_{he\,uv} = -2j\omega\sqrt{\mu_0\varepsilon_0}\, c_{1h\,u} c_{2e\,v}\frac{1}{\gamma_{1h\,u}}\sqrt{\frac{|\gamma_{1h\,u}|}{|\gamma_{2e\,v}|}} \qquad \text{integral 2}$$

$$V_{eh\,uv} = \frac{-2}{j\omega\sqrt{\mu_0\varepsilon_0}} c_{1e\,u} c_{2h\,v}\ \gamma_{2h\,v}\sqrt{\frac{|\gamma_{1e\,u}|}{|\gamma_{2h\,v}|}} \qquad \text{integral 3}$$

$$V_{ee\,uv} = 2c_{1e\,u} c_{2e\,v}\sqrt{\frac{|\gamma_{1e\,u}|}{|\gamma_{2e\,v}|}} \qquad \text{integral 4}$$

$$V_{hh\ vu} = 2c_{1h\ u}c_{2h\ v}\sqrt{\frac{|\gamma_{2h\ v}|}{|\gamma_{1h\ u}|}} \qquad \text{integral 1}$$

$$V_{he\ vu} = \frac{-2}{j\omega\sqrt{\mu_0\varepsilon_0}}c_{1e\ u}c_{2h\ v}\ \gamma_{1e\ u}\sqrt{\frac{|\gamma_{2h\ v}|}{|\gamma_{1e\ u}|}} \qquad \text{integral 3}$$

$$V_{eh\ vu} = -2j\omega\sqrt{\mu_0\varepsilon_0}c_{1h\ u}\ c_{2e\ v}\frac{1}{\gamma_{2e\ v}}\sqrt{\frac{|\gamma_{2e\ v}|}{|\gamma_{1h\ u}|}} \qquad \text{integral 2}$$

$$V_{ee\ vu} = 2c_{1e\ u}c_{2e\ v}\frac{\gamma_{1e\ u}}{\gamma_{2e\ v}}\sqrt{\frac{|\gamma_{2e\ v}|}{|\gamma_{1e\ u}|}} \qquad \text{integral 4}$$

where

$$\text{integral 1} = f(k_1, k_2) \quad \text{with} \quad k_1 = k_{1th\ u} \text{ and } k_2 = k_{2th\ v}$$

$$\text{integral 2} = g(k_1, k_2) \quad \text{with} \quad k_1 = k_{1th\ u} \text{ and } k_2 = k_{2te\ v}$$

$$\text{integral 3} = g(k_1, k_2) \quad \text{with} \quad k_1 = k_{1te\ u} \text{ and } k_2 = k_{2th\ v}$$

$$\text{integral 4} = f(k_1, k_2) \quad \text{with} \quad k_1 = k_{1te\ u} \text{ and } k_2 = k_{2te\ v}$$

and

$$f(k_1, k_2) = \pi\int_0^a r\ k_1k_2J_n'(k_1r)J_n'(k_2r) + r\left(\frac{n}{r}\right)^2 J_n(k_1r)J_n(k_2r)\ dr$$

$$f(k_1, k_2) = \begin{cases} \pi k_1k_2\left[\frac{r}{k_1^2-k_2^2}\{k_1J_2(k_1r)J_1(k_2r) - k_2J_1(k_1r)J_2(k_2r)\}\right]_0^a \\ \qquad n = 0 \\ \pi k_1k_2\left[\frac{r}{k_1^2-k_2^2}\{k_1J_n(k_1r)J_{n-1}(k_2r) - k_2J_{n-1}(k_1r)J_n(k_2r)\}\right]_0^a \\ \qquad -n[J_n(k_1r)J_n(k_2r)]_0^a\ n>0 \end{cases}$$

$$g(k_1, k_2) = \pi\int_0^a n\ k_1J_n'(k_1r)J_n(k_2r) + nk_2J_n(k_1r)J_n'(k_2r)\ dr$$

$$g(k_1, k_2) = \pi n[J_n(k_1r)J_n(k_2r)]_0^a$$

Note that when the u and v modes are such that the n values are different ($n_1 \neq n_2$), functions f and g are both zero. This explains the single n parameter used in the expressions.

APPENDIX 4

TRISECTIONS WITH UNITY INVERTERS ON THE INSIDE OR ON THE OUTSIDE

The trisection (L, jX, K) has the same response as the trisection(C, jB, J) Figure (A4.1). The responses of the two trisections (with unity inverters inside and outside the circuit) are, respectively,

$$S_{21}(s) = \frac{1 - jK(sL + jX)}{(1 + jK)\{(1 - jK)[s(L/2) + j(X/2)] + 1\}}$$

$$S_{21}(s) = \frac{1 + jJ(sC + jB)}{(1 - jJ)\{(1 + jJ)[s(C/2) + j(B/2)] + 1\}}$$

The two responses are the same if

$$K = -J, \qquad L = C, \qquad X = B$$

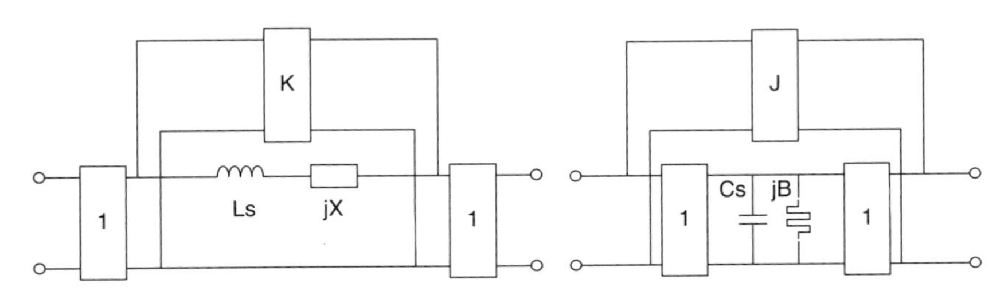

Figure A4.1

Advanced Design Techniques and Realizations of Microwave and RF Filters,
By Pierre Jarry and Jacques Beneat

Consequently, K or J represents the coupling between the two frequencies ω_{10p} and ω_{m0q}.

REFERENCES

Jarry, P., M. Guglielmi, E. Kerhervé, J. M. Pham, O. Roquebrun, and D. Schmitt, Synthesis of dual-mode in-line microwave rectangular filters with higher modes, *Int. J. RF Microwave Comput. Aided Eng.*, pp. 241–248, Mar. 2005.

APPENDIX 5

REFERENCE FIELDS AND SCATTERING MATRICES FOR MULTIMODAL RECTANGULAR WAVEGUIDE FILTERS

A5.1 REFERENCE ELECTRIC AND MAGNETIC FIELDS

The electric and magnetic described later in this section are based on a uniform section of rectangular waveguide, shown in Figure A5.1. The TE modes are given by

$$E_z = 0 \qquad H_z = K \cos \frac{m\pi x}{l_X} \cos \frac{n\pi y}{l_Y} e^{-\gamma z}$$

$$E_x = K \frac{j\omega\mu}{k_c^2} \frac{n\pi}{l_Y} \cos \frac{m\pi x}{l_X} \sin \frac{n\pi y}{l_Y} e^{-\gamma z}$$

$$E_y = -K \frac{j\omega\mu}{k_c^2} \frac{m\pi}{l_X} \sin \frac{m\pi x}{l_X} \cos \frac{n\pi y}{l_Y} e^{-\gamma z}$$

$$H_x = K \frac{\gamma}{k_c^2} \frac{m\pi}{l_X} \sin \frac{m\pi x}{l_X} \cos \frac{n\pi y}{l_Y} e^{-\gamma z}$$

$$H_y = K \frac{\gamma}{k_c^2} \frac{n\pi}{l_Y} \cos \frac{m\pi x}{l_X} \sin \frac{n\pi y}{l_Y} e^{-\gamma z}$$

where $k_c^2 = (m\pi/l_X)^2 + (n\pi/l_Y)^2$ and $\gamma^2 = k_c^2 - \mu\varepsilon\omega^2$. The factor K is often taken to normalize the power to 1. This has the effect of making the fields dependent on frequency and makes the equations more difficult, due to the introduction of square roots. To simply the expressions, we decide to take

Advanced Design Techniques and Realizations of Microwave and RF Filters,
By Pierre Jarry and Jacques Beneat

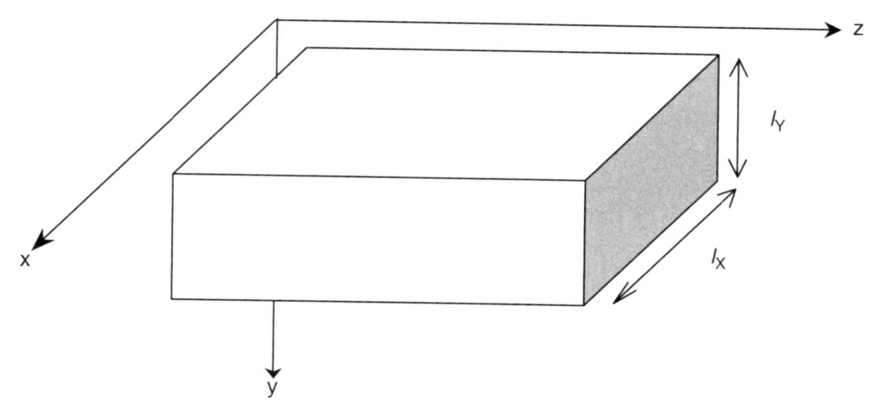

Figure A5.1 Description of a uniform section of rectangular waveguide.

$$K = \frac{k_c^2}{j\omega\mu}$$

In this case, the reference TE fields used in this appendix will be taken as

$$E_{x\ \text{reference1}} = \frac{n\pi}{l_Y} \cos \frac{m\pi x}{l_X} \sin \frac{n\pi y}{l_Y}$$

$$E_{y\ \text{reference1}} = -\frac{m\pi}{l_X} \sin \frac{m\pi x}{l_X} \cos \frac{n\pi y}{l_Y}$$

$$H_{\text{reference1}} = Y_{\text{TE}}\, \vec{z} \wedge E_{\text{reference1}} \qquad \text{with} \quad Y_{\text{TE}} = \frac{\gamma}{j\mu\omega}$$

$$\gamma^2 = k_c^2 - \mu\varepsilon\omega^2 \qquad k_c^2 = \left(\frac{m\pi}{l_X}\right)^2 + \left(\frac{n\pi}{l_Y}\right)^2$$

$$P_{\text{referenceTE}} = \frac{j l_X l_Y k_c^2 \gamma^*}{8\ \mu\omega}(1 + \delta_{0nm})$$

The TM modes are given

$$E_z = K \sin \frac{m\pi x}{l_X} \sin \frac{n\pi y}{l_Y}\, e^{-\gamma z} \qquad H_z = 0$$

$$E_x = -K \frac{\gamma}{k_c^2} \frac{m\pi}{l_X} \cos \frac{m\pi x}{l_X} \sin \frac{n\pi y}{l_Y}\, e^{-\gamma z}$$

$$E_y = -K \frac{\gamma}{k_c^2} \frac{n\pi}{l_Y} \sin \frac{m\pi x}{l_X} \cos \frac{n\pi y}{l_Y} e^{-\gamma z}$$

$$H_x = K \frac{j\omega\varepsilon}{k_c^2} \frac{n\pi}{l_Y} \sin \frac{m\pi x}{l_X} \cos \frac{n\pi y}{l_Y} e^{-\gamma z}$$

$$H_y = -K \frac{j\omega\varepsilon}{k_c^2} \frac{m\pi}{l_X} \cos \frac{m\pi x}{l_X} \sin \frac{n\pi y}{l_Y} e^{-\gamma z}$$

where $k_c^2 = \left(\frac{m\pi}{l_X}\right)^2 + \left(\frac{n\pi}{l_Y}\right)^2$ and $\gamma^2 = k_c^2 - \mu\varepsilon\omega^2$. We decide to take the parameter K such that

$$K = \frac{k_c^2}{\gamma}$$

In this case, the reference TM fields used in this appendix will be taken as

$$E_{x\ \text{reference1}} = -\frac{m\pi}{l_X} \cos \frac{m\pi x}{l_X} \sin \frac{n\pi y}{l_Y}$$

$$E_{y\ \text{reference1}} = -\frac{n\pi}{l_Y} \sin \frac{m\pi x}{l_X} \cos \frac{n\pi y}{l_Y}$$

$$H_{\text{reference1}} = Y_{\text{TM}}\, \vec{z} \wedge E_{\text{reference1}} \qquad \text{with} \quad Y_{\text{TM}} = \frac{j\varepsilon\omega}{\gamma}$$

$$\gamma^2 = k_c^2 - \mu\varepsilon\omega^2 \qquad k_c^2 = \left(\frac{m\pi}{l_X}\right)^2 + \left(\frac{n\pi}{l_Y}\right)^2$$

$$P_{\text{referenceTM}} = \frac{\varepsilon\omega l_X l_Y k_c^2}{8j\gamma^*}$$

A5.2 SCATTERING MATRIX OF MULTIPLE WAVEGUIDE SECTIONS

The scattering matrix of multiple waveguide sections is illustrated in Figure A5.2 for the case of three waveguide sections. This will be referred to as a *trisection*. The scattering matrix of the trisection is defined as

$$\begin{vmatrix} B_1^1 \\ B_1^2 \\ B_1^3 \\ B_2^1 \\ B_2^2 \\ B_2^3 \end{vmatrix} = S_L \cdot \begin{vmatrix} A_1^1 \\ A_1^2 \\ A_1^3 \\ A_2^1 \\ A_2^2 \\ A_2^3 \end{vmatrix}$$

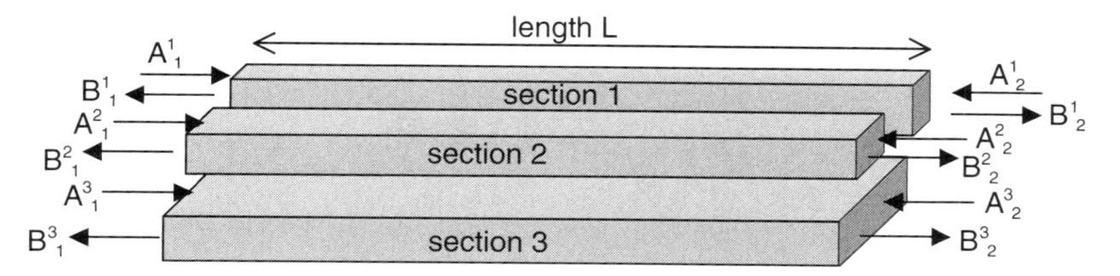

Figure A5.2 Representation of a trisection.

In the trisection it is assumed that there are no interactions between the three sections of waveguides. In addition, inside a section of waveguide, the various modes do not interact between themselves since they are orthogonal.

The incident field on the left of the trisection is given by

$$E_{\text{incident1}} = \sum_{s=1}^{3} \sum_{\text{modes } m} a_1^{s,m} E_{\text{reference1}}^{s,m}$$

The incident field on the right of the trisection is given by

$$E_{\text{incident2}} = \sum_{s=1}^{3} \sum_{\text{modes } m} a_2{}^{s,m} E_{\text{reference2}}^{s,m}$$

where s represents the section and m the mode. Also, the reference fields $E_{\text{reference1}}^{s,m}$ and $E_{\text{reference2}}^{s,m}$ are null outside their waveguide section.

The reflected fields are given by

$$E_{\text{reflected1}} = \sum_{s=1}^{3} \sum_{\text{modes } m} a_2{}^{s,m} E_{\text{reference2}}^{s,m} e^{-\gamma^{s,m} L}$$

$$E_{\text{reflected2}} = \sum_{s=1}^{3} \sum_{\text{modes } m} a_1{}^{s,m} E_{\text{reference1}}^{s,m} e^{-\gamma^{s,m} L}$$

where $e^{-\gamma^{s,m} L}$ corresponds to the field dependence in the direction z. Since in our convention, the reference electric fields do not change sign when the wave changes its direction of travel, we have

$$E_{\text{reflected1}} = \sum_{s=1}^{3} \sum_{\text{modes } m} b_1{}^{s,m} E_{\text{reference1}}^{s,m}$$

$$E_{\text{reflected2}} = \sum_{s=1}^{3} \sum_{\text{modes } m} b_2{}^{s,m} E_{\text{reference2}}^{s,m}$$

Since the waveguide section is assumed homogeneous, we have

$$E^{s,m}_{\text{reference1}} = E^{s,m}_{\text{reference2}}$$

By matching the two expressions for the reflected fields, we have

$$b^{s,m}_1 = e^{-\gamma^{s,m}L} a^{s,m}_2$$
$$b^{s,m}_2 = e^{-\gamma^{s,m}L} a^{s,m}_1$$

and the scattering matrix of the trisection is given by

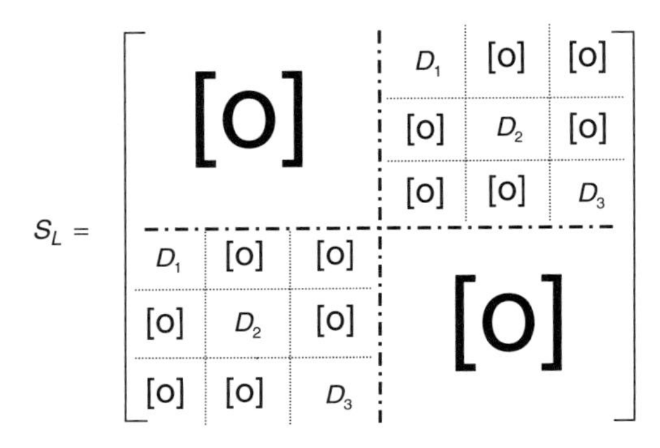

where the matrices D_s, $s = 1, 2, 3$, are diagonal matrices such that their mth row and mth column correspond to the term $e^{-\gamma^{s,m}L}$, where $\gamma^{s,m}$ is the propagation constant of the mth mode in the section s.

A5.3 SCATTERING MATRIX FOR DISCONTINUITY IN DIELECTRIC PERMITTIVITY

The scattering matrix for a discontinuity in dielectric permittivity is illustrated in Figure A5.3 for three waveguide sections. In the case of a discontinuity, the scattering matrix is defined at the junction. In Figure A5.3, the junction is exaggerated since its length should be considered to be zero.

As in the case of the trisection, the incident fields are given by

$$E_{\text{incident1}} = \sum_{s=1}^{3} \sum_{\text{modes } m} a^{s,m}_1 E^{s,m}_{\text{reference1}}$$

$$E_{\text{incident2}} = \sum_{s=1}^{3} \sum_{\text{modes } m} a^{s,m}_2 E^{s,m}_{\text{reference2}}$$

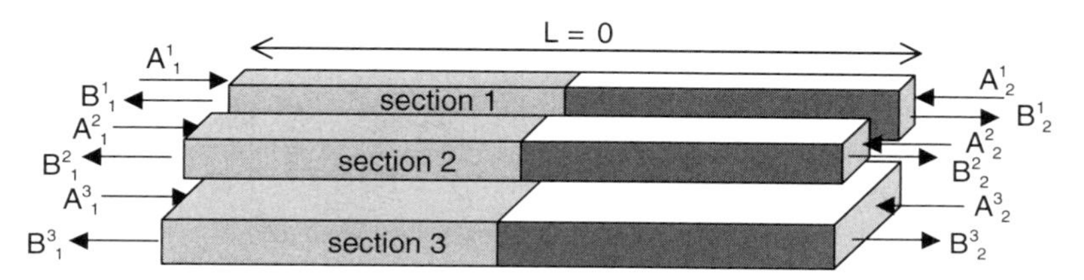

Figure A5.3 Dielectric discontinuity in a trisection.

At the junction, the electric and magnetic tangential fields from both sides of the junction must be equal. In our convention, the reference electric fields do not change sign, whereas the reference magnetic fields do change sign. In summary,

$$\sum_{s=1}^{3} \sum_{\text{modes } m} (a_1^{s,m} + b_1^{s,m}) E_{\text{reference1}}^{s,m} = \sum_{s=1}^{3} \sum_{\text{modes } m} (a_2^{s,m} + b_2^{s,m}) E_{\text{reference2}}^{s,m}$$

$$\sum_{s=1}^{3} \sum_{\text{modes } m} (a_1^{s,m} - b_1^{s,m}) H_{\text{reference1}}^{s,m} = \sum_{s=1}^{3} \sum_{\text{modes } m} (a_2^{s,m} - b_2^{s,m}) H_{\text{reference2}}^{s,m}$$

Since the dimensions of the waveguide sections are the same on both sides of the junction, we have

$$\begin{aligned} E_{\text{reference1}}^{s,m} &= E_{\text{reference2}}^{s,m} \\ Y_2^{s,m} H_{\text{reference1}}^{s,m} &= -Y_1^{s,m} H_{\text{reference2}}^{s,m} \end{aligned}$$

where following the TE or TM nature of the mode,

$$Y_i^{\text{TE},s,m} = \frac{\gamma_i^{s,m}}{j\mu\omega} \qquad Y_i^{\text{TM},s,m} = \frac{j\varepsilon_i^s \omega}{\gamma_i^{s,m}}$$

Using these relations for the reference fields in the field continuity equations, we have

$$\begin{aligned} a_1^{s,m} + b_1^{s,m} &= a_2^{s,m} + b_2^{s,m} \\ Y_1^{s,m} (a_1{}^{s,m} - b_1^{s,m}) &= -Y_2^{s,m} (a_2{}^{s,m} - b_2^{s,m}) \end{aligned}$$

and the scattering matrix of a discontinuity in dielectric permittivity in the trisection is given by

$$S_{\Delta\varepsilon} = \left[\begin{array}{ccc|ccc} R_1 & [o] & [o] & A_1 & [o] & [o] \\ [o] & R_2 & [o] & [o] & A_2 & [o] \\ [o] & [o] & R_3 & [o] & [o] & A_3 \\ \hline B_1 & [o] & [o] & -R_1 & [o] & [o] \\ [o] & B_2 & [o] & [o] & -R_2 & [o] \\ [o] & [o] & B_3 & [o] & [o] & -R_3 \end{array}\right]$$

where the matrices R_s, A_s and B_s, $s = 1, 2, 3$, are diagonal matrices such that their mth row and mth column correspond respectively to the terms

$$\frac{Y_1^{s,m} - Y_2^{s,m}}{Y_1^{s,m} + Y_2^{s,m}} \qquad \frac{2Y_2^{s,m}}{Y_1^{s,m} + Y_2^{s,m}} \qquad \frac{2Y_1^{s,m}}{Y_1^{s,m} + Y_2^{s,m}}$$

A5.4 SCATTERING MATRIX FOR DISCONTINUITY IN WAVEGUIDE DIMENSIONS

The scattering matrix for a discontinuity waveguide dimensions is illustrated in Figure A5.4 for the case of three waveguide sections connected to a larger waveguide. In the case of a discontinuity, the scattering matrix is defined at the junction. In Figure A5.4 the junction is exaggerated, since its length should be considered as to be zero.

The incident field on the left (the large waveguide) is defined as

$$E_{\text{incident1}} = \sum_{m=1}^{N_1} a_1^m E_{\text{reference1}}^m$$

where N_1 is the number of modes considered in the large waveguide (medium 1). The incident field on the right (medium 2, a combination of the three waveguide sections of the trisection) is given by

$$E_{\text{incident2}} = \sum_{s=1}^{N_s} \sum_{n_2=1}^{N_2} a_2^{s,n_2} E_{\text{reference2}}^{s,n_2}$$

where N_s is the number of sections (e.g., $N_s = 3$ for a trisection) and n_2 and N_2 are the index and total number of modes considered in the waveguide sections.

We call S_1 the transversal surface in the large waveguide and S_2 the union of all the transversal surfaces of the waveguide sections S_2^s. The continuity of the electric tangential fields at the surface of separation gives the equation

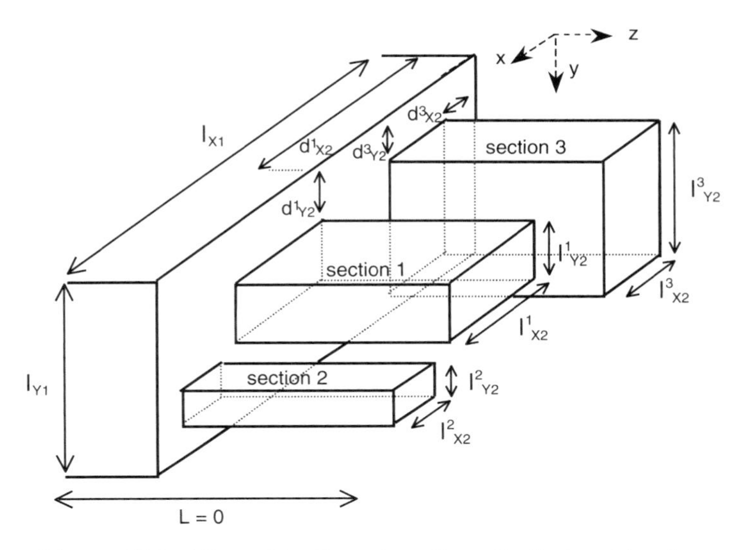

Figure A5.4 Waveguide dimensions discontinuity in a trisection.

$$\sum_{m=1}^{N_1} (a_1^m + b_1^m) E_{\text{reference1}}^m = \sum_{s=1}^{N_s} \sum_{n_2}^{N_2} (a_2^{s,n_2} + b_2^{s,n_2}) E_{\text{reference2}}^{s,n_2} = E_{\text{tangential}}$$

This equation is true over the entire surface S_1 since the reference fields of medium 2 are assumed to be null outside their respective sections. Since the modes are orthogonal in a rectangular waveguide, and hence the reference fields, we have

$$b_1^m = \frac{\iint_{S_2} E_{\text{tangential}} E_{\text{reference1}}^m \, ds}{\iint_{S_1} E_{\text{reference1}}^m E_{\text{reference1}}^m \, ds} - a_1^m \tag{A5.1}$$

$$b_2^{s,n_2} = \frac{\iint_{S_2} E_{\text{tangential}} E_{\text{reference2}}^{s,n_2} \, ds}{\iint_{S_2} E_{\text{reference2}}^{s,n_2} E_{\text{reference2}}^{s,n_2} \, ds} - a_2^{s,n_2} \tag{A5.2}$$

The numerator integral for b_1^m is limited to S_2 since the electric tangential field is null everywhere else on the surface S_1, due to the perfect conducting assumptions of the structure.

To simplify the equations, we define

$$U_1^m = \iint_{S_1} E_{\text{reference1}}^m E_{\text{reference1}}^m \, ds$$

$$U_2^{s,n_2} = \iint_{S_2} E_{\text{reference2}}^{s,n_2} E_{\text{reference2}}^{s,n_2} \, ds$$

The continuity of the magnetic tangential fields at the surface of separation gives the equation

$$\sum_{m=1}^{N_1} (a_1^m - b_1^m) H_{\text{reference1}}^m = \sum_{s=1}^{N_s} \sum_{n_2}^{N_2} (a_2^{s,n_2} - b_2^{s,n_2}) H_{\text{reference2}}^{s,n_2}$$

where following our convention the reference electric fields do not change sign, whereas the reference magnetic fields do change sign when changing the direction of propagation.

Furthermore, we have

$$\begin{aligned} H_{\text{reference1}}^{n_1} &= Y_1^{n_1} \overrightarrow{z} \wedge E_{\text{reference1}}^{n_1} \\ H_{\text{reference2}}^{s,n_2} &= Y_2^{s,n_2} (\overrightarrow{-z}) \wedge E_{\text{reference2}}^{s,n_2} \end{aligned}$$

Therefore, the continuity of the magnetic tangential fields at the junction can be expressed as

$$\sum_{n_1=1}^{N_1} (a_1^{n_1} - b_1^{n_1}) Y_1^{n_1} E_{\text{reference1}}^{n_1} = -\sum_{s=1}^{N_s} \sum_{n_2}^{N_2} (a_2^{s,n_2} - b_2^{s,n_2}) Y_2^{s,n_2} E_{\text{reference2}}^{s,n_2}$$

Replacing $b_1^{n_1}$ and b_2^{s,n_2} by their expressions in equations (A5.1) and (A5.2), we have

$$\begin{aligned} &\sum_{n_1=1}^{N_1} \frac{Y_1^{n_1}}{U_1^{n_1}} \left(\iint_{S_2} E_{\text{tangential}} E_{\text{reference1}}^{n_1} \, ds \right) E_{\text{reference1}}^{n_1} \\ &\quad + \sum_{s=1}^{N_s} \sum_{n_2}^{N_2} \frac{Y_2^{s,n_2}}{U_2^{s,n_2}} \left(\iint_{S_2} E_{\text{tangential}} E_{\text{reference2}}^{s,n_2} \, ds \right) E_{\text{reference2}}^{s,n_2} \\ &\quad = \sum_{n_1=1}^{N_1} 2a_1^{n_1} Y_1^{n_1} E_{\text{reference1}}^{n_1} + \sum_{s=1}^{N_s} \sum_{n_2}^{N_2} 2a_2^{s,n_2} Y_2^{s,n_2} E_{\text{reference2}}^{s,n_2} \end{aligned} \tag{A5.3}$$

Since the incident fields are introduced arbitrarily from $a_1^{n_1}$ and a_2^{s,n_2}, it is convenient to define the electric tangential fields in the form

$$E_{\text{tangential}} = \sum_{t_1=1}^{N_1'} 2a_1^{t_1} Y_1^{t_1} f_1^{t_1} + \sum_{\text{guide}=1}^{N_s} \sum_{t_2=1}^{N_2'} 2a_2^{\text{guide},t_2} Y_2^{\text{guide},t_2} f_2^{\text{guide},t_2}$$

where $f_1^{t_1}$ and f_2^{guide,t_2} are vector functions. These functions are assumed to be null outside the surface S_2. One should note that $N_1' < N_1$ and $N_2' < N_2$. The modes from N_1' to N_1 and from N_2' to N_2 are created by the discontinuity and are not incident. Using the expression of the tangential field above with equation (A5.3) gives

$$E_{\text{reference1}}^{t_1} = \sum_{n_1=1}^{N_1} \frac{Y_1^{n_1}}{U_1^{n_1}} \left(\iint_{S_2} f_1^{t_1} E_{\text{reference1}}^{n_1} \, ds \right) E_{\text{reference1}}^{n_1} + \sum_{s=1}^{N_s} \sum_{n_2}^{N_2} \frac{Y_2^{s,n_2}}{U_2^{s,n_2}} \left(\iint_{S_2} f_1^{t_1} E_{\text{reference2}}^{s,n_2} \, ds \right) E_{\text{reference2}}^{s,n_2}$$

$$E_{\text{reference2}}^{\text{guide},t_2} = \sum_{n_1=1}^{N_1} \frac{Y_1^{n_1}}{U_1^{n_1}} \left(\iint_{S_2} f_2^{\text{guide},t_2} E_{\text{reference1}}^{n_1} \, ds \right) E_{\text{reference1}}^{n_1} + \sum_{s=1}^{N_s} \sum_{n_2}^{N_2} \frac{Y_2^{s,n_2}}{U_2^{s,n_2}} \left(\iint_{S_2} f_2^{\text{guide},t_2} E_{\text{reference2}}^{s,n_2} \, ds \right) E_{\text{reference2}}^{s,n_2}$$

These relations are true for $t_1 = 1, 2, \ldots, N_1'$, $t_2 = 1, 2, \ldots, N_2'$ and guide $= 1, 2, \ldots, N_s$.

At this stage, we make use of the moment method [A5.1,A5.2]. The relations in Equation (A5.4) are integral equations depending on the vector functions $f_1^{t_1}$ and f_2^{guide,t_2}. To simplify the equations, the reference electric fields of medium 2 (waveguide sections) are chosen as base functions. In addition, we follow the conditions of the Galerkin method [A5.2]. In this case the test functions are also taken as the reference electric fields of medium 2.

We use N_2' base and test functions for small cross sections. We are looking for functions $f_1^{t_1}$ and f_2^{guide,t_2} under the form

$$f_1^{t_1} = \sum_{\text{part}=1}^{N_s} \sum_{n_3=1}^{N_2'} \alpha_{\text{part},n_3}^{t_1} E_{\text{reference2}}^{\text{part},n_3}$$

$$f_2^{\text{guide},t_2} = \sum_{\text{part}=1}^{N_s} \sum_{n_3=1}^{N_2'} \beta_{\text{part},n_3}^{\text{guide},t_2} E_{\text{reference2}}^{\text{part},n_3}$$

Using successively the test functions $E_{\text{reference2}}^{\text{piece},n_4}$ for $n_4 = 1, 2, \ldots, N_2'$ and piece $= 1, 2, \ldots, N_s$ in Equation (A5.4), we get

$$\iint_{S_2^{\text{piece}}} E_{\text{reference1}}^{t_1} E_{\text{reference2}}^{\text{piece},n_4} \, ds$$

$$
\begin{aligned}
&= \sum_{\text{part}=1}^{N_s} \sum_{n_3=1}^{N_2'} \alpha_{\text{part},n_3}^{t_1} \left[\sum_{n_1=1}^{N_1} \frac{Y_1^{n_1}}{U_1^{n_1}} \left(\iint_{S_2} E_{\text{reference2}}^{\text{part},n_3} \; E_{\text{reference1}}^{n_1} \, ds \right) \right. \\
&\quad \times \left(\iint_{S_2} E_{\text{reference1}}^{n_1} \; E_{\text{reference2}}^{\text{piece},n_4} \, ds \right) + \sum_{s=1}^{N_s} \sum_{n_2}^{N_2} \frac{Y_2^{s,n_2}}{U_2^{s,n_2}} \left(\iint_{S_2^{\text{part}} \cap S_2^s} E_{\text{reference2}}^{\text{part},n_3} \; E_{\text{reference2}}^{s,n_2} \, ds \right) \\
&\quad \left. \times \left(\iint_{S_2^{\text{piece}} \cap S_2^s} E_{\text{reference2}}^{s,n_2} \; E_{\text{reference2}}^{\text{piece},n_4} \, ds \right) \right]
\end{aligned}
$$

$$
\begin{aligned}
&\iint_{S_2^{\text{guide}} \cap S_2^{\text{piece}}} E_{\text{reference2}}^{\text{guide},t_2} E_{\text{reference2}}^{\text{piece},n_4} \, ds \\
&= \sum_{\text{part}=1}^{N_s} \sum_{n_3=1}^{N_2'} \beta_{\text{part},n_3}^{\text{guide},t_2} \left[\sum_{n_1=1}^{N_1} \frac{Y_1^{n_1}}{U_1^{n_1}} \left(\iint_{S_2^{\text{part}}} E_{\text{reference2}}^{\text{part},n_3} \; E_{\text{reference1}}^{n_1} \, ds \right) \right. \\
&\quad \times \left(\iint_{S_2^{\text{piece}}} E_{\text{reference1}}^{n_1} \; E_{\text{reference2}}^{\text{piece},n_4} \, ds \right) + \sum_{s=1}^{N_s} \sum_{n_2}^{N_2} \frac{Y_2^{s,n_2}}{U_2^{s,n_2}} \left(\iint_{S_2^{\text{part}} \cap S_2^s} E_{\text{reference2}}^{\text{part},n_3} \; E_{\text{reference2}}^{s,n_2} \, ds \right) \\
&\quad \left. \times \left(\iint_{S_2^{\text{piece}} \cap S_2^s} E_{\text{reference2}}^{s,n_2} \; E_{\text{reference2}}^{\text{piece},n_4} \, ds \right) \right]
\end{aligned}
$$

Due to the orthogonal nature of the modes in medium 2, these relations reduce to

$$
\begin{aligned}
&\iint_{S_2^{\text{piece}}} E_{\text{reference1}}^{t_1} E_{\text{reference2}}^{\text{piece},n_4} \, ds \\
&\quad = \sum_{\text{part}=1}^{N_s} \sum_{n_3=1}^{N_2'} \alpha_{\text{part},n_3}^{t_1} \left[\sum_{n_1=1}^{N_1} \frac{Y_1^{n_1}}{U_1^{n_1}} \left(\iint_{S_2} E_{\text{reference2}}^{\text{part},n_3} \; E_{\text{reference1}}^{n_1} \, ds \right) \right. \\
&\qquad \left. \times \left(\iint_{S_2} E_{\text{reference1}}^{n_1} \; E_{\text{reference2}}^{\text{piece},n_4} \, ds \right) + \delta_{n_3,n_4} \delta_{\text{part},piece} Y_2^{\text{part},n_3} U_2^{\text{part},n_3} \right]
\end{aligned}
$$

$$\iint_{S_2^{\text{guide}} \cap S_2^{\text{piece}}} E_{\text{reference2}}^{\text{guide},t_2} E_{\text{reference2}}^{\text{piece},n_4}\, ds$$

$$= \sum_{\text{part}=1}^{N_s} \sum_{n_3=1}^{N_2'} \beta_{\text{part},n_3}^{\text{guide},t_2} \left[\sum_{n_1=1}^{N_1} \frac{Y_1^{n_1}}{U_1^{n_1}} \left(\iint_{S_2^{\text{part}}} E_{\text{reference2}}^{\text{part},n_3}\, E_{\text{reference1}}^{n_1}\, ds \right) \right.$$

$$\left. \times \left(\iint_{S_2^{\text{piece}}} E_{\text{reference1}}^{n_1}\, E_{\text{reference2}}^{\text{piece},n_4}\, ds \right) + \delta_{n_3,n_4} \delta_{\text{part},\,piece} Y_2^{\text{part},n_3} U_2^{\text{part},n_3} \right]$$

$$= \delta_{t_2,n_4} \delta_{\text{guide},\,piece} U_2^{\text{piece},n_4}$$

for $t_1 = 1, 2, \ldots, N_1'$, $t_2 = 1, 2, \ldots, N_2'$, guide $= 1, 2, \ldots, N_s$, piece $= 1, 2, \ldots, N_s$, and $n_4 = 1, 2, \ldots, N_2'$. In this equation, $\delta_{i,j} = 1$ for $i = j$ and 0 otherwise. Solving these equations provides $\alpha_{\text{part},n_3}^{t_1}$ for $t_1 = 1, 2, \ldots, N_1'$, $n_3 = 1, 2, \ldots, N_2'$, part $= 1, 2, \ldots, N_s$ and $\beta_{\text{part},n_3}^{\text{guide},t_2}$ for $t_2 = 1, 2, \ldots, N_2'$, $n_3 = 1, 2, \ldots, N_2'$, part $= 1, 2, \ldots, N_s$. The tangential field can then be expressed as

$$E_{\text{tangential}} = \sum_{t_1=1}^{N_1'} 2a_1^{t_1} Y_1^{t_1} \left(\sum_{\text{part}=1}^{N_s} \sum_{n_3=1}^{N_2'} \alpha_{\text{part},n_3}^{t_1} E_{\text{reference2}}^{\text{part},n_3} \right)$$

$$+ \sum_{\text{guide}=1}^{N_s} \sum_{t_2=1}^{N_2'} 2a_2^{\text{guide},t_2} Y_2^{\text{guide},t_2} \left(\sum_{\text{part}=1}^{N_s} \sum_{n_3=1}^{N_2'} \beta_{\text{part},n_3}^{\text{guide},t_2} E_{\text{reference2}}^{\text{part},n_3} \right)$$

Using equations (A5.1) and (A5.2), equations representative of the scattering matrix can be defined:

$$b_1^{n_1} = \sum_{t_1=1}^{N_1'} 2a_1^{t_1} \frac{Y_1^{t_1}}{U_1^{n_1}} \left(\sum_{\text{part}=1}^{N_s} \sum_{n_3=1}^{N_2'} \alpha_{\text{part},n_3}^{t_1} \iint_{S_2^{\text{part}}} E_{\text{reference2}}^{\text{part},n_3} E_{\text{reference1}}^{n_1}\, ds \right)$$

$$+ \sum_{\text{guide}=1}^{N_s} \sum_{t_2=1}^{N_2'} 2a_2^{\text{guide},t_2} \frac{Y_2^{\text{guide},t_2}}{U_1^{n_1}} \left(\sum_{\text{part}=1}^{N_s} \sum_{n_3=1}^{N_2'} \beta_{\text{part},n_3}^{\text{guide},t_2} \iint_{S_2^{\text{part}}} E_{\text{reference2}}^{\text{part},n_3} E_{\text{reference1}}^{n_1}\, ds \right) - a_1^{n_1}$$

$$b_2^{s,n_2} = \sum_{t_1=1}^{N_1'} 2a_1^{t_1} \frac{Y_1^{t_1}}{U_2^{s,n_2}} \left(\sum_{\text{part}=1}^{N_s} \sum_{n_3=1}^{N_2'} \alpha_{\text{part},n_3}^{t_1} \iint_{S_2^{\text{part}} \cap S_2^{s}} E_{\text{reference2}}^{\text{part},n_3} E_{\text{reference2}}^{s,n_2}\, ds \right)$$

$$+ \sum_{\text{guide}=1}^{N_s} \sum_{t_2=1}^{N_2'} 2a_2^{\text{guide},t_2} \frac{Y_2^{\text{guide},t_2}}{U_2^{s,n_2}} \left(\sum_{\text{part}=1}^{N_s} \sum_{n_3=1}^{N_2'} \beta_{\text{part},n_3}^{\text{guide},t_2} \iint_{S_2^{\text{part}} \cap S_2^s} E_{\text{reference2}}^{\text{part},n_3} E_{\text{references}}^{s,n_2} \, ds \right) - a_2^{s,n_2}$$

These relations are used for $n_1 = 1, 2, \ldots, N_1'$, $n_2 = 1, 2, \ldots, N_2'$, and $s = 1, 2, \ldots, N_s$ to define the scattering matrix parameters. Note that for the transition from small waveguide sections to a large waveguide, the scattering matrix is obtain from the results above using $S_{11}' = S_{22}$, $S_{22}' = S_{11}$, $S_{12}' = S_{21}$, and $S_{21}' = S_{12}$.

REFERENCES

[A5.1] R. F. Harrington, *Field Computation by Moment Methods*, IEEE Press, Piscatoury, NJ, 1993.

[A5.2] J. J. H. Wang, *Generalized Moment Methods in Electromagnetics: Formulation and Computer Solution of Integral Equations*, Wiley, New york, 1991.

INDEX

Advanced Design Techniques and Realizations of Microwave and RF Filters,
By Pierre Jarry and Jacques Beneat